Additive Manufacturing - Present and Sustainable Future, Materials and Applications

Edited by Isabel Montealegre-Meléndez

Published in London, United Kingdom

Additive Manufacturing - Present and Sustainable Future, Materials and Applications
http://dx.doi.org/10.5772/intechopen.1001756
Edited by Isabel Montealegre-Meléndez

Contributors
Alfredo Suárez, Ana R. Reis, Anselmo Thiesen Júnior, Atilla Atli, Babin Dangal, Baiqing Xiong, Changshu He, Cristina Berges, Daniel Figueiredo, Donghua Zhao, Elsa W. Sequeiros, Etienne Lemaire, Fernando Veiga, Gemma Herranz, Hao Deng, Harshith Boppana, Henrique Santos Ferreira, Javier Hidalgo, Jhonattan Gutjahr, Jing Wang, Jose M. Costa, Juliane Ribeiro da Cruz, Jurandir Marcos Sá de Sousa, Kai Wen, Lizhen Yan, Longqing Chen, Lvjun Zhou, Manikandan Nachimuthu, Manuel F. Vieira, Milton Pereira, Pedro Villanueva, Rajesh Pungampalayam Kannappan, Sangjin Jung, Sheng Cao, Shuying Wang, Tomas Ballesteros, Virginia Uralde, Vishvajitsinh Kosamiya, Wei Xiao, Wenhao Liu, Xiwu Li, Yanan Li, Ying Li, Yongan Zhang

First published in London, United Kingdom, 2025 by IntechOpen
IntechOpen is the global imprint of INTECHOPEN LIMITED, registered in England and Wales, registration number: 11086078, 167-169 Great Portland Street, London, W1W 5PF, United Kingdom

For EU product safety concerns: IN TECH d.o.o., Prolaz Marije Krucifikse Kozulić 3, 51000 Rijeka, Croatia, info@intechopen.com or visit our website at intechopen.com.

British Library Cataloguing-in-Publication Data
A catalogue record for this book is available from the British Library

Additive Manufacturing - Present and Sustainable Future, Materials and Applications
Edited by Isabel Montealegre-Meléndez
p. cm.

This title is part of the Industrial Engineering and Management Book Series, Volume 11
Topic: Production Engineering
Series Editor: Fausto Pedro Garcia Marquez
Topic Editor: Orhan Korhan

Print ISBN 978-0-85014-814-5
Online ISBN 978-0-85014-813-8
eBook (PDF) ISBN 978-0-85014-815-2
ISSN 3029-0511

If disposing of this product, please recycle the paper responsibly.

IntechOpen Book Series

Industrial Engineering and Management

Volume 11

Aims and Scope of the Series

Industrial Engineering and Management (IEM) is a discipline that focuses on optimizing complex processes and systems within various industries. It involves the integration of engineering, business, economics, mathematics, and behavioral sciences to improve efficiency, productivity, quality, and overall performance in organizations. Key aspects of Industrial Engineering and Management include: Process Optimization; System Analysis and Design; Quality Control and Management; Supply Chain Management; Operations Management; Human Factors and Ergonomics; Project Management; Cost Analysis and Financial Management; Decision Analysis.

Overall, Industrial Engineering and Management aims to optimize resources, improve processes, enhance productivity, and ensure the effective and efficient utilization of all elements involved in the production or delivery of goods and services. It is crucial in today's competitive business environment for organizations to stay efficient and competitive.

Production Engineering and Operational Excellence are fields of study and practices that focus on optimizing and improving the manufacturing and production processes within an organization. It combines principles from engineering, management, and operational strategies to enhance productivity, efficiency, quality, safety, and sustainability in the production of goods and services.

Here are the key components of Production Engineering and Operational Excellence: Process Optimization; Operational Excellence; Manufacturing Systems Design; Quality Management; Supply Chain Optimization; Production Planning and Scheduling; Automation and Technology Integration; Health, Safety, and Environmental Management; Cost Management; Performance Measurement and Key Performance Indicators (KPIs); Continuous Improvement and Innovation.

Production Engineering and Operational Excellence are crucial for organizations aiming to stay competitive in the global market by achieving high levels of efficiency, quality, and customer satisfaction while optimizing resources and minimizing waste. It is a multidisciplinary approach that encompasses engineering principles, management strategies, and the effective use of technology to drive operational success.

Meet the Series Editor

Fausto Pedro Garcia Marquez is a Full Professor at UCLM, Spain, with accreditation since 2013. He also holds the position of Honorary Senior Research Fellow at Birmingham University, UK, and serves as a Lecturer at the Postgraduate European Institute. In addition to these roles, Fausto has experience as a Senior Manager at Accenture from 2013 to 2014. He earned his European Ph.D. with the highest distinction. Throughout his career, Fausto has received numerous awards and honors. These include the Nominate Prize (2022), Gran Maestre (2022), Grand Prize (2021), Runner Prize (2020), and Advancement Prize (2018), as well as Runner (2015), Advancement (2013), and Silver (2012) by the International Society of Management Science and Engineering Management (ICMSEM). He was also the recipient of the First International Business Ideas Competition 2017 Award. Fausto's contributions extend to academic publishing, with over 242 papers to his name. Notably, his work has been recognized in journals like "Applied Energy" (Q1, IF 9.746, Best Paper 2020) and "Renewable Energy" (Q1, IF 8.001, Best Paper 2014). His affiliations include the editorial and authorship roles in more than 50 books, with publications through respected publishers such as Elsevier, Springer, Pearson, Mc-GrawHill, IntechOpen, IGI, Marcombo, and AlfaOmega. He has authored over 100 international chapters and holds 6 patents. Fausto serves as the Editor of 5 International Journals and is a Committee Member for more than 70 International Conferences. His research portfolio encompasses being the Principal Investigator in 4 European Projects, 8 National Projects, and participating in over 150 projects involving universities and companies. His areas of expertise and research interests span Artificial Intelligence, Maintenance, Management, Renewable Energy, Transport, Advanced Analytics, and Data Science. Fausto is a recognized Expert in the European Union in AI4People (EISMD) and ESF. He also serves as the Director of www.ingeniumgroup.eu, holds the status of Senior Member at IEEE since 2021, and has been honored as an Honorary Member of the Research Council of the Indian Institute of Finance since 2021. Fausto is also the Committee Chair of The International Society for Management Science and Engineering Management (ISMSEM) since 2020.

Meet the Volume Editor

Dr. Isabel Montealegre-Meléndez is an associate professor at the University of Seville. She is Head of the Department of Materials Engineering and Science and Transportation. She received her bachelor's degree in Chemical Engineering from the University of Granada in Spain and her M.D. at the Technische Universität Wien in Austria before becoming a Doctor at the Austrian Research Centre in 2009. As an industrial doctoral student at ARCs and TUWien, she developed investigations in powder metallurgy and metal-matrix composites, researching titanium reinforced by CNTs, boron, nanodiamonds, etc. After, she returned to Spain. She currently investigates in-situ reinforced composites and additive manufacturing. She works and collaborates with companies in the metallurgy industry, focusing on developing components using additive manufacturing techniques.

Contents

Preface

Since its emergence in the 20th century, additive manufacturing has profoundly transformed. Initially conceived as a set of experimental techniques aimed at innovating prototyping processes, additive manufacturing has since evolved into a fundamental pillar of modern industrial production. Simultaneous advances in materials science, process optimization, computational modeling, and digital design tools have driven this transformation. Together, these developments have enabled the shift from conceptual prototypes to fully functional, high-performance components across various industries and applications.

In today's rapidly evolving society, there is a growing demand for increasingly specialized products with complex, customized geometries. In response, the manufacturing sector continuously adapts to faster and more precise production processes. The inherent flexibility and sustainability of additive techniques directly address these evolving demands, enabling the production of components with minimal material waste tailored to highly specific requirements. This adaptability has made additive manufacturing particularly valuable in sectors such as biomedical engineering, aerospace, and automotive manufacturing, among others. Consequently, additive techniques are increasingly recognized as essential for overcoming the contemporary challenges of sustainable manufacturing, resource optimization, and personalized production.

This volume brings together contributions from researchers, engineers, and professionals from diverse disciplines whose collective expertise reflects the inherently interdisciplinary nature of additive manufacturing.

By combining fundamental research with applied case studies, this book not only offers a comprehensive understanding of the current capabilities of additive manufacturing but also provides a critical perspective on how these technologies can continue to contribute to technological and industrial progress in the years to come.

The first chapter, *Advancements and Methodologies in Directed Energy Deposition (DED-Arc) Manufacturing: Design Strategies, Material Hybridization, Process Optimization and Artificial Intelligence*, presents a detailed analysis of innovative design strategies, hybrid material applications, process control techniques, and the increasing role of artificial intelligence in optimizing DED-Arc processes. This is followed by *High-k, Low-Loss Ceramic-Thermoplastic Composite Feedstock Filaments for Fused Deposition Modeling of Microwave and mm-Wave Devices*, a chapter that highlights the intersection of advanced materials engineering and additive manufacturing, with particular emphasis on applications in microwave and millimeter-wave technologies.

The third chapter, *Integrating Aesthetics and Robotic Additive Manufacturing in Artistic Design*, offers a distinctive perspective, demonstrating how additive technologies can transcend industrial and scientific applications, contributing to artistic creation and expanding these processes' cultural and design potential.

The book also features a comprehensive chapter *A Review of Solid-State Additive Manufacturing: Method, Microstructural Evolution, Mechanical Properties, Applications and Challenges*, thoroughly examining methods, microstructural evolution, mechanical properties, and potential application scenarios for these promising techniques. This is complemented by the chapter *Topology Optimization for Metal Additive Manufacturing: Effect of Build Orientation and Overhang Constraint*, which underscores the critical importance of design methodologies tailored specifically for additive processes, highlighting how build orientation and overhang limitations influence the final performance of parts.

Innovation in additive manufacturing continues to expand the boundaries of what is possible. Chapter *4D Printing-3D Printing of Smart Materials* introduces time-responsive materials and structures, offering a forward-looking glimpse into the future of adaptive and self-transforming components. This volume also includes industry-relevant research, such as *Mechanical Properties of AISI 410L Stainless Steel Parts Manufactured by Laser Directed Energy Deposition in the as-Built and Heat-Treated States*, presenting valuable insights into the mechanical behavior of additively manufactured components, both in their as-built state and after specific heat treatments.

Sustainability, an increasingly central theme in modern manufacturing, is addressed in the chapter *Additive Manufacturing of Ferroelectric and Piezoelectric Composites Based on Eco-Circular Salts*, illustrating how additive manufacturing can contribute to developing environmentally responsible materials and circular production cycles. The importance of design for additive manufacturing (DfAM) is further emphasized in *Optimizing Metal AM Potential through DfAM: Design, Performance, and Industrial Impact*, underscoring the need to rethink conventional design processes to fully exploit the unique capabilities of additive techniques.

One crucial aspect of additive manufacturing—often overlooked despite its importance—is post-processing. This is addressed comprehensively in *Post-Process Treatments for Additive Manufactured Titanium Alloys*, a chapter that examines how post-processing techniques influence the mechanical properties and microstructural characteristics of additively manufactured titanium components. Finally, the volume concludes with a broad review of *Material Extrusion Additive Technologies: Benefits, Challenges, and Niche Applications*, which synthesizes the main benefits, challenges, and niche applications of this widely adopted and continuously evolving technique.

This book would not have been possible without the dedicated research efforts of all contributing authors whose work and expertise have enriched each chapter and whose knowledge continues to drive innovation in the field of additive manufacturing.

I would like to express my deepest and most sincere gratitude to Dr. Cristina Arévalo Mora (reviewer of "Material Extrusion Additive Technologies: Benefits, Challenges, and Niche Applications" and Dr. Eva M. Pérez Soriano (reviewer of "A Review of Solid-State Additive Manufacturing: Method, Microstructural Evolution, Mechanical Properties, Applications and Challenges"), who served as reviewers for chapters. Their careful attention to detail, constructive and insightful feedback, and deep expertise have significantly enhanced the academic quality of this work.

Thanks to ongoing technological advances and the continuous resolution of challenges in the additive manufacturing landscape, it is clear that integrating these techniques into industrial practice will only deepen in the future. The unique synergy between sustainability and flexibility, coupled with significant cost savings enabled by additive manufacturing, ensures that this technology will remain at the forefront of innovation throughout the 21st century.

I sincerely hope this book serves as a valuable reference for researchers, engineers, and students and a source of inspiration for future developments that will continue to enhance the transformative potential of additive manufacturing for generations to come.

Isabel Montealegre-Meléndez,
University of Seville,
Seville, Spain

Chapter 1

Advancements and Methodologies in Directed Energy Deposition (DED-Arc) Manufacturing: Design Strategies, Material Hybridization, Process Optimization and Artificial Intelligence

Virginia Uralde, Alfredo Suárez, Fernando Veiga, Pedro Villanueva and Tomas Ballesteros

Abstract

This chapter explores the latest advancements and methodologies in directed energy deposition (DED-arc) manufacturing. The introduction sets the stage for understanding the significance of these developments in the context of modern manufacturing needs. The discussion includes design strategies for DED-arc, emphasizing topological optimization, functional design, and generative design, alongside the application of artificial intelligence (AI) in enhancing design processes. Innovative approaches to material hybridization are detailed, focusing on both multi-layer and *in situ* techniques for combining different materials to optimize component performance. The paper also covers slicing and pathing, examining slicing strategies, the use of lattice structures, and the implementation of 2D and 3D patterns to improve manufacturing efficiency and product quality. The conclusion summarizes key findings, discusses their implications for the additive manufacturing industry, and suggests potential future research directions in DED-arc technology, highlighting the emerging trends and innovations that are shaping the field.

Keywords: wire arc additive manufacturing (WAAM), near-net shape, artificial intelligence, material hybridization, topological optimisation

1. Introduction

Directed energy deposition (DED) technology encompasses various techniques where the energy source melts material as it is fed [1]. Its primary advantage over powder bed systems lies in its capability to produce larger parts with a greater working volume and repair damaged components [2]. The amount of material needing machining is significantly reduced, and the preform closely matches the final part geometry. DED includes wire-fed processes [3] and powder-fed processes [4]. The potential use of wire reduces costs per kilogram, enhances material efficiency, and diminishes demand for powder recycling systems [5, 6].

Energy sources for melting materials in DED include lasers, electron beams, or electric arcs. Machines with specific kinematics incorporate feed heads and material-feeding equipment, delivering material to designated locations [7]. Consequently, these technologies lack closed-system machine providers. Within the DED category, processes differ by energy source type and input material format, encompassing electron beam DED (DED-EB), laser beam DED (DED-LB), and arc DED (DED-Arc) [8].

1.1 Directed energy deposition arc (DED-arc)

Arc-directed energy deposition technologies (DED-Arc), also known as wire arc additive manufacturing (WAAM), represent a significant advancement in advanced manufacturing as it is summarized in **Figure 1**. These technologies combine energy input through an electric arc with material addition via metal wire, enabling efficient creation of complex, customized parts [9].

DED-Arc employs specialized welding equipment including power sources, torches, and wire feeding systems. Movement and precision during manufacturing are ensured through robotic systems or numerically controlled gantries (CNC), guaranteeing repeatability and precision in material deposition.

This technology includes several welding processes, notably gas metal arc welding (GMAW), plasma arc welding (PAW), and gas tungsten arc welding (GTAW), each with specific foundations, key differences, and diverse applications in industry.

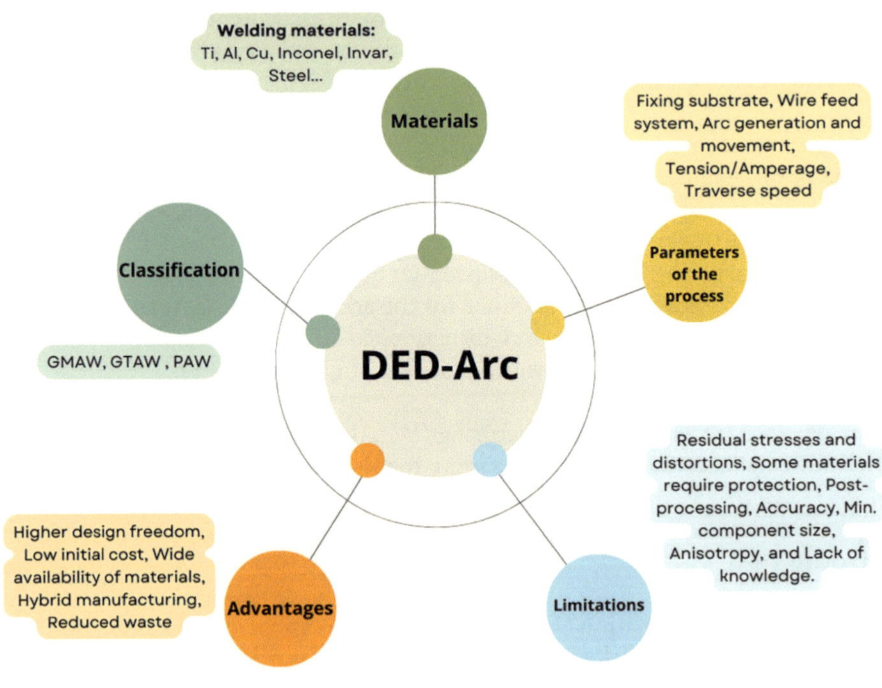

Figure 1.
Main characteristics of DED-Arc technology.

Gas metal arc welding (GMAW), or metal inert gas (MIG) welding, utilizes a continuous wire electrode fed through a torch and melted by an electric arc between the electrode and the workpiece. An inert or mixed gas shields the arc and weld pool from atmospheric contamination [10]. Consumable electrodes facilitate high deposition rates, making GMAW suitable for various materials in automotive, aerospace, and shipbuilding.

Plasma arc welding (PAW) employs an electric arc formed between a tungsten electrode and the workpiece, confined and concentrated by a nozzle to create a narrow, precise plasma arc [11]. Shielded by inert gas, PAW ensures high precision and quality welds, particularly beneficial in aerospace, instrumentation, and electronics.

Gas tungsten arc welding (GTAW), or tungsten inert gas (TIG) welding, utilizes an electric arc between a non-consumable tungsten electrode and the workpiece [12]. The arc and weld pool are shielded by inert gas, facilitating precise welding of non-ferrous metals and thin materials in applications requiring high-quality welds.

Additional advanced techniques like cold metal transfer (CMT), an advanced variant of GMAW, control short-circuit transfer for high material deposition rates with a low heat input [13].

This chapter provides an overview of new trends in DED-Arc technology, focusing on application and significance in modern manufacturing processes, including design for additive manufacturing (DfAM), bead-layer geometry determination, slicing and pathing, and material hybridization. **Figure 2** illustrates the main features and applications of the new trends in DED-Arc and gives an overview of the topics covered by this chapter.

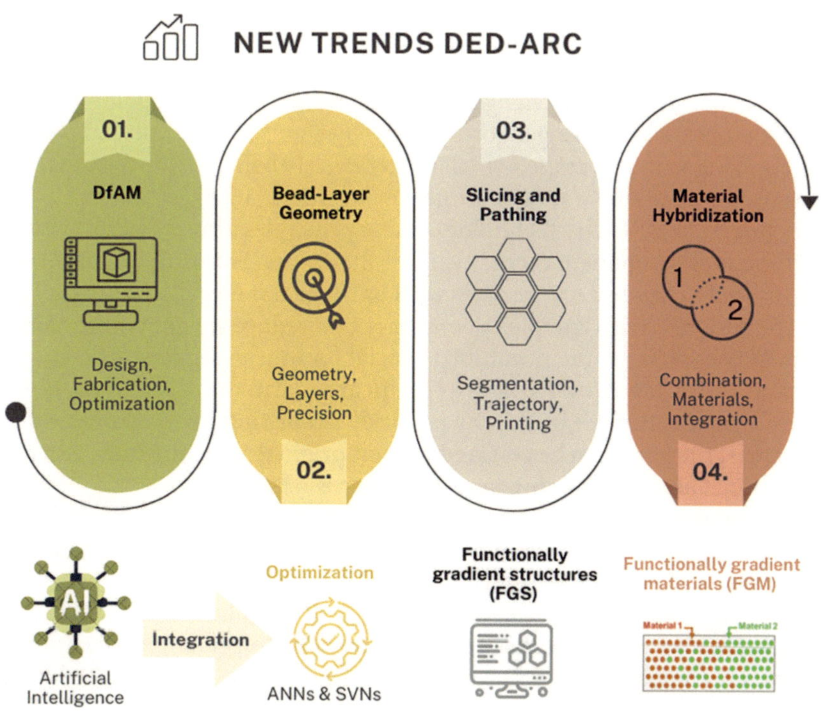

Figure 2.
New trends in DED-Arc technology.

This comprehensive exploration sets the stage for detailed examination of DED-Arc's operational principles, technological advancements, and prospects in the realm of additive manufacturing.

2. Design for additive manufacturing

Design for additive manufacturing (DfAM) involves creating designs tailored specifically for the capabilities and constraints of additive manufacturing (AM) technologies. The fundamental principles of DfAM include leveraging AM's ability to produce intricate and complex shapes that are challenging to achieve with traditional methods. This capability allows designers to explore complex geometries and internal lattice structures, which enhance the strength-to-weight ratio of parts. Additionally, AM reduces material waste through precise deposition, as material is added layer by layer, only where necessary [14]. This contrasts with traditional subtractive methods, where significant material is often wasted [15]. In this chapter, traditional methods refer to conventional subtractive manufacturing techniques such as machining, milling, and turning. These methods involve removing material from a larger block to create the desired shape, which often involves significant material waste. In addition, traditional methods include casting and forging, where material is poured into molds or molded under high pressure to shape parts. These techniques, while well established, often have limitations in producing complex geometries and internal structures, which DED-Arc and other additive manufacturing processes can achieve more efficiently and with greater design flexibility.

Another crucial principle of DfAM is integrating multiple functions into single parts to minimize assembly needs. This approach reduces the number of components, thereby simplifying supply chains and improving product reliability by reducing the potential points of failure. Moreover, DfAM enables mass customization, allowing for the efficient production of bespoke parts tailored to individual requirements without the need for expensive tooling changes. Computational tools for topology optimization are also employed, utilizing algorithms to redesign parts for optimal performance while minimizing material usage, which is particularly beneficial in aerospace and automotive industries where weight reduction is critical [16, 17].

In the context of DED-Arc technology, these principles are applied to create high-performance components across various industries. For instance, in aerospace, DED-Arc is used to fabricate large parts with optimized internal structures. These parts can be designed with complex geometries that reduce weight while maintaining or enhancing strength and durability, crucial for improving fuel efficiency and performance [18]. DED-Arc also plays a significant role in the repair of high-value components, such as turbine blades. By precisely depositing material onto worn or damaged areas, the parts can be restored to their original specifications, extending their operational life and reducing maintenance costs [19].

DED-Arc technology is also utilized for creating customized tooling and fixtures. These tools often require complex geometries that traditional manufacturing methods struggle to produce. By leveraging DfAM principles, companies can design and fabricate tools that not only are tailored to specific applications but also exhibit improved performance characteristics. This customization enhances the manufacturing process's efficiency, as the tools can be optimized for specific tasks, reducing cycle times and improving overall productivity [20]. Additionally, DED-Arc allows for the creation of hybrid parts that combine multiple materials, offering new possibilities for designing parts with varied properties within a single component [21].

Another approach is given by Ref. [22–24] with a functional design based on DED-Arc constraints of an aeronautical mounting turret made of 5356 series aluminum alloy, inspired by vault-based architectural design, showing remarkable results compared to the topologically optimized design (**Figure 3**). The work presented by the authors aims to lighten three different parts of the fixturing system using DfAM techniques for additive manufacturing (AM) with topological optimization. This work considers the solution as provided by the topological optimization software [22]. The article focuses on the aluminum aeronautical turret from the previous work, which is the one used in **Figure 3**, aiming at the actual fabrication of the part by WAAM, thus not achieving a significant weight reduction [23]. This article complements previous studies by considering and observing the limitations of WAAM for topological optimization, adopting a vault structure design to lighten the part's volume. Its mechanical behavior is tested and compared with the FEM model [24]. This approach has resulted in a reduction of the part weight to 2.2 kg, which represents a 63% decrease compared to the original design. Despite the weight reduction, the DED-Arc design withstands a maximum stress of 4.86 MPa, slightly higher than that of the optimized design, which is 4.65 MPa. However, the yield strength of the material has been adjusted to 109 MPa, which is a significant reduction compared to the 145 MPa of the original design. The safety factor of the DED-Arc design is maintained at 18.6, which, although lower than the optimized design of 23.5, still provides adequate robustness for many applications. This design demonstrates how it is possible to achieve a balance between weight, strength, and safety factors by implementing advanced design techniques specific to additive manufacturing, taking full advantage of the capabilities of DED-Arc.

The benefits of DfAM in DED-Arc include flexibility in design, material efficiency, the ability to repair high-value components, and the production of large, complex parts that are difficult to manufacture with other technologies. One

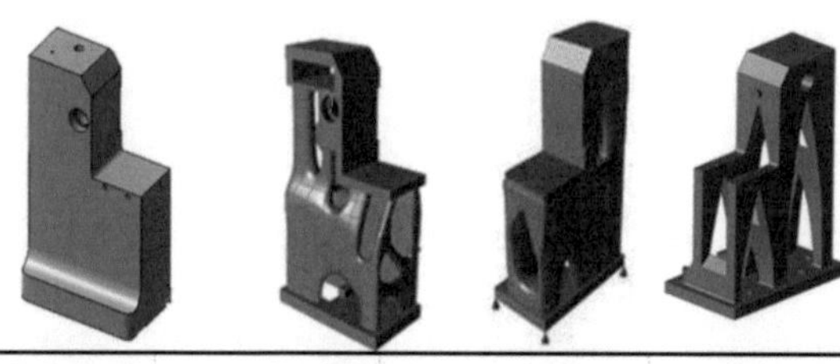

	Original Design	**Topologically optimized design [22]**	**DED-arc Topologically optimized design [23]**	**DED-Arc oriented design [24]**
Weight (Kg)	6	2,3 (↓61%)	4,1 Kg (↓31%)	2,2 Kg
Maximum Stress (MPa)	1,99	4,83	4,65	4,86
Yield Strength (MPa)	145	145	109	109
Factor of safety	73	30	23.5	18.6

Figure 3.
Aeronautical turret use case with different approaches [22–24] summarizing work on optimized design.

significant advantage is the ability to produce complex shapes and customized parts, which reduces the need for multiple manufacturing processes. This flexibility allows manufacturers to respond quickly to design changes and market demands, enhancing their competitive edge [15]. Material efficiency is another key benefit, as AM processes typically use less material compared to traditional methods, leading to cost savings and reduced environmental impact [16].

However, there are limitations to consider. DED-Arc often results in rough surface finishes, which require additional post-processing to meet quality standards. This additional step can increase production time and costs. Achieving high dimensional accuracy can also be challenging due to thermal effects and material properties during the deposition process. Variations in temperature can cause warping or residual stresses, affecting the final part's dimensions and mechanical properties [25]. Ensuring consistent quality and performance across different batches requires sophisticated control systems and a high level of expertise, which can be a barrier for smaller manufacturers [26].

Furthermore, the initial investment in DED-Arc equipment can be high, potentially limiting its adoption to larger organizations or specific high-value applications. While the long-term benefits of reduced material costs and enhanced design flexibility are significant, the upfront costs can be prohibitive for some companies [27]. Additionally, the development of comprehensive computer-aided design (CAD) and computer-aided manufacturing (CAM) software tailored for DED-Arc processes is still evolving, and further advancements are needed to fully exploit the technology's potential [28]. Despite these challenges, ongoing research and technological improvements continue to expand the capabilities and applications of DED-Arc in additive manufacturing [29].

DfAM in the context of DED-Arc offers significant advantages in terms of flexibility, efficiency, and the ability to produce large, complex, and customized parts. By understanding and addressing the benefits and limitations, manufacturers can effectively integrate DED-Arc into advanced manufacturing processes to achieve superior outcomes. This approach not only enhances the design and functionality of parts but also drives innovation and sustainability in the manufacturing industry [30, 31].

3. Determination of bead-layer geometry

In the domain of direct energy deposition (DED) utilizing arc-based methods, the precise determination and control of bead-layer geometry stand as fundamental pillars for achieving desired part properties and ensuring high-quality manufacturing outcomes. This section explores the foundational concepts, various measurement methodologies, and the intricate influence of process parameters on bead geometry within DED manufacturing, with a special focus on DED-Arc (DED-Arc) or wire arc additive manufacturing (WAAM).

Fundamentally, the geometry of the bead, encompassing its width, height, and overall shape, holds significant implications for the structural integrity, dimensional accuracy, and surface finish of fabricated parts. **Figure 4** shows some of the key parameters in the geometry of a weld bead. This aspect has garnered considerable attention within the scientific community, underscoring its pivotal role in ensuring the mechanical properties and functional reliability of additively manufactured components [32]. **Figure 4** illustrates the bead geometry and wall geometry in a welded structure of mild steel ER70. The bead geometry includes measurements

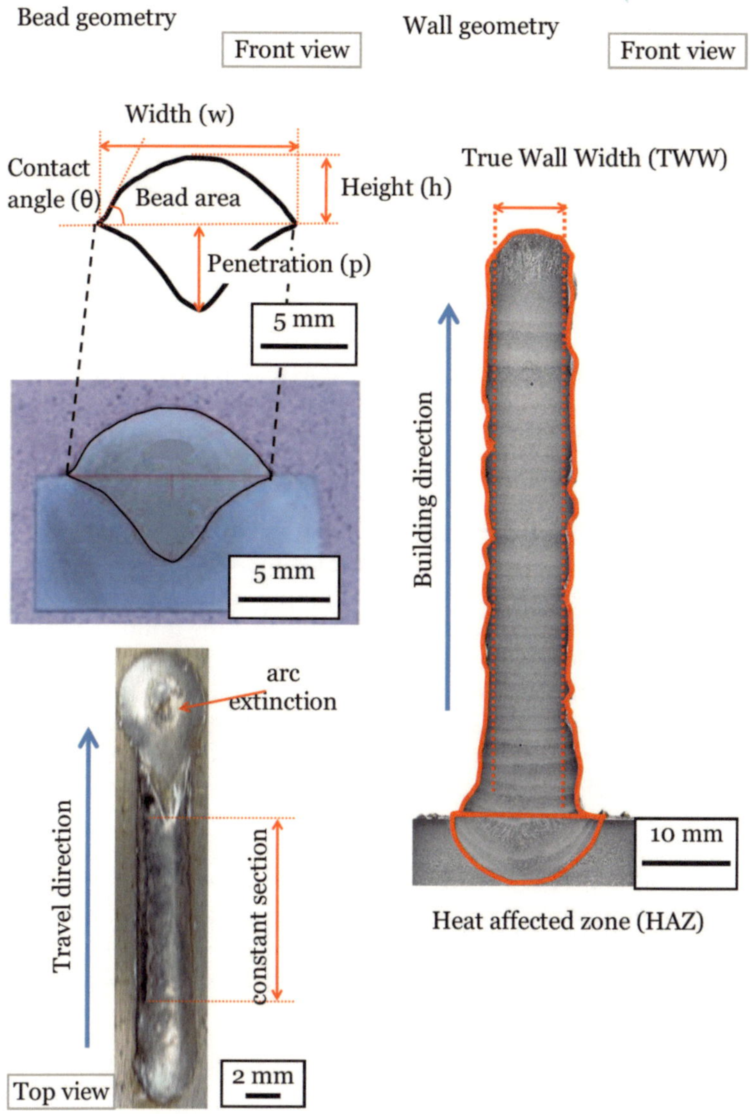

Figure 4.
Key parameters in weld bead and wall geometry in DED-Arc.

such as width (w), height (h), bead area, and penetration (p). The contact angle (θ) is the angle formed between the bead's surface and the base material, indicating how well the bead wets the base material. The wall geometry includes the true wall width (TWW), which is the actual width of the wall formed by multiple layers of beads, and the heat affected zone (HAZ), the area around the weld where the microstructure and properties have been altered by the welding heat. The top view of the bead shows the arc extinction point and a constant section of the bead, with the material tending to solidify backward, demonstrating the flow of the molten metal during the welding process. The travel direction of the welding torch or electrode is also indicated, showing how the bead is formed. These details provide a comprehensive understanding of the critical dimensions and features in the welding process.

Advanced measurement techniques play a crucial role in the comprehensive analysis of bead geometry. Techniques such as 3D laser scanning, coordinate measuring machines (CMM), and microscopy-based image analysis enable precise quantification of bead dimensions and profiles. For instance, Ding et al. [33] have integrated machine learning approaches to automate bead modeling, demonstrating advancements in real-time quality control and process optimization. Similarly, teaching–learning-based optimization techniques are utilized to enhance bead geometry and reduce energy consumption, reflecting ongoing efforts to refine manufacturing processes [34].

The influence of process parameters on bead geometry is multifaceted and profound. Variables such as arc voltage, wire feed speed, deposition angle, and travel speed exert critical influences on bead dimensions and morphology. Optimizing these parameters is crucial for achieving consistent bead characteristics and enhancing manufacturing efficiency. Research indicates that adjustments in these parameters can lead to significant variations in bead geometry, thereby affecting the mechanical properties and surface quality of additively manufactured parts [35].

In-depth analyses of bead geometry provide valuable insights into process robustness and quality assurance. Studies by Tang et al. on WAAM highlight detailed investigations into the geometry at both the initiation and termination zones of the bead, crucial for minimizing geometric errors and ensuring uniformity throughout the deposition process [36]. Moreover, similarly, geometric terms such as the ratio between width and height (RWTH) are introduced as indicators for assessing bead uniformity and cross-sectional stability [37, 38]. In bead geometry, symmetry plays a critical role in ensuring the quality and consistency of additive manufacturing processes, particularly in direct energy deposition (DED) methods like DED-Arc [39–41].

The meticulous determination and control of bead-layer geometry in DED-Arc processes are indispensable for advancing additive manufacturing capabilities [42]. Through continuous refinement of measurement methodologies and a deeper understanding of parameter interactions, researchers and industry practitioners aim to optimize bead quality, thereby pushing the boundaries of manufacturing precision and reliability in the production of complex, near-net-shape components.

4. Slicing and pathing

Path strategies in WAAM are pivotal for guiding the deposition nozzle or torch along trajectories that build up the desired part geometry layer by layer [43]. These strategies not only influence the structural integrity and surface finish of the final component but also impact process efficiency and material usage.

This commonly used strategy involves the back-and-forth movement of the welding torch during material deposition. By oscillating along the weld axis, heat distribution becomes more uniform, minimizing the risk of thermal distortion and ensuring consistent bead geometry [44]. This method is particularly effective in enhancing the quality and reliability of deposited layers. Initially proposed by Ref. [45], contouring strategies utilize geometrically defined paths such as circular arcs or straight-line segments to efficiently fill the designated deposition area. These methods optimize material usage and contribute to the precision of the manufacturing process.

Strategies involving fractal patterns, proposed by Ref. [46, 47], explore the use of intricate, self-similar patterns to achieve comprehensive area coverage while minimizing deposition time and ensuring structural robustness. Recent advancements, as highlighted by Ref. [23, 48], advocate for the integration of polynurbs—non-uniform rational B-spline curves—to define complex geometries. These methods combine contouring with zig-zagging or other oscillatory patterns to address the challenges of manufacturing intricate parts efficiently.

In the domain of additive manufacturing, lattice structures represent a significant advancement, offering enhanced mechanical properties, reduced material consumption, and improved weight-to-strength ratios. These structures, characterized by their repetitive geometric patterns, are particularly advantageous in applications requiring lightweight components with high structural integrity.

These consist of planar configurations such as honeycomb, diamond, or triangular patterns. They are widely utilized in applications where weight reduction and material efficiency are critical, such as in aerospace and automotive industries [49].

Extending the concept into three dimensions, 3D lattice structures incorporate intricate volumetric arrangements like gyroids, octet trusses, or periodic minimal surfaces. These structures offer superior mechanical properties due to their optimized load distribution and are increasingly employed in biomedical implants and advanced structural components [50].

By replacing solid volumes with geometrically optimized patterns, lattice structures significantly reduce overall component weight without compromising strength or functionality. The geometric complexity of lattice structures enhances stiffness, fatigue resistance, and energy absorption capabilities compared to traditional solid structures. Additive manufacturing facilitates the fabrication of lattice structures with intricate geometries that are challenging or impossible to produce using conventional methods. This design freedom enables the creation of customized components tailored to specific performance requirements.

The integration of optimized pathing strategies and lattice structures in DED-Arc holds promising implications for the future of manufacturing. Continued research aims to refine these techniques, leveraging advancements in computational modeling, material science, and process optimization to further enhance the performance, efficiency, and versatility of additive manufacturing technologies across various industrial sectors.

Pathing strategies in DED-Arc manufacturing and the utilization of lattice structures represent critical avenues for advancing additive manufacturing capabilities. By optimizing deposition paths and harnessing the potential of geometrically optimized structures, researchers and industry practitioners strive to push the boundaries of manufacturing efficiency and performance in the production of complex, functional components.

5. Material hybridization

In the realm of additive manufacturing, material hybridization, particularly through bi-metallic structures (BMS), is a burgeoning area of research aiming to optimize material usage and enhance component performance across diverse industries. This approach involves combining different materials within a single part to leverage their respective strengths and properties where needed most. In additive manufacturing using DED, materials can be hybridized in two primary ways as it can be seen in **Figure 5**. One method is layer-by-layer deposition, where different metals are deposited in distinct layers to create a multi-material structure [51], allowing for precise control of material distribution and tailoring properties in specific regions. The

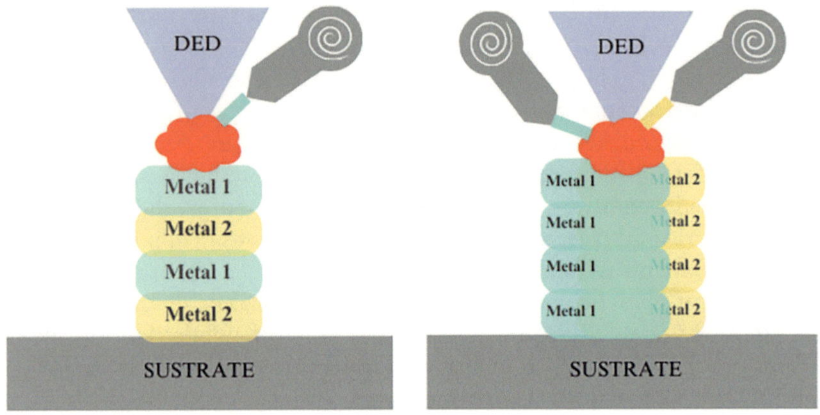

Figure 5.
Schematic of the hybridization of two materials: layer-by-layer and "In-situ" deposition.

other method is *in-situ* mixing [52], where different metals are simultaneously fed into the same melt pool, creating a composite material directly in the fusion process. This enables the creation of functionally graded materials with a smooth transition between different metals, enhancing the component's performance characteristics.

Bi-metallic structures are pivotal in applications where specific zones of a component require tailored material properties such as strength, corrosion resistance, or conductivity. By strategically integrating materials like steels, nickel-based alloys, aluminum, or titanium, manufacturers can achieve significant benefits. These include maximizing performance in critical areas while reducing costs by using expensive materials only where necessary [53, 54].

Various material combinations cater to specific application needs:

- Copper-Stainless Steel (Cu-SS): This combination blends copper's high thermal conductivity with stainless steel's corrosion resistance, making it ideal for applications like fusion reactors, heat exchangers, and cookware [55].

- Aluminum-Steel: Combining aluminum's corrosion resistance with the structural strength of steel, this hybrid is commonly used in adapters for oxygen regenerators and electrolytic refining equipment [56].

- Nickel-Steel: Despite challenges in reconciling their differing properties, successful combinations yield materials with excellent mechanical strength, corrosion resistance, and high-temperature creep resistance, suitable for demanding applications such as nuclear reactors and gas turbines [54].

- Inconel718-Austenitic Stainless Steel: Used in aerospace and nuclear applications, this pairing combines the robustness of Inconel with the corrosion resistance of austenitic stainless steel [56, 57]

Challenges in material hybridization include managing differences in coefficient of thermal expansion (CTE), ensuring chemical compatibility, and mitigating the formation of brittle or unstable interfaces within the material matrix [36, 58, 59]. These factors influence the overall structural integrity and performance of the fabricated components.

Additive manufacturing facilitates the creation of functionally gradient structures (FGS), where the composition of materials varies layer-by-layer or within specific regions of a component [60, 61]. This capability allows for the seamless transition between materials, optimizing mechanical properties such as hardness, flexibility, and wear resistance along the gradient [53].

The ability to customize material properties according to specific design needs is a hallmark of material hybridization. This tailoring enhances the performance and longevity of components in demanding applications across sectors such as aerospace, naval, and automotive industries. For instance, in aerospace, the integration of lightweight materials with high-strength alloys improves fuel efficiency and structural integrity, while in automotive manufacturing, hybrid structures enhance vehicle performance and safety through optimized material combinations [62].

Moreover, DED-Arc's *in situ* material hybridization offers significant advantages in sectors where weight reduction, durability, and performance optimization are critical. This approach enables cost-effective production by reducing material waste and simplifying manufacturing processes, thereby driving innovation in component design and fabrication [63].

By leveraging additive manufacturing for material hybridization, industries can achieve not only performance enhancements but also substantial cost savings and environmental benefits. Future research endeavors aim to further refine deposition techniques, optimize material combinations, and deepen our understanding of material interfaces to unlock the full potential of hybrid structures in advancing modern manufacturing capabilities [64].

Additive manufacturing techniques like DED-Arc facilitate both multilayer and *in situ* material hybridization approaches. These methods allow for the deposition of distinct materials layer-by-layer or in a graded fashion, creating functionally gradient structures (FGS). FGS transition smoothly from one material composition to another, optimizing mechanical properties such as hardness, flexibility, and wear resistance along the gradient [65–67].

The ongoing evolution of material hybridization through additive manufacturing promises to redefine manufacturing capabilities across aerospace, automotive, and other industries. Future research aims to refine deposition techniques, optimize material combinations, and enhance understanding of material interfaces to further harness the potential of hybrid structures for superior performance and efficiency in component manufacturing.

6. Use of artificial intelligence (AI) in DED-arc

Artificial intelligence (AI) has become indispensable in enhancing the capabilities of DED-Arc, a key subset of additive manufacturing. This technology harnesses AI to optimize processes, ensure quality, and enable novel material applications [68].

AI's integration in DED-Arc has revolutionized how manufacturing processes are controlled and optimized. By employing machine learning algorithms, AI models analyze vast amounts of data collected from sensors and simulations [69]. This analysis allows for predictive modeling of parameters such as wire feed speed, travel speed, arc voltage, and nozzle-to-plate distance. Such predictive capabilities enable real-time adjustments during fabrication, ensuring consistent and precise material deposition. This capability is particularly beneficial in achieving complex geometries and meeting stringent dimensional tolerances required in industries like aerospace and automotive.

Process control in DED-Arc benefits significantly from AI technologies. AI algorithms, including artificial neural networks (ANNs) [70, 71], support vector machines (SVMs) [72], and deep learning models [73], continuously monitor and adjust manufacturing variables [74]. For instance, ANNs model bead geometry to optimize parameters like bead width, height, and overlap, ensuring uniform material distribution and structural integrity [75]. SVMs are employed for automation and quality control by analyzing molten pool dynamics and ensuring optimal energy input [33]. Deep learning models excel in path planning for intricate structures, minimizing material waste and optimizing deposition efficiency [76]. These applications collectively enhance productivity and reduce manufacturing costs by mitigating errors and optimizing energy consumption.

AI-driven parameter optimization plays a pivotal role in enhancing the performance of DED-Arc processes. By learning from historical data and real-time feedback, AI models adjust parameters such as weaving amplitude, interlayer temperature, and contact angle to achieve desired material properties and geometric accuracy. This adaptive optimization not only improves the mechanical properties of fabricated components but also reduces trial-and-error iterations, thereby accelerating the development cycle of new products. AI's ability to predict and optimize parameters ensures that manufacturers can consistently produce parts with precise dimensions and superior mechanical characteristics [77, 78].

Ensuring the quality of additively manufactured components is critical in industrial applications, and AI significantly enhances quality assurance in DED-Arc. AI-based systems employ advanced inspection techniques, such as CNNs for image analysis and Bayesian models for defect detection, to identify surface imperfections, porosity, cracks, and dimensional inaccuracies in real time [79]. By detecting defects early in the manufacturing process, AI minimizes scrap and rework, thereby optimizing yield and reducing production costs. Furthermore, AI-driven quality assurance ensures that components meet stringent industry standards and customer specifications, bolstering reliability and performance in demanding environments [80].

The application of AI in DED-Arc enables the production of functionally gradient structures (FGS) and hybrid materials with tailored properties. FGS seamlessly transition between different materials, optimizing mechanical properties such as hardness, flexibility, and wear resistance along the component's gradient [81]. This capability is particularly advantageous in aerospace, naval, and automotive industries where lightweight components with superior strength and durability are essential. By customizing material properties according to specific design requirements, AI-driven DED-Arc facilitates innovation and product differentiation in highly competitive markets [82].

AI's integration into DED-Arc represents a transformative leap in additive manufacturing capabilities. As AI technologies continue to evolve, their application in optimizing processes, ensuring quality, and expanding material possibilities will drive further advancements in manufacturing efficiency, product performance, and industrial innovation [83]. Future research efforts will focus on refining AI algorithms, enhancing predictive capabilities, and expanding the scope of AI.

7. Conclusions and future directions

This study has examined various critical aspects of directed energy deposition using arc welding (DED-Arc), emphasizing its integration with advanced manufacturing principles and artificial intelligence (AI). Design for additive manufacturing with DED-Arc enables the creation of complex geometries and functionally gradient structures (FGS), optimizing mechanical properties such as strength, flexibility, and wear resistance through tailored material properties layer-by-layer.

Determining bead-layer geometry in DED-Arc is pivotal for achieving precise deposition outcomes. AI technologies like artificial neural networks (ANNs) and support vector machines (SVMs) enhance real-time control and optimization of welding parameters such as bead width, height, and interlayer bonding. This adaptive control ensures consistent quality and minimizes material waste.

Slicing and pathing strategies have evolved significantly to support DED-Arc's capability in efficient fabrication of large-scale components. AI-driven path planning algorithms optimize tool paths for complex structures, enhancing manufacturing speed, accuracy, and cost-effectiveness.

Material hybridization in DED-Arc leverages AI-enhanced process control to seamlessly integrate multiple materials. This approach combines metals and ceramics with complementary properties, benefiting industries like aerospace, automotive, and medical sectors with enhanced component performance.

The use of artificial intelligence (AI) in DED-Arc represents a transformative shift in additive manufacturing. AI algorithms enable predictive modeling, adaptive control, and real-time defect detection, enhancing process reliability and quality assurance. Optimization of parameters such as wire feed speed, travel speed, and energy input through AI enhances manufacturing efficiency and product consistency.

Implications for the industry are profound. By integrating DED-Arc with AI technologies, manufacturers can achieve higher productivity, reduced lead times, and improved product performance. Sectors requiring customized, high-strength components, such as aerospace and medical industries, are poised to benefit significantly from these innovations. Moreover, AI-driven quality assurance mitigates risks associated with manufacturing defects, ensuring components meet stringent standards and bolstering confidence in additive manufacturing processes.

Future research in DED-Arc should focus on advancing AI applications tailored for specific processes, exploring novel material combinations for optimized performance, and establishing industry standards and certification protocols. Sustainable manufacturing practices, human-machine collaboration, and efficiency enhancement strategies are also critical areas for exploration. By addressing these research areas, the additive manufacturing community can unlock new opportunities for innovation, efficiency, and sustainability, reshaping global manufacturing industries.

Author details

Virginia Uralde[1], Alfredo Suárez[2], Fernando Veiga[1*], Pedro Villanueva[1] and Tomas Ballesteros[1]

1 Department of Engineering, Public University of Navarre, Pamplona, Spain

2 TECNALIA, Basque Research and Technology Alliance (BRTA), Donostia-San Sebastián, Spain

*Address all correspondence to: fernando.veiga@unavarra.es

References

[1] Suárez A, Ramiro P, Veiga F, Ballesteros T, Villanueva P. Benefits of aeronautical preform manufacturing through arc-directed energy deposition manufacturing. Materials. 2023;**16**(22):7177

[2] Chen D, Chen M, Wu C. Effects of phase difference on the behavior of arc and weld pool in tandem P-GMAW. Journal of Materials Processing Technology. 2015;**225**:45-55

[3] Ding D, Pan Z, Cuiuri D, Li H. A multi-bead overlapping model for robotic wire and arc additive manufacturing (WAAM). Robotics and Computer-Integrated Manufacturing. 2015;**31**:101-110

[4] Hu D, Kovacevic R. Sensing, modeling and control for laser-based additive manufacturing. International Journal of Machine Tools and Manufacture. 2003;**43**(1):51-60

[5] Liu ZY, Li C, Fang XY, Guo YB. Energy consumption in additive manufacturing of metal parts. Procedia Manufacturing. 2018;**26**:834-845

[6] Jackson MA, Van Asten A, Morrow JD, Min S, Pfefferkorn FE. A comparison of energy consumption in wire-based and powder-based additive-subtractive manufacturing. Procedia Manufacturing. 2016;**5**:989-1005

[7] Herzog D, Seyda V, Wycisk E, Emmelmann C. Additive manufacturing of metals. Acta Materialia. 2016;**117**:371-392

[8] Svetlizky D, Das M, Zheng B, Vyatskikh AL, Bose S, Bandyopadhyay A, et al. Directed energy deposition (DED) additive manufacturing: Physical characteristics, defects, challenges and applications. Materials Today. 2021;**49**:271-295

[9] Taşdemir A, Nohut S. An overview of wire arc additive manufacturing (WAAM) in shipbuilding industry. Ships and Offshore Structures. 2020;0(0):1-18

[10] Moore KL, Naidu DS, Yender R, Tyler J. Gas metal arc welding control: Part I: Modeling and analysis. Nonlinear Analysis, Theory, Methods and Applications. 1997;**30**(5):3101-3111

[11] Miao Y, Li C, Yin C, Wei C, Lin Z. Joint characteristics of carbon steel bypass-current PAW on additive manufacturing. Journal of Manufacturing Processes. 2021;**61**:408-416

[12] Fahimpour V, Sadrnezhaad SK, Karimzadeh F. Microstructure and mechanical property change during FSW and GTAW of Al6061 alloy. Metallurgical and Materials Transactions A: Physical Metallurgy and Materials Science. 2013;**44**(5):2187-2195

[13] González H, Arrizubieta I, Calleja A, Ruiz JE, Lamikiz A. CAM development for additive manufacturing in turbo-machinery components. Procedia Manufacturing. 2017;**13**:802-809

[14] Kruth JP. Material Incress manufacturing by rapid prototyping techniques. CIRP Annals. 1991;**40**(2):603-614

[15] Plocher J, Panesar A. Review on design and structural optimisation in additive manufacturing: Towards next-generation lightweight structures. Materials & Design. 2019;**183**:108164

[16] Sauerwein M, Doubrovski E, Balkenende R, Bakker C. Exploring the potential of additive manufacturing for product design in a circular economy. Journal of Cleaner Production. 2019;**226**:1138-1149

[17] Gibson I, Rosen D, Stucker B, Khorasani M. Design for Additive Manufacturing. In: Gibson I, Rosen D, Stucker B, Khorasani M, editors. Additive Manufacturing Technologies [Internet]. Cham: Springer International Publishing; 2021. pp. 555-607. DOI: 10.1007/978-3-030-56127-7_19

[18] Ahmed N. Direct metal fabrication in rapid prototyping: A review. Journal of Manufacturing Processes. 2019;**42**:167-191

[19] Leary M. Design for Additive Manufacturing. 1st ed. Amsterdam Kidlington, Oxford Cambridge, MA: Elsevier; 2019. p. 358

[20] Gebisa AW, Lemu HG. Design for manufacturing to design for additive manufacturing: Analysis of implications for design optimality and product sustainability. Procedia Manufacturing. 2017;**13**:724-731

[21] Großmann A, Weis P, Clemen C, Mittelstedt C. Optimization and re-design of a metallic riveting tool for additive manufacturing—A case study. Additive Manufacturing. 2020;**31**:100892

[22] Suárez A, Veiga F, Bhujangrao T, Aldalur E. Study of the Mechanical Behavior of Topologically Optimized Arc Wire Direct Energy Deposition Aerospace Fixtures. Journal of Materials Engineering and Performance. 1 Aug 2022;**31**(8):6270-6282. DOI: 10.1007/s11665-022-06702-x

[23] Veiga F, Suárez A, Aldalur E, Goenaga I, Amondarain J. Wire arc additive manufacturing process for topologically optimized aeronautical fixtures. 3D Printing and Additive Manufacturing. 2023;**10**(1):23-33

[24] Veiga F, Bhujangrao T, Suárez A, Aldalur E, Goenaga I, Gil-Hernandez D. Validation of the mechanical behavior of an aeronautical fixing turret produced by a Design for Additive Manufacturing (DfAM). Polymers. 2022;**14**(11):2177

[25] Boschetto A, Bottini L, Eugeni M, Cardini V, Nisi GG, Veniali F, et al. Selective laser melting of a 1U CubeSat structure. Design for Additive Manufacturing and assembly. Acta Astronautica. 2019;**159**:377-384

[26] Oyesola MO, Mpofu K, Mathe NR, Daniyan I. Development of an integrated design methodology model for quality and throughput of additive manufacturing processes. Procedia CIRP. 2019;**84**:688-693

[27] Walton D, Moztarzadeh H. Design and development of an additive manufactured component by topology optimisation. Procedia CIRP. 2017;**60**:205-210

[28] Greer C, Nycz A, Noakes M, Richardson B, Post B, Kurfess T, et al. Introduction to the design rules for metal big area additive manufacturing. Additive Manufacturing. 2019;**27**:159-166

[29] Nguyen L, Buhl J, Bambach M. Continuous Eulerian tool path strategies for wire-arc additive manufacturing of rib-web structures with machine-learning-based adaptive void filling. Additive Manufacturing. 2020;**35**:101265

[30] Bi G, Sun CN, Gasser A. Study on influential factors for process monitoring and control in laser aided additive manufacturing. Journal of

Materials Processing Technology. 2013;**213**(3):463-468

[31] Fleischer J, Denkena B, Winfough B, Mori M. Workpiece and tool handling in metal cutting machines. CIRP Annals. 2006;**55**(2):817-839

[32] Murray PE. Selecting parameters for GMAW using dimensional analysis. Welding Journal-New York. 2002;**81**(7):125-12S

[33] Ding D, He F, Yuan L, Pan Z, Wang L, Ros M. The first step towards intelligent wire arc additive manufacturing: An automatic bead modelling system using machine learning through industrial information integration. Journal of Industrial Information Integration. 2021;**23**:100218

[34] Venkatarao K. The use of teaching-learning based optimization technique for optimizing weld bead geometry as well as power consumption in additive manufacturing. Journal of Cleaner Production. 2021;**279**:123891

[35] Dinovitzer M, Chen X, Laliberte J, Huang X, Frei H. Effect of wire and arc additive manufacturing (WAAM) process parameters on bead geometry and microstructure. Additive Manufacturing. 2019;**26**:138-146

[36] Tang ZJ, Liu WW, Wang YW, Saleheen KM, Liu ZC, Peng ST, et al. A review on in situ monitoring technology for directed energy deposition of metals. International Journal of Advanced Manufacturing Technology. 2020;**108**(11-12):3437-3463

[37] Li Y, Sun Y, Han Q, Zhang G, Horváth I. Enhanced beads overlapping model for wire and arc additive manufacturing of multi-layer multi-bead metallic parts. Journal of Materials Processing Technology. 2018;**252**:838-848

[38] Li F, Chen S, Shi J, Zhao Y, Tian H. Thermoelectric cooling-aided bead geometry regulation in wire and arc-based additive manufacturing of thin-walled structures. Applied Sciences. 2018;**8**(2):207

[39] Uralde V, Veiga F, Aldalur E, Suarez A, Ballesteros T. Symmetry and its application in metal additive manufacturing (MAM). Symmetry. 2022;**14**(9):1810

[40] Uralde V, Veiga F, Suarez A, Aldalur E, Ballesteros T. Symmetry Analysis in Wire Arc Direct Energy Deposition for Overlapping and Oscillatory Strategies in Mild Steel. Symmetry. Jun 2023;**15**(6):1231

[41] Curiel D, Veiga F, Suarez A, Villanueva P. Methodology for the Path Definition in Multi-Layer Gas Metal Arc Welding (GMAW). Symmetry. Feb 2023;**15**(2):268

[42] Wang C, Bai H, Ren C, Fang X, Lu B. A comprehensive prediction model of bead geometry in wire and arc additive manufacturing. Journal of Physics: Conference Series. 2020;**1624**(2):022018

[43] Veiga F, Arizmendi M, Suarez A, Bilbao J, Uralde V. Different path strategies for directed energy deposition of crossing intersections from stainless steel SS316L-Si. Journal of Manufacturing Processes. 2022;**84**:953-964

[44] Uyen TMT, Minh PS, Nguyen VT, Do TT, Nguyen VT, Le MT, et al. Trajectory strategy effects on the material characteristics in the WAAM technique. Micromachines. 2023;**14**(4):827

[45] Farouki RT, Koenig T, Tarabanis KA, Korein JU, Batchelder JS. Path planning with offset curves for layered fabrication processes. Journal of Manufacturing Systems. 1995;**14**(5):355-368

[46] Chiu WK, Yeung YC, Yu KM. Toolpath generation for layer manufacturing of fractal objects. Rapid Prototyping Journal. 2006;**12**(4):214-221

[47] Wasser T, Jayal AD, Pistor C. Implementation and Evaluation of Novel Buildstyles in Fused Deposition Modeling (FDM). Solid Freeform Fabrication Symposium Proceedings. Salt Lake City: University of Utah; Jan 1999. pp. 95-102

[48] Jin W, Zhang C, Jin S, Tian Y, Wellmann D, Liu W. Wire Arc Additive Manufacturing of Stainless Steels: A Review. Applied Sciences. Jan 2020;**10**(5):1563

[49] Jancar L, Pagac M, Mesicek J, Stefek P. Design procedure of a topologically optimized scooter frame part. Symmetry. 2020;**12**(5):755

[50] Ueno A, Guo H, Takezawa A, Moritoyo R, Kitamura M. Temperature distribution design based on variable lattice density optimization and metal additive manufacturing. Symmetry. 2021;**13**(7):1194

[51] Uralde V, Suarez A, Aldalur E, Veiga F, Ballesteros T. Wall fabrication by direct energy deposition (DED) combining mild steel (ER70) and stainless steel (SS 316L): Microstructure and mechanical properties. Materials. 2022;**15**(17):5828

[52] Bandyopadhyay A, Traxel KD, Lang M, Juhasz M, Eliaz N, Bose S. Alloy design via additive manufacturing: Advantages, challenges, applications and perspectives. Materials Today. 2022;**52**:207-224

[53] Rajesh Kannan A, Mohan Kumar S, Pramod R, Pravin Kumar N, Siva Shanmugam N, Palguna Y. Microstructure and mechanical properties of wire arc additive manufactured bi-metallic structure. Science and Technology of Welding and Joining. 2021;**26**(1):47-57

[54] Wu B, Qiu Z, Pan Z, Carpenter K, Wang T, Ding D, et al. Enhanced interface strength in steel-nickel bimetallic component fabricated using wire arc additive manufacturing with interweaving deposition strategy. Journal of Materials Science & Technology. 2020;**52**:226-234

[55] Kotoban D, Aramov A, Tarasova T. Possibility of multi-material laser cladding fabrication of nickel alloy and stainless steel. Physics Procedia. 2016;**83**:634-646

[56] Devendranath Ramkumar K et al. Influence of filler metals and welding techniques on the structure-property relationships of Inconel 718 and AISI 316L dissimilar weldments. Materials and Design. 2014;**62**:175-188

[57] Hinojos A et al. Joining of Inconel 718 and 316 stainless steel using electron beam melting additive manufacturing technology. Materials and Design. 2016;**94**:17-27

[58] Moritz T et al. Hybridization of materials and processes by additive Manufacturing. In: Chua CK, Yeong WY, Tan MJ, Liu E, Tor SB, editors. Proceedings of the 3rd International Conference on Progress in Additive Manufacturing (Pro-AM 2018). Nanyang Technological University. 2018. ISSN: 2424-8967. DOI: 10.25341/D48W20

[59] Subramaniyan M, Arasappan R, Nachimuthu PK, Pramod R, Shanmugam N, Vishnu AS, et al. Microstructural Features and Mechanical Integrity of Wire Arc Additive Manufactured SS321/Inconel 625 Functionally Gradient Material.

Journal of Materials Engineering and Performance. 12 Mar 2021

[60] Nazir A, Gokcekaya O, Md Masum Billah K, Ertugrul O, Jiang J, Sun J, et al. Multi-material additive manufacturing: A systematic review of design, properties, applications, challenges, and 3D printing of materials and cellular metamaterials. Materials & Design. 2023;**226**:111661

[61] Carroll BE, Otis RA, Borgonia JP, Suh JO, Dillon RP, Shapiro AA, et al. Functionally graded material of 304L stainless steel and inconel 625 fabricated by directed energy deposition: Characterization and thermodynamic modeling. Acta Materialia. 2016;**108**:46-54

[62] Ahsan MRU, Tanvir ANM, Ross T, Elsawy A, Oh MS, Kim DB. Fabrication of bimetallic additively manufactured structure (BAMS) of low carbon steel and 316L austenitic stainless steel with wire + arc additive manufacturing. Rapid Prototyping Journal. 2020;**26**(3):519-530

[63] Karunakaran KP. Hybrid manufacturing. In: Pei E, Bernard A, Gu D, Klahn C, Monzón M, Petersen M, et al., editors. Springer Handbook of Additive Manufacturing [Internet]. Cham: Springer International Publishing; 2023. pp. 425-441. Available from:. DOI: 10.1007/978-3-031-20752-5_26

[64] Spierings AB, Klahn C. Benefits using additive manufacturing. In: Pei E, Bernard A, Gu D, Klahn C, Monzón M, Petersen M, et al., editors. Springer Handbook of Additive Manufacturing [Internet]. Cham: Springer International Publishing; 2023. pp. 115-124. Available from:. DOI: 10.1007/978-3-031-20752-5_8

[65] Weflen E, Frank MC. Hybrid additive and subtractive manufacturing of multi-material objects. Rapid Prototyping Journal. 1 Jan 2021;**27**(10):1860-1871

[66] Nyamuchiwa K, Palad R, Panlican J, Tian Y, Aranas C. Recent Progress in hybrid additive manufacturing of metallic materials. Applied Sciences. 2023;**13**(14):8383

[67] Singh HV, Verma GC. Optimizing time in additive manufacturing: A hybrid fused deposition modeling and photopolymerization approach. Rapid Prototyping Journal. 1 Jan 2024;**30**(7):1269-1286

[68] He F, Yuan L, Mu H, Ros M, Ding D, Pan Z, et al. Research and application of artificial intelligence techniques for wire arc additive manufacturing: A state-of-the-art review. Robotics and Computer-Integrated Manufacturing. 2023;**82**:102525

[69] Fernández-Zabalza A, Veiga F, Suárez A, López JRA. The use of virtual sensors for bead size measurements in wire-arc directed energy deposition. Applied Sciences. 2024;**14**(5):1972

[70] Teti R, Kumara SRT. Intelligent computing methods for manufacturing systems. CIRP Annals. 1997;**46**(2):629-652

[71] Veiga F, Suarez A, Aldalur E, Artaza T. Wire arc additive manufacturing of invar parts: Bead geometry and melt pool monitoring. Measurement. 2022;**189**:110452

[72] Wang Q, Jiao W, Wang P, Zhang Y. Digital twin for human-robot interactive welding and welder behavior analysis. IEEE/CAA Journal of Automatica Sinica. 2021;**8**(2):334-343

[73] Li Y, Mu H, Polden J, Li H, Wang L, Xia C, et al. Towards intelligent monitoring system in wire arc additive

manufacturing: A surface anomaly detector on a small dataset. International Journal of Advanced Manufacturing Technology. 2022;**120**(7):5225-5242

[74] Bacioiu D, Melton G, Papaelias M, Shaw R. Automated defect classification of aluminium 5083 TIG welding using HDR camera and neural networks. Journal of Manufacturing Processes. 2019;**45**:603-613

[75] Reisch R et al. Distance-Based Multivariate Anomaly Detection in Wire Arc Additive Manufacturing. 19th IEEE International Conference on Machine Learning and Applications (ICMLA), Miami, FL, USA; Dec 2020. pp. 659-664. DOI: 10.1109/ICMLA51294.2020.00109

[76] Kwon O, Kim HG, Ham MJ, Kim W, Kim GH, Cho JH, et al. A deep neural network for classification of melt-pool images in metal additive manufacturing. Journal of Intelligent Manufacturing. 2020;**31**(2):375-386

[77] Kumar S, Gopi T, Harikeerthana N, Gupta MK, Gaur V, Krolczyk GM, et al. Machine learning techniques in additive manufacturing: A state of the art review on design, processes and production control. Journal of Intelligent Manufacturing. 2023;**34**(1):21-55

[78] Era IZ, Farahani MA, Wuest T, Liu Z. Machine learning in directed energy deposition (DED) additive manufacturing: A state-of-the-art review. Manufacturing Letters. 2023;**35**:689-700

[79] Carpanzano E, Knüttel D. Advances in artificial intelligence methods applications in industrial control systems: Towards cognitive self-optimizing manufacturing systems. Applied Sciences. 2022;**12**(21):10962

[80] Qin J, Hu F, Liu Y, Witherell P, Wang CCL, Rosen DW, et al. Research and application of machine learning for additive manufacturing. Additive Manufacturing. 2022;**52**:102691

[81] Reimann J, Hammer S, Henckell P, Rohe M, Ali Y, Rauch A, et al. Directed energy deposition-arc (DED-arc) and numerical welding simulation as a hybrid data source for future machine learning applications. Applied Sciences. 2021;**11**(15):7075

[82] Ahn DG. Directed energy deposition (DED) process: State of the art. International Journal of Precision Engineering and Manufacturing-Green Technology. 2021;**8**(2):703-742

[83] Lund H, Penttilä S, Skriko T. Extended reality implementation possibilities in direct energy deposition-arc. Frontiers in Sustainability. 19 Jun 2024;(5):1408604

Chapter 2

High-k, Low-Loss Ceramic-Thermoplastic Composite Feedstock Filaments for Fused Deposition Modeling of Microwave and mm-Wave Devices

Vishvajitsinh Kosamiya and Jing Wang

Abstract

Maturing of additive manufacturing (AM) techniques has increased their utilization for fabricating radio frequency (RF) and microwave devices. Solid composites used in material extrusion AM have experienced considerable expansion over the past decade, incorporating functional properties into 3D-printed objects. There are encouraging indications from AM material research that electrically efficient AM materials can be discovered. These materials would be useful for producing microwave components in the future. One of the enabling techniques for fabricating these materials is to incorporate nano/microparticles or fillers into thermoplastic material. Composite material 3D printing is a novel approach to managing materials' microwave properties. While extrinsic qualities (effective permittivity) can be controlled by shape and porosity management, intrinsic attributes are tied to the composition of composites. Furthermore, combining various materials to increase the spectrum of available microwave characteristics is made possible by multi-material 3D printing. In this chapter, we explore different methodologies to fabricate ceramic/thermoplastic composites for fused deposition modeling (FDM) of RF and microwave devices. Analytical models for predicting effective permittivity of the composite are discussed and application examples of FDM printed RF, microwave and mm-wave devices employing composites are presented.

Keywords: ceramic-thermoplastic composite, fused deposition modeling, high-k low-loss dielectrics, dielectric constant, dielectric loss tangent

1. Introduction

Research on electromagnetic (EM) devices and microwave circuits has historically mostly focused on two-dimensional designs that complement established printed circuit board (PCB) technologies [1, 2]. Additive manufacturing (AM) has gained a lot of traction for creating small quantities of highly specialized devices, whereas printed circuit board (PCB) technology has been thoroughly refined and optimized for mass-producing affordable consumer electronics products. Though it cannot fully replace traditional manufacturing for mass production, additive manufacturing (AM) presents a disruptive alternative in businesses that require high levels of design customization and small production volumes [3, 4].

In recent years, there has been a rise in study on AM in the field of electrical engineering [5–7]. By removing the limitations on possible shapes that were previously imposed by traditional production techniques, the third dimension made possible by AM provides substantial benefits [8–13]. Using a graded material (easier to produce using AM) to achieve a required EM response allows the geometry to be optimized for other reasons, such as aerodynamics. Some of the main types of 3D printing methods used for fabricating radio frequency (RF) and microwave components include fused deposition modeling (FDM) [14], poly-jet printing, stereolithographic apparatus (SLA) [15], selective laser sintering (SLS) [16] and selective laser melting (SLM) [17].

One popular AM method that is well-known for its ease of use and adaptability is FDM [18, 19]. This technique is fundamental to the field of 3D printing since it provides a useful and affordable way to bring complex ideas to life. In FDM process, the printer heats a material (thermoplastic or composites made from thermoplastic) above its glass transition temperature and then deposits it through a nozzle, building an object layer by layer, as shown in **Figure 1**. Although FDM has comparatively poor resolution control, it is compatibility with a wider range of available material and their composites makes it suitable for prototyping and experimenting with different varieties of recipes [5, 20, 21]. Furthermore, FDM is normally a single step process, not requiring and post-processing steps. Depending on the quality of the printer mechanics and the use of the proper printer settings, FDM can achieve a degree of precision of about 0.1 to 0.2 mm.

The FDM printers can have more than one extrusion nozzle to support printing of multi-material parts [22]. FDM can be considered as the most versatile AM technology for microwave component manufacturing. High design flexibility is made possible by

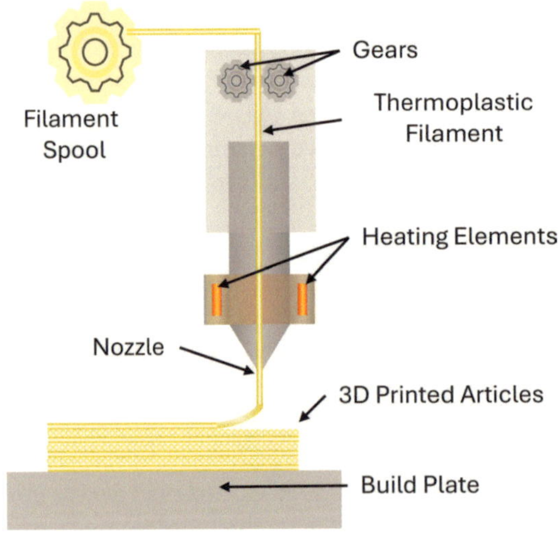

Figure 1.
Schematic representation of fused deposition modeling (FDM) process.

the substantial size of the FDM material catalog. Additionally, FDM prototyping is a popular option for RF engineers because of its inexpensive cost [23].

However, the possibility of creating extremely small devices is limited by the moderate dielectric constant (ε_r~2–3) displayed by the most popular printable filaments, as well as the extra reduction brought on by the inevitable interstitial air gaps between the rods extruded during the printing process [24, 25]. For electronics engineering, especially RF and microwave devices permittivity of the material matters significantly because of its influence on guided wavelength inside the material, as seen in the following equation,

$$\lambda_g = \frac{c_0}{f\sqrt{\varepsilon_r}} \tag{1}$$

where λ_g is the guided wavelength, C_0 is the speed of light in vacuum. As Eq. (1) suggests that for a signal with a given frequency, higher permittivity material will cause it to have a smaller wavelength. This relation is widely adopted by antenna engineers to miniaturize antennas for RF and microwave applications [20, 26, 27]. At the microscopic level, several dielectric mechanisms can contribute to the dielectric constant of a given material [28]. Dipole orientation and ionic conduction interact strongly at microwave frequencies, as shown in **Figure 2**. For instance, water molecules rotate in response to an alternating electric field because they are permanent dipoles. Food warms in a microwave oven because these mechanisms are quite lossy. Over the microwave area, atomic and electronic mechanisms are typically constant and rather weak.

To increase low permittivity of common thermoplastic materials, researchers have come up with ways to disperse ceramic particles in thermoplastic matrix and utilize it to fabricate RF and microwave devices [5, 20, 21, 26, 29, 30]. This chapter is subdivided into five sections. Section I introduces relative permittivity of materials and why it matters to RF and microwave engineers as well as FDM process. Different fabrication methods of ceramic/thermoplastic composite filaments for FDM of RF and

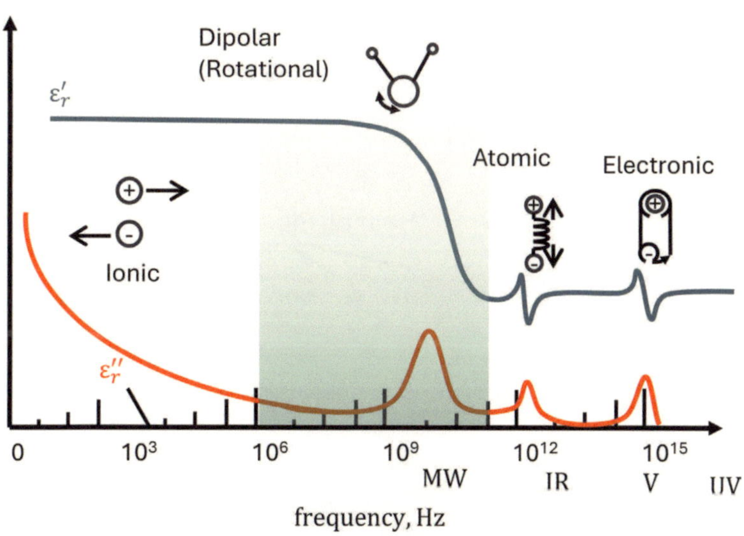

Figure 2.
Frequency responses of dielectric mechanisms [28]. The highlighted green area represents RF, microwave, and mm-wave frequency range.

microwave devices in literature are discussed in section II. Section III explores different analytical models available for prediction of the effective permittivity of a composite material based on permittivity values and volume fraction of constituent materials. Experimental demonstrations of microwave and mm-wave devices 3D printed using ceramic/thermoplastic feedstock filaments are discussed in section IV. Finally, section V presents concluding remarks and an outlook on future research in this topic.

2. Synthesis methods

Polymer-based composites can be used to combine the processability of the polymer matrix with the EM characteristics of discrete filler particles using predetermined mixing methods [31, 32]. For composites to act as a homogenous medium in terms of electromagnetic energy transmission, the wavelength inside the medium must be substantially larger than the scale of the particles or particle agglomerates [33, 34]. This condition can be easily met with well-dispersed particles of 10^{-6} to 10^{-4} m exposed to microwaves with free-space wavelengths higher than 10^{-2} m, resulting in non-resonant bulk materials appropriate for broadband applications.

Synthesis methods can be characterized based on the method of mixing the filler particles inside the thermoplastic matrix: (1) melt mixing (MM) and (2) solution-assisted dispersion (SD) [35, 36].

2.1 Melt mixing

This technique involves heating the thermoplastic matrix to its melting point, followed by the addition of filler particles to the semi-liquid thermoplastic and stirring with a mechanical screw [37], as illustrated in **Figure 3**. With an increase in filler particle volume, it becomes more difficult to distribute the filler particles in the melted polymer due to a decrease in the solution's viscosity. Melt mixing can produce consistent results for a filler volume fraction of 0.3 or less [31, 38–40]. Nevertheless, no additional chemicals are needed for this method, and process adjustments can be made as thermoplastic is being extruded in the FDM filament.

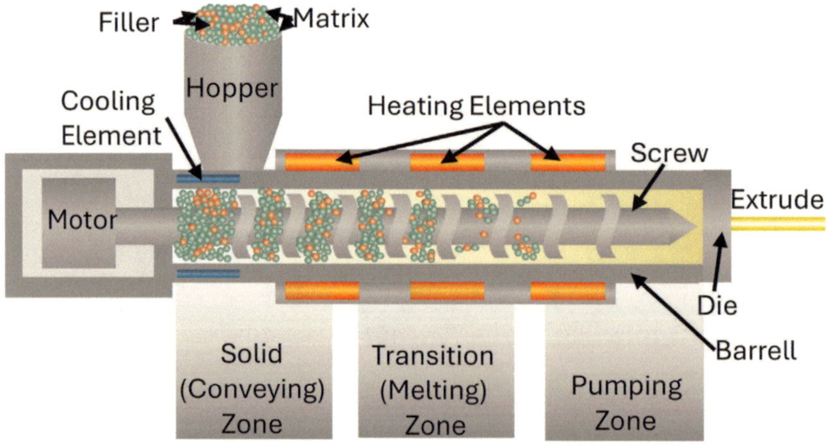

Figure 3.
Schematic representation of melt mixing process using a single screw extruder.

An extruder essentially consists of a steel screw that rotates inside a heated barrel, like an auger. Thermoplastic resin, in the form of pellets, is fed to the screw through a hopper, and the screw compacts, melts and pressurizes the polymer. After the tip of the screw, a die is attached to the barrel, and the high-pressure melt is forced through the die and extruded into desired shape. The screw has many functions, which include compressing the solid pellets, heating the resin through mechanical deformation, pressurizing the melt, and mixing. Using special screw geometries, mixing can be achieved on a single screw extruder, but it is generally recognized that the best mixing is achieved on twin screw extruders [37].

2.2 Solution assisted dispersion

The SD method is broken into multiple sub-steps: (i) agitation to disperse filler particles in a solvent, (ii) agitation to mix filler and polymer solutions, and (iii) controlled solvent evaporation and/or nanocomposite precipitation. The process diagram of the SD method is shown in **Figure 4**. In this method, the thermoplastic is first dissolved using a chemical solvent at room temperature, and then filler particles are added to liquid solution and stirred. After a homogeneous dispersion is achieved, the solvent is evaporated at 60°C to leave behind just the thermoplastic polymer and filler

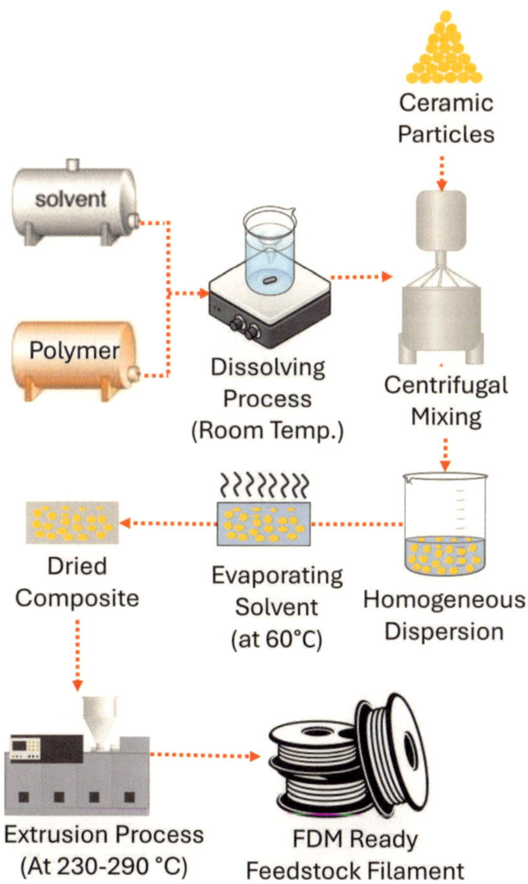

Figure 4.
Process flow diagram of solution-assisted dispersion (SD) method.

particles [21, 30, 41, 42]. This process used by authors in [43] with extrusion temperatures in the range of 230–290°C, can achieve repeatable results up to 50% filler volume concentration.

However, this process requires additional processing steps, and if the solvent is not reused, it could become expensive. Residual solvent inside final composite can also have a negative impact on the dielectric and loss properties of the composite. A key challenge is to keep the filler particles from agglomeration during the process of evaporating the solvent.

Both MM and SD methods could be modified by introducing a surfactant either during or before the mixing process [5, 21, 31, 43]. The surfactant could perform a dual function of improving the adhesion between filler and matrix as well as hindering the filler particles from agglomerating. Care must be taken to optimize the quantity of the surfactant to maintain both mechanical properties required for 3D printing as well as dielectric properties for EM applications [40, 41].

2.3 Examples of works in literature

Castles et al. [21] used the SD method to disperse $BaTiO_3$ (BT) particles in acrylonitrile butadiene styrene (ABS) matrix up to 29 vol% to achieve relative permittivity values in the range of 2.6–8.7 and loss tangents in the range 0.005–0.027. They also fabricated 1D and 2D periodic structures to demonstrate the feasibility of 3D printing; however, the EM functionalities of those prototypes were not discussed.

In a sequential study, Wu et al. [41] described the production of a better polymer-ceramic composite feedstock filament for utilization in FDM 3D printing. The preceding formulation was enhanced by adding a surfactant and plasticizer. The optimum ratio of surfactant and plasticizer was found by doing a parametric study of their effects on dielectric permittivity. Filament with a maximum solid loading of 32 vol% was manufactured; its dielectric permittivity and loss tangents measured at 15 GHz using a split post-dielectric resonator were found to be 11 and 0.0289, respectively.

Khatri et al. [40] also fabricated ABS-BT composite; however, they utilized the MM method to disperse BT particles in the ABS matrix. They also utilized 1.1 wt % steric acid as a surfactant to improve adhesion between inorganic filler and organic matrix. They obtained a relative permittivity of 11.5 for the composite containing 35 vol% BT. This study's findings were consistent with the Maxwell-Garnett [32, 44] and Jayasundere-Smith [45] mixing models.

To improve the overall dielectric loss tangent of composite feedstock, Castro et al. [29] used cyclo-olefin polymer (COP) due to its superb electromagnetic properties with $\varepsilon_r \sim 2.3$ and $\tan(\delta_d) < 0.0005$ at 17 GHz. They employed the MM method to disperse TiO_2, $MgCaTiO_2$ (MCT), and $Ba_{0.55}Sr_{0.45}TiO_3$ (BST) particles into COP matrix. They were able to achieve a relative permittivity value of 4.88 for dielectric composite with 25% volume loading of BST in COP with a low-loss tangent of 0.007 at 17 GHz. For composite with 30% volume loading of TiO_2, they were able to achieve a low-loss tangent of 0.0014 and permittivity of 4.57. One of the key differences was that they also performed co-firing of ceramic particles before dispersing them in COP matrix. This helped in increasing crystallinity of ceramic particles and hence improving loss performance of composite.

Czarny et al. [46] used cyclic olefin copolymer (COC) polymer ($\varepsilon_r \sim 2.3$, $\tan(\delta_d) \sim 1.10^{-3}$) and loaded by titanium oxide (TiO_2) ($\varepsilon_r \sim 100$, $\tan(\delta_d) < 1.10^{-4}$), at 36 vol%. This innovative thermoplastic composite material has a high permittivity of up to 9.2 and a low-loss tangent of 1.10^{-3} at 9.3GHz. They used spray-drying to produce titanium oxide grains on a micrometer size. They also sintered (TiO_2) particles at the optimized temperature of 1100°C. However, they fail to mention the method that was used to disperse ceramic filler in thermoplastic matrix.

Mazzei Capote et al. [5] used the SD method to create ABS and BT-based composites, which included the use of a surfactant and a plasticizer to aid processing. The ABS-based composite contained 75% BT (nominally 35% by volume, based on density), 5% dibutyl phthalate, and 1% octyl gallate. They were able to obtain a record high permittivity value of 12.85 in the 10 to 12 GHz band. The composite had a high-loss tangent (tan (δ_d) = 0.046), perhaps due to ABS's poor loss attributes. They implemented the use of ABS-based composite in the manufacturing process of microwave photonic devices. They 3D printed a 2D photonic crystal and experimentally verified its transmission performance. Furthermore, a prototype 3D photonic crystal was also 3D printed; however, it had insufficient unit cells to function properly.

To the best of our knowledge, authors [30] used SD method for the first time to disperse BST, MCT and TiO_2 particles into COP matrix. We also employed silane surfactant to improve adhesion between inorganic filler and organic polymer. The maximum measured ceramic particle filler volume concentration of 50% resulted in relative permittivity values of up to 12.35 and a low-loss tangent of 0.003. Ceramic particles were co-fired before surfactant-assisted ball milling to improve their dielectric and loss properties. Two functional dielectric waveguides were fabricated and characterized using developed composite and commercial counterpart [43].

Table 1 summarizes the state-of-the-art works in literature for developing ceramic-thermoplastic feedstock filament with high-k, low-loss properties. As can be seen, that SD method generally allows for higher permittivity values due to homogeneous dispersion of ceramic particles in matrix compared to MM method.

3. Analytical model predictions

The emergence of RF and microwave communication necessitated the development of substrate materials with tailorable dielectric and mechanical properties. Traditional dielectric materials cannot meet new requirements, such as balanced dielectric characteristics and good mechanical flexibility, as electronics become more multi-functional, integrated, and miniaturized. As a result of their excellent processibility, flexibility, and dielectric qualities, high dielectric polymer composites have sparked widespread interest. As the interaction between EM energy and matter are heavily influenced by the compositions and internal structure of the material, the dielectric properties of the composite have a direct impact on all applications in electronics. Hence there has been much research throughout history toward tackling the problem of rationalizing and predicting electrical properties of composite materials.

Maxwell and Faraday first studied formulae for the permittivity of heterostructures in the mid-nineteenth century as part of their comprehensive electric and magnetic research [47]. The models currently in literature can be broadly characterized in two categories: (1) mixing rule and (2) effective medium theory (EMT). This chapter focuses on the various cases in which composite regions are large enough to contain macroscopic fields. The advanced models based on percolation theory and complex electrodynamic-based models are not discussed here, but interested readers can find relevant information in the references.

Ref.	Matrix	Filler	Filler Fraction	Mixing Method	Surface Modification Applied.	Permittivity @GHz	Loss tangent @GHz
[27]	COC	TiO_2	36 vol %	n/a	No	7.7@37.1GHz	0.001 @37.1GHz
[5]	ABS	$BaTiO_3$	35 vol %	SD	Yes	12.85@12 GHz	0.046 @12GHz
[42]	ABS	$BaTiO_3$ $SrTiO_3$ $Ba_{0.67}Sr_{0.33}TiO_3$	30 vol %	SD	Yes	8.39 8.18 8.02 @18-40GHz	0.016 <0.01 0.069 @18-40GHz
[38]	ABS	$CaCu_3Ti_4O_{12}$	40 wt %	MM	No	9.6@18GHz	0.0057 @18GHz
[43]	COP	TiO_2 (Fired 1100°C)	50 vol %	SD	Yes	12@16GHz	0.003 @16GHz
[30]	COP	$Ba_{0.55}Sr_{0.45}TiO_3$ (Fired 1100°C) $MgCaTiO_2$ (Fired 1100°C)	50 vol %	SD	Yes	8.46@16GHz 12.35@12GHz	0.0099 @16GHz 0.0033 @12GHz
[41]	ABS	$BaTiO_3$	32 vol %	SD	Yes	11@15GHz	0.03 @15GHz
[29]	COP	$MgCaTiO_2$ (Fired 1200°C) TiO_2 (Fired 1200°C)	30 vol %	MM	No	4.78 4.82 @17GHz	0.0012 0.0018 @17GHz
[21]	ABS	$BaTiO_3$	29 vol %	SD	No	8.72@14.13 GHz	0.027 @14.13 GHz
[26]	ABS	$BaSrTiO_3$	15 vol %	SD	No	6.05@5GHz	0.007 @5GHz
[39]	PLA	GNP CNT	6%wt	MM & SD	No	12.73 (ME) @26–37 GHz 11.84(SB) @26–37 GHz	n/a

Table 1.
Comparison of literature results on ceramic-thermoplastic feedstock filaments used for FDM printing of RF and microwave devices.

To begin with mixing rules, think of a homogeneous medium (matrix) with relative dielectric permittivity ε_m and spheroids with relative permittivity ε_f. The volume fraction of filler spheroid particles is denoted as Θ_f, while the arising volume fraction of the matrix is $\Theta_m = 1 - \Theta_f$. ε_{eff} represents the effective permittivity of composite material. It is assumed that both filler and matrix have no dielectric loss and that they do not interact with one another chemically.

3.1 Mixing rules-based models

The analytical bounds of the effective permittivity of a binary lossless sandwich composite in terms of its constituent parts can be computed by considering two extreme cases of interchanging layers: the first case with the layers being normal to the applied field (**Figure 5b**) and the second case with the layers being tangential to the applied field (**Figure 5a**). The upper bound, which represents parallel model, can be calculated using,

$$\varepsilon_{eff,max} = \varepsilon_m \Theta_m \varepsilon_f \Theta_f \tag{2}$$

The lower bound, which represents series model, can be calculated using,

$$\varepsilon_{eff,min} = \frac{\varepsilon_m \varepsilon_f}{\varepsilon_f \Theta_m + \varepsilon_m \Theta_f} \tag{3}$$

3.1.1 Maxwell-Garnett model

A more accurate equation for predicting dielectric properties can be achieved by the following equation known as Maxwell-Garnett equation [32, 44],

$$\varepsilon_{eff} = \varepsilon_m \left[1 + \frac{3\Theta_f (\varepsilon_f - \varepsilon_m)}{(1 - \Theta_f)(\varepsilon_f - \varepsilon_m) + 3\varepsilon_m} \right] \tag{4}$$

This equation considers the dielectric permittivity of spherical fillers scattered in a medium, as illustrated in **Figure 5c**. In the literature, the Maxwell-Garnett equation is represented in a variety of ways and referred to by various other names, including Maxwell-Wagner [48], Rayleigh [49], and Lorentz-Lorenz [50]. One of such Maxwel-Garnett equation forms is given below,

$$\frac{\varepsilon_{eff} - \varepsilon_m}{\varepsilon_{eff} + 2\varepsilon_m} = \Theta_f \left[\frac{\varepsilon_f - \varepsilon_m}{\varepsilon_f + 2\varepsilon_m} \right] \tag{5}$$

This equation applies solely to spherical particles. For dispersed particles that are not spherical in shape, Eq. (5) must be changed. A popular method for including information about the geometry of scattered particles is to incorporate a depolarization factor that is proportional to their divergence from sphericity. Thus, the Maxwell-Garnett model equation (Eq. (4)) could be altered to this form to make it more general [51],

$$\varepsilon_{eff} = \varepsilon_m \left[1 + \frac{\Theta_f (\varepsilon_f - \varepsilon_m)}{D(1 - \Theta_f)(\varepsilon_f - \varepsilon_m) + \varepsilon_m} \right] \tag{6}$$

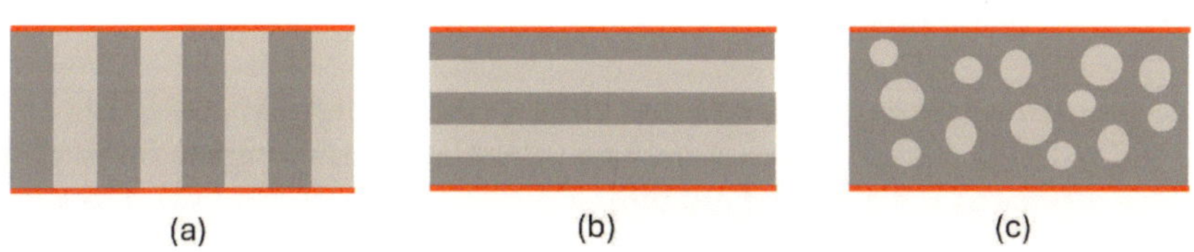

Figure 5.
(a) Parallel model, (b) series model, and (c) mixing model. Red lines are for electrodes.

where parameter D is depolarization factor. For D = 1/3, Eq. (6) simplifies as Eq. (4). Value of D could be computed or can be found in the literature for different geometrical shapes [51].

3.1.2 Lichtenker model

According to Weiner's theory, the lower and upper boundaries are given by eqs. (2) and (3). The higher bound for the effective dielectric permittivity is attained by a system made up of parallel layers arranged tangentially to the applied electric field. The lower bound is achieved by system with the layers that are normal to the applied electrical field. Cross-sectional analysis yields similar results. Starting with eqs. (2) and (3), Lichtenker assumed that the effective dielectric function of the considered composite satisfies the equation:

$$\varepsilon_{eff}^{\alpha} = \varepsilon_f^{\alpha}\Theta_f + \varepsilon_m^{\alpha}\Theta_m \tag{7}$$

where parameter α varies from −1 to 1. The value of −1 for represents anisotropy and value of 1 for α represents isotropy. Each value of α represents a distinct micro-geometrical structure of the composite. One can thus predict that the applicability of Eq. (7) is broader than that of the well-known Maxwell-Garnett equation (Eq. (4)), which, in general, has no free parameters connected to the topology.

3.1.3 Raleigh model

An important advance to the subject matter was made by Lord Rayleigh [52]. The standard assessments of the effective dielectric constant of binary mixes are derived from the Raleigh model. It assumes cylindrical particles mixed in a matrix with effective dielectric constant calculated from the following equation:

$$\frac{\varepsilon_{eff} - \varepsilon_m}{\varepsilon_{eff} + \varepsilon_m} = \Theta_f\left[\frac{\varepsilon_f - \varepsilon_m}{\varepsilon_f + \varepsilon_m}\right] \tag{8}$$

This is quite like Eq. (5). Rayleigh demonstrated that Maxwell's model provides an accurate first-order estimation for cubically ordered spherical and cylindrical fillers in a host matrix. Three things are worth noting: (1) Lord Rayleigh further developed Maxwell's dipole theory [53] of the dispersion of spheres within a medium by considering the contribution of induced octopole moments to conductivity, (2) Rayleigh's model is simpler, but it is equivalent to the Maxwell-Garnett model, and (3) Rayleigh's model revealed an equivalence of electrical conduction and acoustic propagation.

3.1.4 Jaysundere: Smith equation

At greater particle concentrations, the interaction among particles is more important due to the shorter distance between them, as seen in **Figure 6b**. Jaysundere-Smith [45] estimated the electric field with a dielectric sphere contained in a continuous dielectric matrix, considering the polarization of consecutive particles. According to Jaysundere and Smith, effective relative permittivity can be given *via* equation,

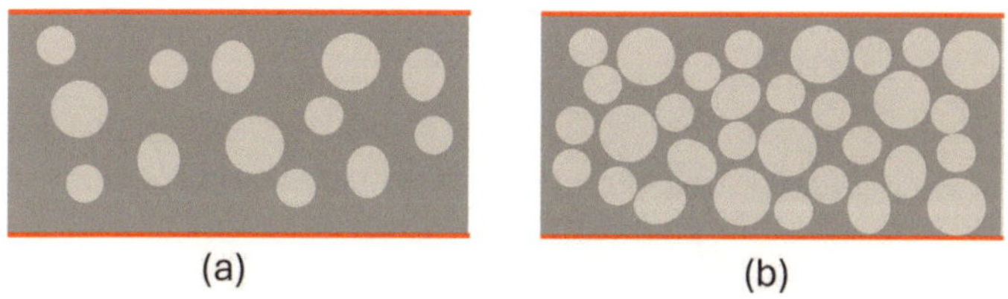

Figure 6.
(a) Composite with lower filler volume fraction and (b) composite with higher filler volume fraction.

$$\varepsilon_{eff} = \frac{\varepsilon_m \Theta_m + \varepsilon_f \Theta_f \left(\frac{3\varepsilon_m}{2\varepsilon_m + \varepsilon_f}\right)\left[1 + 3\Theta_f \frac{(\varepsilon_f - \varepsilon_m)}{2\varepsilon_m + \varepsilon_f}\right]}{\Theta_m + \Theta_f \left(\frac{3\varepsilon_m}{2\varepsilon_m + \varepsilon_f}\right)\left[1 + 3\Theta_f \frac{(\varepsilon_f - \varepsilon_m)}{2\varepsilon_m + \varepsilon_f}\right]} \tag{9}$$

3.2 Effective medium theory-based models

Effective medium theories (EMT) apply when the fields are quasistatic, meaning that the wavelength of the electromagnetic radiation probing the system is substantially larger than the typical scale of inhomogeneities. The frequency also defines the nature of the microscopic mechanisms within a material that govern the bulk material's macroscopic polarization (and magnetization) response. Polarization in materials can be caused by a variety of mechanisms, including electronic, ionic, atomic, dipolar, and interfacial, as detailed in the introduction.

3.2.1 Hanai-Bruggman model

The Bruggeman model [54], which was also developed by Hanai [55], is a popular model for calculating the relative permittivity of EM composites. It is cited as the Hanai-Bruggeman model [56]. One advantage of this model is that it assumes filler particles to be very close to each other; this is indicated by the following equation,

$$\frac{\varepsilon_f - \varepsilon_{eff}}{\varepsilon_f - \varepsilon_m}\left(\frac{\varepsilon_m}{\varepsilon_{eff}}\right)^{\frac{1}{3}} 1 - \Theta_f \tag{10}$$

The Hanai-Bruggeman model can accurately predict permittivities of composites with volume fractions of particles Θ_f up to around 50%, under the assumption that the scattered filler particles in the composite do not create a percolative path across the matrix.

3.2.2 Stroud-Pan model

Stroud and Pan [57] expanded the EMT to finite frequencies, accounting for particle scattering with a full multipole expansion. In their model, composite is thought of as being made of spherical particles for both matrix and filler as shown in **Figure 7**. Effective permittivity of the composite can be calculated using the following equation,

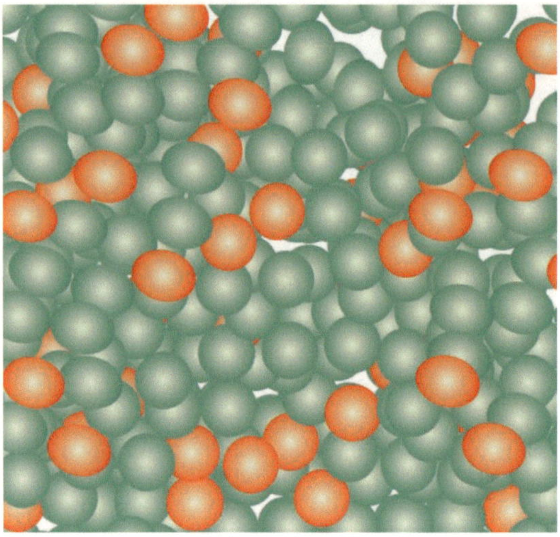

Figure 7.
Composite geometry utilized by Stroud and Pan for computing frequency-dependent permittivity. Red spheres represent filler while green spheres represent matrix representation of fused deposition modeling (FDM) process.

$$\sum_i \Theta_i \left[\frac{\varepsilon_i - \varepsilon_{eff}}{\varepsilon_i + 2\varepsilon_{eff}} + \frac{1}{30} \left(\frac{\omega R_i}{c} \right)^2 \left(\varepsilon_i - \varepsilon_{eff} \right) \right] = 0 \tag{11}$$

where Θ_i is the volume fraction of ith material, ε_i is relative permittivity of ith material, R_i is the radius of ith material particles and ω is angular frequency of EM wave and c is speed of light in vacuum.

For the sake of illustration, assume that the composite is made up of only two types of dielectric particles, ε_1 and ε_2, with volume fractions Θ_i and $\Theta_2 = 1 - \Theta_1$, and that each type of particle is formed into approximately spherical shapes of the same radius R. Then eq. (11) assumes the simple form.,

$$\Theta_1 \left[\frac{\varepsilon_1 - \varepsilon_{eff}}{\varepsilon_1 + 2\varepsilon_{eff}} + \gamma \left(\varepsilon_1 - \varepsilon_{eff} \right) \right] + \Theta_2 \left[\frac{\varepsilon_2 - \varepsilon_{eff}}{\varepsilon_2 + 2\varepsilon_{eff}} + \gamma \left(\varepsilon_2 - \varepsilon_{eff} \right) \right] = 0 \tag{12}$$

where γ equals to $\frac{1}{30}\omega^2 R^2 / c^2$. The above equation looks to be a cubic equation for ε_{eff} which simplifies to a prominent quadratic Eq. (5), in the limit (γ = 0).

Different models for predicting the effective dielectric constants of composite materials are compiled and presented to help researchers in developing new high performance ceramic-thermoplastic composite feedstock filaments for fused deposition modeling of RF and microwave devices.

4. Application of FDM printed ceramic-thermoplastic feedstock filaments for RF and microwave applications

As the FDM technology matured and became more available, more researchers in RF and microwave field started using it to 3D print passive RF and microwave devices with complex geometry [58]. The majority of works focused on all-dielectric RF devices, as it is challenging to fabricate smooth conductor surfaces using FDM method [7].

However, there have been advances in research for providing smooth FDM printed conductive surfaces, recently [59, 60].

Given FDM technology's obvious proclivity for printing dielectric materials, most scientific endeavors that use this technique are mainly focused on manufacturing pure dielectric devices such as reflect arrays, dielectric resonator antennas (DRAs), lenses, and dielectric rod antennas (DRAs). Companies such as Preperm [61] and Zetamix [62] are beginning to cater to the specialty material market for FDM printing of RF devices, which will improve the momentum of research publication and potential commercial application.

Although not exhaustive, authors have tried to compile a list of known RF and microwave device applications presented in literature that use ceramic-thermoplastic feedstock filaments for fused deposition modeling. These works are broadly characterized based on whether the material was developed in lab or commercially purchased.

4.1 Laboratory-made ceramic-thermoplastic composites

Castro et al. [31] demonstrated for the first time 3D printed microwave devices employing cyclo-olefin-polymer (COP)-based composites *via* an FDM process. They also observed the highest relative dielectric permittivity value of $\sim$5 and an ultra-low dielectric loss tangent tan (δ_d) < 0.0014 for COP-based composites. **Figure 8a** illustrates how they used 3D printing to create rectangular edge-fed patch antennas operating at around 17 GHz. The substrates for these antennas were 3D printed using 25% v/v COP-$MgCaTiO_2$ ($\varepsilon_r \sim 4.47$) and pure COP ($\varepsilon_r \sim 2.27$). A 49.35% reduction in size was reported for antenna fabricated with 25% v/v COP-$MgCaTiO_2$ compared to edge-fed patch antenna made of pure COP.

Goulas et al. [26] used $BaSrTiO_2$ in acrylonitrile butadiene styrene (ABS) composite ($\varepsilon_r \sim 6.05$ and tan (δ_d) $\sim$ 0.007 at 5 GHz) with 15.26 vol% filler to 3D print a prototype microstrip patch 5G antenna and a hemispherical dielectric lens, as shown

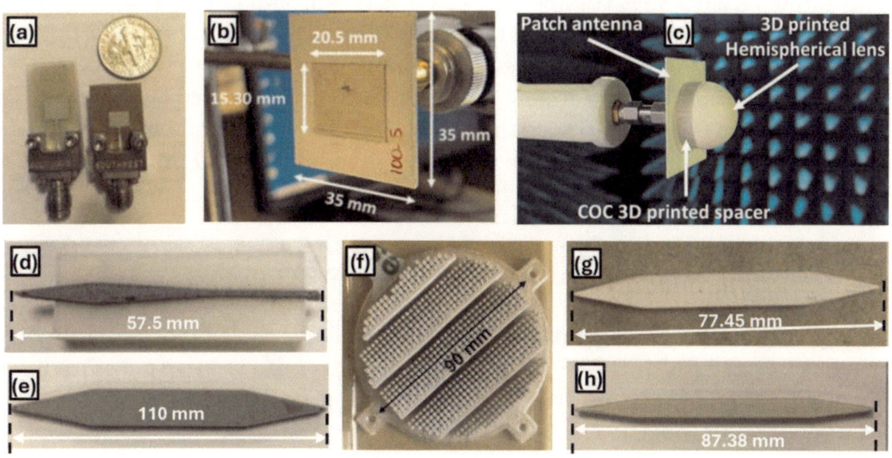

Figure 8.
(a) 3D printed patch antennas over pure COP and 25%v/v COP-$MgCaTiO_2$ substrate [20] (b) microstrip patch 5G antenna on 3D printed 15.26%v/v ABS-$BaSrTiO_2$ substrate [26] (c) hemispherical dielectric lens 3D printed using 15.26%v/v ABS-$BaSrTiO_2$ [26](d) FDM printed DRA using 30%v/v COP- $MgCaTiO_2$ composite filament along with ABS cladding [63](e) FDM printed DRW using 30%v/v COP- $MgCaTiO_2$ composite [64](f) 3D printed subwavelength deflector using 36%v/v COC-TiO_2 composite [27](g)DRW FDM printed using 30%v/v COP- TiO_2 composite filament [30] (h) DRW FDM printed using 40%v/v COP- TiO_2 composite filament [43].

in **Figure 8b** and **c**, respectively. The rectangular microstrip patch was fabricated *via* fused filament fabrication (FFF) 3D printing. A hemispherical dielectric lens with a radius of 25 mm (which fed with a different microstrip patch antenna) was also fabricated using FFF 3D printing using the same composite and a 100% infill setting.

Lugo et al. [63] used COP and $MgCaTiO_2$ (MCT) based composite (ε_r~4.34 and tan (δ_d) ~ 0.0060 at 44 GHz), with 30% volume of ceramic filers to FDM print dielectric rod antenna operating in mm-wave frequencies. The peak gain of the multilayer antenna is over 22 dBi for the entire frequency range of 30–40 GHz. **Figure 8d** shows the DRA along with low permittivity (ABS) cladding used to enhance the antenna performance.

In a follow-up paper, they used the same 30 vol. % COP-$MgCaTiO_2$ composite (ε_r ~4.82 and tan (δ_d) ~ 0.0018 at 17 GHz) to fabricate a multilayer dielectric rod waveguide (DRW) for operation at an extended Ku-band from 10 to 18 GHz [64]. The insertion loss of the dielectric rod waveguide stays within 0.01 dB/mm and 0.012 dB/mm, which is comparable to the previously reported multilayer DRW that was constructed from a commercial high-k material not compatible with FDM [64]. **Figure 8e** presents the picture of the 3D-printed DRW made of composite material.

Hoang et al. [27] fabricated a subwavelength deflector using cyclo-olefin copolymer (COC) and TiO_2 composite and compared it to a counterpart 3D printed in ABS. To obtain a high-permittivity and low-loss composite, the TiO_2 (ε_r ~ 100, tan (δ_d) $< 10^{-4}$) was loaded at 36% volume in the low-loss polymer matrix COC (tan (δ_d) $< 10^{-3}$). The deflectors were intended to refract an incoming plane wave with 0° incidence by 30° at 30 GHz. Using this high-permittivity printed material, the thickness of the 90 mm-diameter subwavelength deflector is reduced by more than twice, the realized gain is increased by 1 dB across the full frequency range of 27.5–31 GHz, and the side lobe level is reduced by 7 dB. **Figure 8f** depicts the 3D manufactured subwavelength deflector. Following laboratory research, the disclosed COC-TiO2 filament has been manufactured on an industrial scale and has been sold since the end of 2022 [62].

The authors used a 30% volume COP-TiO_2 (ε_r ~5.37 and tan (δ_d) ~ 0.003 at 16 GHz) composite to 3D print dielectric rod waveguide operating in the frequency range of 12–18 GHz [30]. FDM printed DWG exhibited an impressive low loss of 0.017 dB/mm at 17.5 GHz. In a follow-up paper, authors fabricated DWGs made from 30% volume COP-TiO_2 and 40% volume COP-TiO_2 (ε_r ~6.86 and tan (δ_d) ~ 0.0019 at 16 GHz) for mm-wave frequencies (Ka and V-bands) [43]. 3D printed DWGs for Ka band and V band are shown in **Figure 8g** and **h**, respectively.

4.2 Commercially available ceramic-thermoplastic composites

Huang et al. took advantage of commercially available dielectric composite filament named PREPERM 3-D ABS DK 10.0 from Premix with relative permittivity ε_r of 10.0, and loss tangent tan (δ_d) of 0.003, to fabricate different kinds of antenna applications, from dielectric resonator antenna to dielectric rod antenna [65–68]. In [66, 68], they fabricated low-profile end-fire dielectric antennas with wideband performance consisting of four and five dielectric sections designed with tapered permittivities and sizes to achieve impedance-matched surface-wave launching over a wide bandwidth, as shown in **Figure 9a** and **b**. In several of their works, they used 3-D printing technology to regulate the effective permittivity of dielectric blocks in the 2.0 to 8.3 range by altering the fill factor. They also 3-D-printed a dielectric rod antenna

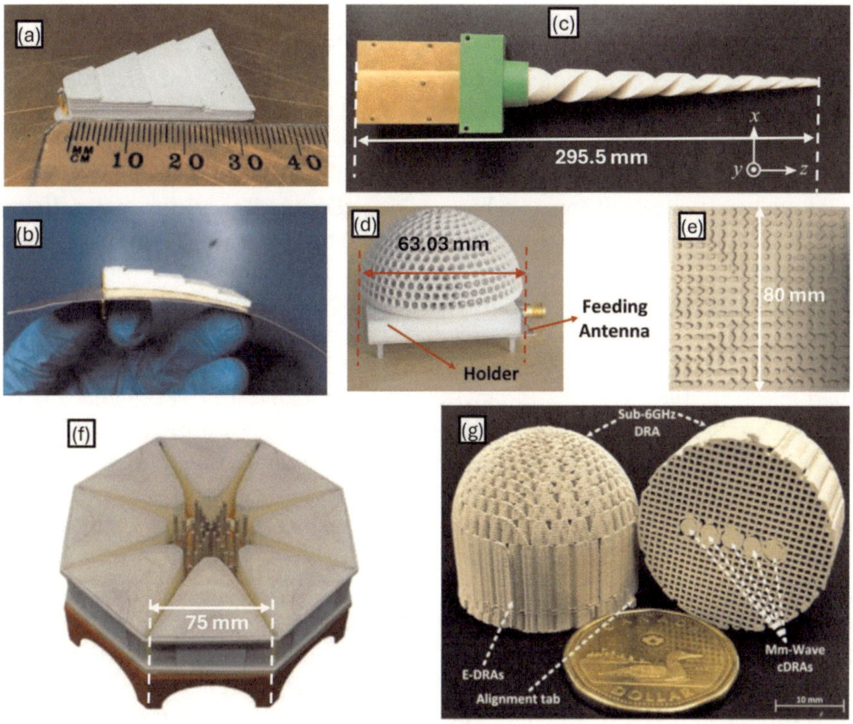

Figure 9.
(a) Low-profile end-fire dielectric antenna [68], (b) conformal low-profile end-fire dielectric antenna [66], (c) dielectric rod antenna with wideband circular polarization using helical grooves in tapered dielectric rod [65], (d) hemispherical lens (18–24 GHz) with circular corrugated matching layer and perforated matching layer regions [69], (e) mm-wave dual-polarized dielectric resonator reflect array [70], (f) gradient-index lens antenna fed by a switched monopole feed array [71], and (g) dual wideband encapsulated dielectric resonator antennas [71].

with wideband circular polarization which used a tapered dielectric rod with two helical grooves (**Figure 9c**) for generating circular polarization.

Alçep and Tokan [69] employed Preperm ABS1000 filament ($\boldsymbol{\varepsilon}_r \sim$10 & tan ($\delta_d$) $\sim$ 0.003) from Premix for 3D printing of hemispherical lens (18–24 GHz) prototype consisting of circular corrugated matching layer and perforated matching layer regions (**Figure 9d**). Instead of the standard broadband matching procedure, which requires the machining of three additional layers of different materials on the lens surface, they employed an alternate matching technique that is feasible to implement with 3D printing technology. The reported performance in terms of beam scanning capability of the perforated lens with 70° scan angle and 2.7 dB scan loss is promising for next-generation automotive radar sensors.

Sun et al. [70] presented a 3D printed millimeter-wave dual-polarized dielectric resonator (DR) reflect array (16 × 16 elements) working in the Ka band. PREPERM 3-D ABS DK 10.0 from Premix materials ($\boldsymbol{\varepsilon}_r \sim$ 10, tan (δ_d) $\sim$ 0.003) was used as the dielectric filament for FDM process. **Figure 9e** shows a picture of the 3D-printed reflect array. The unit cell of the reflect array is composed of two elliptical DRs orthogonally placed in the center. A recorded peak antenna gains of 22.53 dBi was reached, resulting in a gain improvement of 7.43 dB above the feed horn alone.

Giddens and Hao [73] also utilized commercially available standard white ABS from Verbatim and ABS-400 filament from PREPERM with permittivities of 2.69 and 3.99, respectively. They designed and fabricated a gradient-index lens antenna

(**Figure 9f**) designed to radiate with a 45° beamwidth across eight different sectors fabricated by multi-material FDM 3D printing. At 5.8 GHz, the lens antenna was supplied by a switched monopole feed array, resulting in 360° pattern diversity. The lens antenna had a graded dielectric profile from 3.6 down to 1.3.

Malfajani et al. [71] proposed the concept of encapsulated dielectric resonator antennas (EDRAs). E-DRAs are constructed by embedding smaller DRAs with a specified dielectric constant inside a bigger DRA (**Figure 9g**) with a lower dielectric constant, facilitating efficient radiation at two widely spaced and heavily utilized frequency bands. The FFF 3D printing technique is used to fabricate the whole E-DRA structure with a single material (commercial filament PREPERM 3D ABS DK 12.0 from company Avient based in Belgium) but with different infill, percentages to achieve the required permittivity. An array of five elements was employed in a switched mode configuration to add switching beam capability to the antenna in the mm-wave band. The measurement findings depicted a maximum gain of 7.2 dBi at 3.2 GHz and 18 dBi at 31.5 GHz. The calculated efficiency in the sub-6-GHz and mm-wave bands is better than 95% and 80%, respectively. The dielectric constant of the smaller DRA was chosen to be 9, because this is the greatest dielectric constant that can be obtained by ABS1200 through the FFF 3D printing method at mm-wave frequencies.

de Oliveira Neto et al. [72] introduced a technique to produce RF and microwave devices with relative permittivity ($\boldsymbol{\varepsilon_r}$) variations using two or more filaments, with different permittivities, which they introduced into the print head and mixed with designed proportions. They designed and built a dielectric planar lens, with varying permittivity, to work along with a planar antipodal Vivaldi antenna (AVA). Yellow ABS filaments ($\boldsymbol{\varepsilon_r} \sim 2.6$ & tan $(\delta_d) \sim 0.032$) and graphite-filled ABS filaments ($\boldsymbol{\varepsilon_r} \sim 6.5$ & tan $(\delta_d) \sim 0.36$) were used as two feeding materials. As seen in **Figure 10**, they chose the palm tree as the reference antenna for the experiments. Their building process differs from others in the literature in that it creates materials by combining filaments with varying electrical permittivity values.

FDM is foremost and widely utilized AM methods for prototyping EM devices. It's being economical and easy to use, overcome the disadvantages such as lack of thermoplastic materials with extremely high dielectric constant, and the post-processing requirements associated with deposition of (electrically conductive) metallic surfaces.

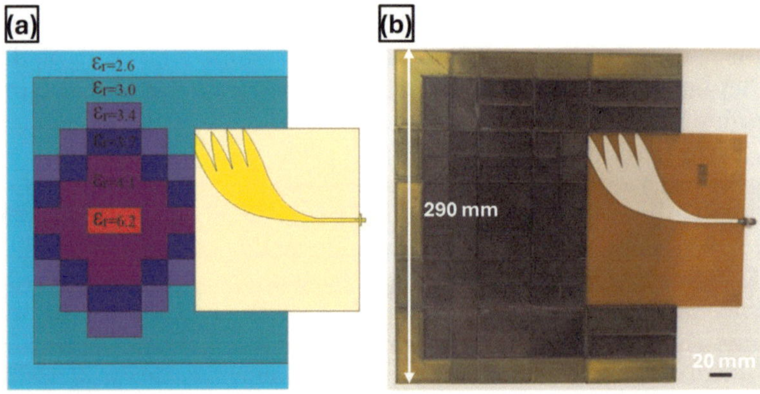

Figure 10.
(a) Simulation model of the lens antipodal Vivaldi antenna with variations of relative permittivity indicated with different colors and (b) picture of 3D printed lens antenna prototype with outermost part composed of yellow ABS and innermost part composed of 95% of graphite-filled ABS [72].

5. Conclusion

Additive manufacturing, especially fused deposition modeling offers feasibility of complex geometries for RF and microwave devices which cannot be achieved or are expensive to achieve using traditional fabrication methods. However, commercially offered 3D printable materials are not optimal for RF and microwave frequencies as they lack high relative permittivity and low-loss-tangents offered by traditional PCB laminates.

Incorporating ceramic filler particles in thermoplastic matrix makes it feasible to combine processibility of thermoplastics with the unique dielectrical properties offered by ceramic materials. Melt mixing is a simpler and more cost-effective way to manufacture ceramic-thermoplastic composites. However, it can only achieve repeatable dispersion of ceramic fillers for composites with up to 30% filler volume fraction. Solution-assisted dispersion can achieve repeatable dispersion of ceramic fillers for higher volume fraction compared to melt mixing. However, it requires expensive and sometimes hazardous chemical solvents, which makes it more expensive. Both above-mentioned methods can be enhanced by incorporating a surface treatment of ceramic fillers before dispersing them in thermoplastic matrix. Surfactant can serve the dual purpose of hindering filler particle agglomeration and improving the interface between filler and matrix.

In this chapter, different methodologies of fabricating ceramic-thermoplastic feedstock filaments are reviewed, and their application for FDM printing of RF and microwave devices is presented. Widely used analytical prediction models for predicting the effective permittivity of the composite materials are listed. Given the limitless and high potential applications of 3-D printing, ongoing research, particularly FDM printed composites, should be encouraged and pursued. The use of microparticles has been shown to generate and build more functional products. It is critical to continue the development of new materials that are precisely tuned for certain electromagnetic applications.

Acknowledgements

This work was supported by NSF grant award number 2329207, entitled “Collaborative Research: FuSe: Thermal Co-Design for Heterogeneous Integration of Low Loss Electromagnetic and RF Systems (The CHILLERS)”.

Author details

Vishvajitsinh Kosamiya* and Jing Wang
Department of Electrical Engineering, University of South Florida (USF), Tampa, United States

*Address all correspondence to: vkkosamiya@ieee.org

References

[1] Clark RH. Handbook of Printed Circuit Manufacturing. Netherlands: Springer; 2012

[2] Ferrando-Rocher M, Herranz-Herruzo JI, Valero-Nogueira A, Bernardo-Clemente B. Performance assessment of gap-waveguide Array antennas: CNC milling versus three-dimensional printing. IEEE Antennas and Wireless Propagation Letters. 2018; **17**:2056-2060

[3] Koul SK, Swapna S, Karthikeya GS. Additive manufacturing in antenna development. In: Koul SK, Swapna S, Karthikeya GS, editors. Antenna Systems for Modern Wireless Devices. Singapore: Springer Nature Singapore; 2024. pp. 283-319

[4] Karthikeyan SS, Mellita RA. Additive manufacturing of MTM-FSS. In: Narayan S, Kesavan A, editors. Handbook of Metamaterial-Derived Frequency Selective Surfaces. Singapore: Springer Nature Singapore; 2022. pp. 695-720

[5] Mazzei, Capote GA, Montoya-Ospina MC, Liu Z, Mattei MS, Liu B, Delgado AP, et al. Compounding a high-permittivity thermoplastic material and its applicability in manufacturing of microwave photonic crystals. Materials (Basel). 2022;**15**(7):1-15

[6] Wang Y, Zhang X, Su R, Chen M, Shen C, Xu H, et al. 3D printed antennas for 5G communication: Current progress and future challenges. Chinese Journal of Mechanical Engineering: Additive Manufacturing Frontiers. 2023;**2**(1): 100065

[7] Chietera FP. Additively manufactured antennas and electromagnetic devices. Hardware. 2024;**2**(2):85-105

[8] Liu R, Mumcu G, Wang J, editors. Towards additive manufacturing based packaging of mm-wave antenna arrays and Beamformer ICs. In: 2024 IEEE Wireless and Microwave Technology Conference (WAMICON). Clearwater, FL, USA: IEEE; 2024. pp. 15-16

[9] Firat OF, Wang J, Weller T. Characterization of additively manufactured suspended finite ground CPW interconnects enhanced by femtosecond laser micromachining. IEEE Transactions on Components, Packaging and Manufacturing Technology. 2024;**14**(6):965-972

[10] Liu R, Braun J, Mitchell G, Wang J, Mumcu G, editors. Packaging of a beamforming IC by laser enhanced direct print additive manufacturing (LE-DPAM). In: 2022 3rd URSI Atlantic and Asia Pacific Radio Science Meeting (AT-AP-RASC); 30 May-4 June 2022. Gran Canaria, Spain: IEEE; 2022

[11] Firat OF, Wang J, Weller T. Additively manufactured metal-insulator-metal capacitors using a high-K dielectric paste. In: 2022 IEEE International Symposium on Antennas and Propagation and USNC-URSI Radio Science Meeting (AP-S/URSI); 10–15 July 2022. Denver, CO, USA: IEEE; 2022

[12] Firat O, Wang J, Weller T, editors. Laser-assisted and additively manufactured multilayer metal-insulator-metal capacitors. In: 2022 IEEE 22nd Annual Wireless and Microwave Technology Conference (WAMICON). Clearwater, FL, USA: IEEE; 2022. pp. 27-28

[13] Abdin MM, Johnson WJD, Wang J, Weller TM. W-band MMIC Chip assembly using laser-enhanced direct print additive manufacturing. IEEE

Transactions on Microwave Theory and Techniques. 2021;**69**(12):5381-5392

[14] Ahmadloo M. Design and fabrication of geometrically complicated multiband microwave devices using a novel integrated 3D printing technique. In: 2013 IEEE 22nd Conference on Electrical Performance of Electronic Packaging and Systems. San Jose, CA, USA: IEEE; 2013

[15] Peverini OA, Addamo G, Lumia M, Virone G, Calignano F, Lorusso M, et al. Additive manufacturing of Ku/K-band waveguide filters: A comparative analysis among selective-laser melting and stereo-lithography. IET Microwaves, Antennas & Propagation. 2017;**11**(14): 1936-1942

[16] Zacharatos F, Theodorakos I, Karvounis P, Tuohy S, Braz N, Melamed S, et al. Selective laser sintering of laser printed Ag nanoparticle micropatterns at high repetition rates. Materials. 2018;**11**(11):2142

[17] Shamvedi D. Design of Microwave Components Using Direct Metal Laser. Sintering: Waterford Institute of Technology; 2018

[18] Rodrigues ASL, Pires ACB, Barbosa PA, Silveira ZC. Polymeric composites in extrusion-based additive manufacturing: A systematic review. Polymer Composites. 2024;**45**(8): 6741-6770

[19] Mogan J, Harun WSW, Kadirgama K, Ramasamy D, Foudzi FM, Sulong AB, et al. Fused deposition modelling of polymer composite: A progress. Polymers [Internet]. 2023;**15**(1):1-27

[20] Castro J, Rojas E, Ross A, Weller T, Wang J. High-k and low-loss thermoplastic composites for fused deposition modeling and their application to 3D-printed Ku-band antennas. In: 2016 IEEE MTT-S International Microwave Symposium (IMS); 22–27 May 2016. San Francisco, CA, USA: IEEE; 2016

[21] Castles F, Isakov D, Lui A, Lei Q, Dancer CE, Wang Y, et al. Microwave dielectric characterisation of 3D-printed $BaTiO_3$/ABS polymer composites. Scientific Reports. 2016;**6**:22714

[22] Mirzavand R, Honari MM, Aslanzadeh S, Saghlatoon H, Mousavi P. Evaluation of one-staged 3-D printed frequency selective surface using carbon-fiber-reinforced thermoplastic composite. IEEE Transactions on Components, Packaging and Manufacturing Technology. 2019;**9**(11): 2298-2304

[23] Li M. Additively Manufactured Electronic Devices and Antennas for Microwave and Millimeter-Wave 5G Applications. Ultimo, NSW, Australia: University of Technology Sydney; 2023

[24] Lumia M, Addamo G, Virone G, Peverini OA, Paonessa F, Rabitsch J, et al. RF characterization of 3D-printed dielectric materials. In: 2023 IEEE Conference on Antenna Measurements and Applications (CAMA); 15–17 Nov. 2023. Genoa, Italy: IEEE; 2023

[25] García CS, Bedia B, Palczynski G, Godja N, Seidler K, Gorsche C. Polymer-based additive manufacturing of a complex RF front-end for new space applications. In: 2024 18th European Conference on Antennas and Propagation (EuCAP); 17–22 March 2024. Glasgow, United Kingdom: IEEE; 2024

[26] Goulas A, McGhee JR, Whittaker T, Ossai D, Mistry E, Whittow W, et al. Synthesis and dielectric characterisation of a low loss $BaSrTiO_3$/ABS ceramic/ polymer composite for fused filament

fabrication additive manufacturing. Additive Manufacturing. 2022;**55**: 102844-102852

[27] Hoang TQV, Vandelle E, Czarny R, Loiseaux B, Fournier T, Penin N, et al. Benefits of using a high-permittivity and low-loss printable material for sub-wavelength deflectors. In: 2023 IEEE Conference on Antenna Measurements and Applications (CAMA) 2023. Genoa, Italy: IEEE; 2023. pp. 492-496

[28] Packard H. Basics of Measuring the Dielectric Properties of Materials. Application Note 1217–1 [Internet]. Palo Alto, CA: Hewlett Packard; 1992

[29] Castro J, Rojas-Nastrucci EA, Ross A, Weller TM, Wang J. Fabrication, modeling, and application of ceramic-thermoplastic composites for fused deposition modeling of microwave components. IEEE Transactions on Microwave Theory and Techniques. 2017;**65**(6):2073-2084

[30] Kosamiya V, Wang J, editors. High-k and low-loss dielectric composite feedstock filaments, tailored for additive manufacturing of microwave devices. In: 2020 IEEE/MTT-S International Microwave Symposium (IMS); 4–6 Aug 2020. Los Angeles, CA, USA: IEEE; 2020

[31] Castro JDD. Engineered Nanocomposite Materials for Microwave/Millimeter-Wave Applications of Fused Deposition Modeling. Tampa, FL, USA: University of South Florida; 2017

[32] Dang Z-M, Yuan J-K, Zha J-W, Zhou T, Li S-T, Hu G-H. Fundamentals, processes and applications of high-permittivity polymer–matrix composites. Progress in Materials Science. 2012;**57**(4):660-723

[33] Todd MG, Shi FG. Complex permittivity of composite systems: A comprehensive interphase approach. IEEE Transactions on Dielectrics and Electrical Insulation. 2005;**12**(3):601-611

[34] Vo HT, Shi FG. Towards model-based engineering of optoelectronic packaging materials: Dielectric constant modeling. Microelectronics Journal. 2002;**33**(5–6):409-415

[35] Kim H, Abdala AA, Macosko CW. Graphene/polymer nanocomposites. Macromolecules. 2010;**43**(16):6515-6530

[36] Lee J-Y, An J, Chua CK. Fundamentals and applications of 3D printing for novel materials. Applied Materials Today. 2017;**7**:120-133

[37] Martin C. Twin screw extruders as continuous mixers for thermal processing: A technical and historical perspective. AAPS PharmSciTech. 2016; **17**(1):3-19

[38] Kattel B, Ayan U, Mohoppu M, Villacorta B, Hutchcraft WE. Enhancing permittivity of 3D printing filaments via Nanocompounding for electromagnetic applications. In: SoutheastCon 2024; 15–24 March 2024. Atlanta, GA, USA: IEEE; 2024

[39] Adami R, Lamberti P, Bychanok D, Kuzhir P, Tucci V. Electromagentic Properties of Filaments Containing Nanofillers for 3D Printing. Milano, Italy: Italian Association of Chemical Engineering; 2021

[40] Khatri B, Lappe K, Habedank M, Mueller T, Megnin C, Hanemann T. Fused deposition modeling of ABS-barium titanate composites: A simple route towards tailored dielectric devices. Polymers (Basel). 2018;**10**(6):1-18

[41] Wu Y, Isakov D, Grant PS. Fabrication of composite filaments with high dielectric permittivity for fused

deposition 3D printing. Materials (Basel). 2017;**10**(10):1-11

[42] Parsons P, Larimore Z, Mirotznik M, Mitchell G. Composite materials development for fused filament fabrication of RF systems. In: 2020 International Applied Computational Electromagnetics Society Symposium (ACES); 27–31 July 2020. Basel, Switzerland: MDPI; 2017

[43] Kosamiya V, Wang J. Composite feedstock filaments for fused deposition modeling (FDM) of microwave interconnects. IEEE Transactions on Components, Packaging and Manufacturing Technology. 2023;**13**(2): 249-256

[44] Garnett JCM, XII LJ. Colours in metal glasses and in metallic films. Philosophical Transactions of the Royal Society of London Series A, Containing Papers of a Mathematical or Physical Character. 1997;**203**(359–371):385-420

[45] Jayasundere N, Smith BV. Dielectric constant for binary piezoelectric 0-3 composites. Journal of Applied Physics. 1993;**73**(5):2462-2466

[46] Czarny R, Hoang TQVTQ, Loiseaux B, Bellomonte G, Lebourgeois R, Leuliet A, et al. High permittivity, low loss, and printable thermoplastic composite material for RF and microwave applications. In: 2018 IEEE Conference on Antenna Measurements & Applications (CAMA); 3–6 Sept. 2018. Västerås, Sweden: IEEE; 2018

[47] Brosseau C. Modelling and simulation of dielectric heterostructures: A physical survey from an historical perspective. Journal of Physics D: Applied Physics. 2006;**39**(7):1277-1294

[48] Wagner K. The after effect in dielectrics. Archiv fur Elektrotechnik. 1914;**2**(378):e394

[49] Nelson SO, You TS. Relationships between microwave permittivities of solid and pulverised plastics. Journal of Physics D: Applied Physics. 1990;**23**(3): 346

[50] Shen LC, Savre WC, Price JM, Athavale K. Dielectric properties of reservoir rocks at ultra-high frequencies. Geophysics. 1985;**50**(4):692-704

[51] Nan C-W. Physics of inhomogeneous inorganic materials. Progress in Materials Science. 1993;**37**(1):1-116

[52] Rayleigh L. On the influence of obstacles arranged in rectangular order upon the properties of a medium. Phiiris Mag. 1892;**34**(481):507

[53] Maxwell JC. A Treatise on Electricity and Magnetism. Oxford, UK: Clarendon Press; 1873

[54] Burggeman D. Calculation of various physics constants in heterogeneous substances I. Dielectricity constants and conductivity of mixed bodies from isotropic substances. Annalen der Physik. 1935;**24**:626-664

[55] Hanai T. Theory of the dielectric dispersion due to the interfacial polarization and its application to emulsions. Kolloid-Zeitschrift. 1960; **171**(1):23-31

[56] Zhou W, Hinojosa BB, Nino JC, Petrovsky V. Applicability of the Bruggeman equation for analyzing dielectric slurries containing ceramic powders with high permittivity. Journal of the American Ceramic Society. 2011; **95**(2):457-460

[57] Stroud D, Pan FP. Self-consistent approach to electromagnetic wave propagation in composite media: Application to model granular metals. Physical Review B. 1978;**17**(4):1602-1610

[58] Munina I, Grigoriev I, O'Donnell G, Trimble D. A review of 3D printed gradient refractive index lens antennas. IEEE Access. 2023;**11**:8790-8809

[59] Colella R, Chietera FP, Michel A, Muntoni G, Casula G, Montisci G, et al. Electromagnetic characterisation of conductive 3D-printable filaments for designing fully 3D-printed antennas. IET Microwaves, Antennas & Propagation. 2022;**16**(11):687-698

[60] Colella R, Chietera FP, Muntoni G, Casula GA, Montisci G, Catarinucci L. Evaluating the effectiveness of planar and waveguide 3D-printed antennas manufactured using dielectric and conductive filaments. IEEE Access. 2023; **11**:34891-34898

[61] Preperm. [Online]. 2019 [Preperm webpage on 3D filaments; [Online]. Available from: https://www.preperm.com/webshop/product-category/3dfilaments

[62] [Online]. Online: Nanoe; 2024 [zetamix]. Available from: https://zetamix.fr/use-case/metasurface-for-flat-antennas/

[63] Lugo DC, Ramirez RA, Castro J, Wang J, Weller TM. 3D printed multilayer mm-wave dielectric rod antenna with enhanced gain. In: 2017 IEEE International Symposium on Antennas and Propagation & USNC/URSI National Radio Science Meeting; 9–14 July 2017. San Diego, CA, USA: IEEE; 2017

[64] Lugo DC, Ramirez RA, Castro J, Wang J, Weller TM. Ku-band additive manufactured multilayer dielectric rod waveguide. In: 2017 IEEE 18th Wireless and Microwave Technology Conference (WAMICON); 24–25 April 2017. Cocoa Beach, FL, USA: IEEE; 2017

[65] Huang J, Chen SJ, Xue Z, Withayachumnankul W, Fumeaux C. Wideband circularly polarized 3-D printed dielectric rod antenna. IEEE Transactions on Antennas and Propagation. 2020;**68**(2):745-753

[66] Huang J, Chen SJ, Xue Z, Withayachumnankul W, Fumeaux C. Wideband 3D printed conformal dielectric antenna with end-fire radiation. In: 2019 IEEE International Symposium on Antennas and Propagation and USNC-URSI Radio Science Meeting; 7–12 July 2019. Atlanta, GA, USA: IEEE; 2019

[67] Huang J, Chen SJ, Xue Z, Withayachumnankul W, Fumeaux C. Impact of infill pattern on 3D printed dielectric resonator antennas. In: 2018 IEEE Asia-Pacific Conference on Antennas and Propagation (APCAP); 5–8 Aug 2018. Auckland, New Zealand: IEEE; 2018

[68] Huang J, Chen SJ, Xue Z, Withayachumnankul W, Fumeaux C. Wideband endfire 3-D-printed dielectric antenna with designable permittivity. IEEE Antennas and Wireless Propagation Letters. 2018;**17**(11): 2085-2089

[69] Alçep M, Tokan F. Impedance matching technique with perforated, inhomogeneous layers for broadband dielectric lenses. IEEE Sensors Journal. 2021;**21**(18):20018-20026

[70] Sun YX, Wu D, Ren J. Millimeter-wave dual-polarized dielectric resonator reflectarray fabricated by 3D printing with high relative permittivity material. IEEE Access. 2021;**9**: 103795-103803

[71] Malfajani RS, Niknam H, Bodkhe S, Therriault D, Laurin JJ, Sharawi MS. A 3D-printed encapsulated dual wide-band

dielectric resonator antenna with beam switching capability. IEEE Open Journal of Antennas and Propagation. 2023;**4**: 492-505

[72] de Oliveira Neto AM, Justo JF, Beccaro W, de Oliveira AM. Designing and building radio frequency devices with tailored dielectric properties using additive manufacturing. Microwave and Optical Technology Letters. 2023;**65**(3): 777-784

[73] Giddens H, Hao Y. Multibeam graded dielectric lens antenna from multimaterial 3-D printing. IEEE Transactions on Antennas and Propagation. 2020;**68**(9):6832-6837

Chapter 3

Integrating Aesthetics and Robotic Additive Manufacturing in Artistic Design

Shuying Wang and Donghua Zhao

Abstract

Robotic additive manufacturing (RAM) is emerging as a transformative technology in artistic design, offering efficiency, scalability, and mobility. This study examines the intersection of technology and aesthetics, exploring how RAM can revolutionize artistic creation. It reviews the fundamental technological aspects of RAM, including process principles, topology optimization, and equipment specifications, while introducing an innovative esthetic perspective. The study discusses how integrating esthetic principles with RAM technology enhances artistic value, enabling complex, visually appealing designs. Key findings highlight RAM's ability to produce intricate designs with high precision and consistency, fostering new creative possibilities in artistic expression. The study also addresses critical technological challenges, such as multi-process integration and digital twin technologies, proposing solutions considering functional and esthetic outcomes. Furthermore, the potential for human-machine collaborative processes in RAM is explored, suggesting a new paradigm that blends artistic creativity with precision engineering. This interdisciplinary approach aims to provide a comprehensive understanding of RAM's role in artistic design and serve as a reference for future research and applications in this field.

Keywords: artistic design, robotic additive manufacturing, aesthetics in technology, digital twin technology, human-machine collaboration

1. Introduction

Integrating RAM with aesthetics can overcome the creative limitations imposed by traditional manual manufacturing techniques, thereby significantly enhancing the dimensions of artistic design. This combination allows unprecedented complexity, customization, and precision in art and design. Artists can engage in bold and unique creative endeavors by utilizing innovative materials to achieve high levels of detail that surpass those of traditional manual manufacturing methods.

In modern society, advocacy for reducing waste and energy consumption and supporting sustainable development is paramount. RAM's single-step forming technology embodies these contemporary values. Modern artistic creation reflects these values, as artists use their work to convey concepts such as ecological sustainability, minimalism, and technological intelligence, thereby influencing the era's cultural landscape. With the increasing variety of RAM equipment and the diverse materials they support, this field has shown significant potential for widespread adoption in recent years. This development can enable a broader creative community to learn and utilize these technologies, facilitating the intersection of art, engineering, and materials science. Such cross-disciplinary innovations foster the incubation of cutting-edge ideas.

RAM has expanded the possibilities for new artistic forms in the digital age and enhanced the functionality and value that artistic designs can embody. RAM strengthens contemporary art's cultural and emotional resonances by employing advanced manufacturing techniques to reflect contemporary themes and interpret technological elements. This integration elevates the overall impact of artistic creation, promoting high-level development and exploration of modern art and design. Simultaneously, art can inspire technological advancements, providing feedback from practical and usage experiences that inform the iterative development of RAM hardware and software, thus ensuring continuous progress.

RAM is an innovative process that employs robots as the primary motion system, revolutionizing technology and artistic design. This technology enables efficient and integrated formation of complex structures, providing unprecedented freedom in component design, which is crucial for artistic endeavors (**Figure 1**) [1–5]. Compared with traditional manual manufacturing methods, RAM offers significant advantages, including larger forming spaces, higher geometric accuracy, and flexible printing paths. These benefits not only enhance the technical capabilities of RAM but also significantly expand the possibilities for creating complex, visually appealing designs that meet stringent service conditions and diverse functional requirements. This synergy of RAM with artistic design principles facilitates the creation of intricate, high-quality artwork that blends esthetic value with technological precision.

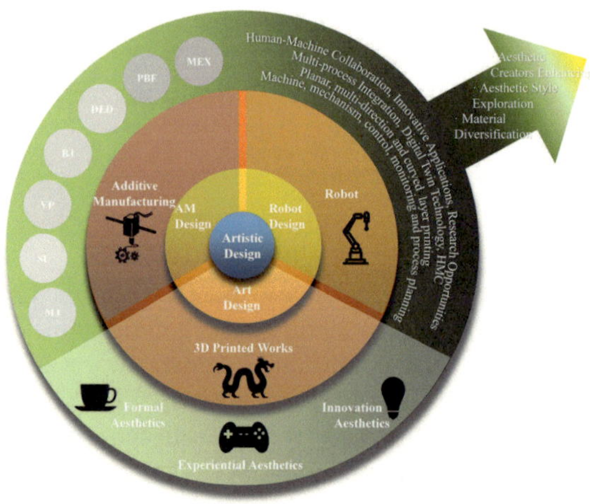

Figure 1.
The framework of integrating aesthetics and RAM in artistic design.

2. Literature review

2.1 Historical development of RAM

Additive Manufacturing (AM) technology, which emerged in the 1980s [6, 7] and was officially named during the inaugural meeting of the ASTM F42 technical committee in 2009 [8], has been widely recognized for its application in the small-batch production of complex parts. AM has continually evolved, developing various processes, such as material extrusion, powder bed fusion, and directed energy deposition [7]. These advancements have paved the way for the integration of robotics, leading to the development of RAM.

Robotic Additive Manufacturing leverages robots' high flexibility, reconfigurability, and adaptability to construct three-dimensional objects through a layer-by-layer material deposition technique [9]. Compared with traditional manual manufacturing, RAM offers superior positioning accuracy, rapid programming, and operational dexterity, significantly enhancing the manufacturing process's efficiency, scalability, and mobility [6, 10].

As highlighted by Attaran et al., robotic manufacturing equipment provides outstanding benefits, including precise positioning, quick reprogramming capabilities, and operational agility, thus endowing RAM with high efficiency, large-format production, and mobility advantages [6, 10]. Wang Zhuoqun et al. further noted that RAM enhances operational accessibility, manufacturing flexibility, and multi-material handling capability. This technological progression has notably improved efficiency, establishing RAM as a cutting-edge trend for producing significant, complex components of high quality [11].

In the context of artistic design, RAM's historical development illustrates a shift from traditional manual manufacturing constraints to the realm in which technology and aesthetics intersect seamlessly. Integrating robotic precision and flexibility with AM processes has opened new possibilities for creating intricate and visually captivating designs that were previously unattainable. This evolution supports the theme of this paper by showcasing how RAM can revolutionize artistic creation and offer novel opportunities for complex, esthetically pleasing, and high-quality artworks.

2.2 Intersection of RAM and art

Progression from traditional manual manufacturing to advanced RAM has led to significant technological and artistic advancements. Initially, artistic creation relied heavily on manual craftsmanship, with artists using tools like chisels, brushes, and lathes. This method allowed for personal expression but was limited by the physical capabilities of the artist, the equipment, and the materials available.

Robot-assisted manufacturing has emerged with technological advancements, introducing greater precision and repeatability. Robots began assisting in subtractive manufacturing processes, such as Computer Numerical Control (CNC) machining, marking the beginning of technology integration into the artistic process. This advancement has allowed for more complex designs and reduced production times.

The advent of AM has marked a significant shift. Techniques such as Stereolithography (SLA) and Selective Laser Sintering (SLS) have enabled the layer-by-layer creation of intricate layers, expanding the creative possibilities for artists.

The integration of robotics with additive manufacturing led to the development of RAM, which combines the flexibility of AM with the precision and speed of robotics. This combination allows the creation of highly detailed and esthetically pleasing artwork.

In artistic design, RAM opens new avenues by enabling artists to achieve intricate details and complex geometries, pushing the boundaries of what can be achieved. The combination of robotic precision and AM processes ensures the artwork is functional and visually compelling. This interdisciplinary approach enhances the quality and complexity of artistic creations, showcasing RAM's potential in artistic design.

2.3 Aesthetics in technology

Integrating robotic fabrication into the design process can significantly enhance creativity and reveal new esthetic potential in architectural design through innovative techniques such as laser cutting and robotic thermoforming [12, 13]. By examining how different additive manufacturing process parameters affect the esthetic properties of 3D-printed parts, designers can achieve greater control over their products' visual and physical qualities [14]. This integration is exemplified by the development of a humanoid-legged robot, where the fusion of visual appeal and functional design demonstrates a seamless blend of artistic creativity and technical innovation [13].

The convergence of aesthetics and robotic additive manufacturing in design processes fosters enhanced creativity and precise control over esthetic properties, as demonstrated in various applications ranging from additive manufacturing to robotics. This synergy between artistic design and technological innovation paves the way for groundbreaking advancements in multiple fields, proving that esthetic considerations are essential for the success and appeal of modern design.

3. Core technologies in RAM

RAM integrates advanced robotics with additive manufacturing processes, enabling the creation of complex and esthetically appealing designs. The following sections detail the core technologies underlying RAM and emphasize their application in artistic design.

3.1 Process principles

Process principles in RAM encompass a comprehensive understanding of the technologies and methods utilized to achieve precise and innovative artistic creation. RAM leverages several AM techniques, each offering unique capabilities contributing to the artistic process.

Material extrusion (MEX): This technique involves layer-by-layer deposition of materials, typically thermoplastic polymers, through a heated nozzle. The precision of robotic arms allows for intricate and complex designs ideal for detailed artistic works. The flexibility of material extrusion supports a wide range of materials and enables artists to experiment with different textures and finishes.

Powder bed fusion (PBF): PBF techniques, such as SLS and Direct Metal Laser Sintering (DMLS), use a laser to fuse powdered materials into solid structures. This method is highly valued for producing fine details and smooth surfaces, crucial for achieving high-quality artistic pieces. The robotic systems' precision enhances this process's accuracy and repeatability, ensuring consistent results across multiple iterations.

Directed energy deposition (DED): DED involves the deposition of a material, usually metal, through a nozzle while simultaneously melting it with a laser or electron beam. This method is particularly suited for creating large, complex structures such as sculptures or architectural elements. The versatility of DED allows the integration of multiple materials within a single piece, adding depth and complexity to artistic design.

Binder jetting (BJ): This process involves the selective deposition of a binding agent onto a powder bed to create a solid object. Binder jetting can produce full-color models, making it an excellent choice for creating vibrant and detailed artistic pieces. Robotics ensures precise control of the deposition process and enhances the quality and consistency of the final artwork.

Vat photopolymerization (VP): Techniques such as SLA and Digital Light Processing (DLP) cure liquid photopolymer resin layer-by-layer using a light source. This process can achieve high-resolution details and smooth finishes for intricate artistic designs. Integrating robotic systems allows for a more complex and large-scale creation, pushing the boundaries of traditional photopolymerization.

Sheet lamination (SL): This technique involves bonding sheets of material to form an object that can be used to create layered, textured artistic designs. Robotic precision in cutting and assembling the sheets enables the production of detailed and complex structures that are difficult to achieve manually.

Material jetting (MJ): This technique involves layer-by-layer deposition of photopolymers or other materials from a print head cured by UV light. It produces high-resolution objects with smooth finishes and complex geometries, which makes it ideal for detailed and vibrant artistic pieces. The ability to simultaneously jet multiple materials allows color mixing and gradients. Robotic precision ensures accurate droplet placement and enhances the quality and consistency of artwork.

The key RAM technologies of RAM involve [4, 15, 16]:

Advanced robotics [16]: Multi-axis robots enhance the dexterity and precision of the manufacturing process, allowing for the creation of highly detailed and complex designs.

Real-time monitoring and control [17, 18]: Integration of sensors and feedback systems ensure continuous monitoring of the manufacturing process, maintains high quality, and minimizes defects, which are crucial for the esthetic integrity of the artwork.

Hybrid manufacturing [19]: Combining additive and subtractive processes, such as milling or grinding, allows for fine-tuning and refining the final product, ensuring functional and esthetic optimization.

Comparison with traditional artistic creation: Human-centered artistic creation relies heavily on an artist's manual skills, creativity, and time. Each piece is often unique, with variations that reflect the artist's hands. Although this method is cherished for its personal touch and originality, it can be time-consuming and limited by human precision.

In contrast, RAM offers the following:

Speed and efficiency: Robotic systems can work continuously and at speeds far beyond human capability, drastically reducing production times.

Precision and consistency: Advanced robotics ensures high precision and repeatability, producing consistently high-quality results.

Complexity and customization: RAM can create intricate designs and complex structures that are challenging or impossible to achieve manually.

Scalability: RAM can easily scale production from single unique pieces to multiple identical copies, thus offering flexibility in artistic production.

By understanding and leveraging these principles, artists and designers can fully exploit the potential of RAM. This technology expands creative possibilities and ensures that the resulting artworks meet the highest standards of quality and precision, pushing the boundaries of what is achievable in artistic design.

3.2 Equipment and tools

RAM has significant advantages in artistic design, such as conformal design for multifunctional structures, integrated manufacturing of large complex components, and creating intricate topologies with composite materials.

In artistic creation, high-end equipment that uses multi-axis robots for additive manufacturing of diverse materials, such as polymers and metals, has become a cutting-edge technology. General-purpose robots, primarily categorized into gantry and articulated types, can be equipped with additive manufacturing heads. These setups, including gantry-type robots [20] and powder bed equipment, have already been widely used to create complex artistic works. Handheld devices also highlight the collaborative nature of human-machine artistic creation. These tools allow artists to engage directly in creating and translating their ideas into physical work in real-time. By integrating robotic technology, handheld devices improve production accuracy and expand the possibilities for artistic creation. For instance, handheld 3D printing pens enable artists to draw three-dimensional structures freely, enhancing the spontaneity and flexibility of their designs. Additionally, handheld devices with integrated sensors and feedback systems can dynamically adjust the printing parameters to accommodate different materials and complex geometries, broadening artistic possibilities and enabling more intricate and adaptive designs [21].

When dealing with cylindrical and irregular thin-walled structures, complex lattice designs, and topologically optimized nonstandard shapes, three-axis equipment is limited in handling the nuanced requirements of artistic designs. The 2.5D layering method often introduces stair-step errors, micro-defects, and stress concentrations, which can compromise artistic pieces' visual and structural integrity. Zhao et al. indicated that this method might reduce the horizontal shear strength of thin-walled structures when processing anisotropic materials, thereby affecting the isotropy of the material [2, 22].

To overcome these limitations, articulated robot-based additive manufacturing provides a highly flexible and precisely controlled method. For example, Ding et al. [23] utilized an 8-axis robotic laser direct metal deposition system combined with a planar hybrid slicing method to optimize the overhanging structures of complex rotating parts such as sculptures and intricate decorative elements. Gibson et al. [24] demonstrated significant capabilities in creating unsupported 45-degree overhangs and large complex components using 5-axis laser hot-wire directed energy deposition technology, integrating a 6-axis industrial robot with a 2-axis rotary platform. These advancements in RAM address the limitations of traditional artistic manufacturing techniques and greatly enhance the production efficiency and quality of high-performance complex artistic structures.

To expand the capabilities of RAM in artistic design, including enhancing the creative workspace, improving equipment flexibility, offering versatile task spaces, and achieving high precision in artistic expression, research and development are advancing toward robotic systems with more axes and many degrees of freedom [25]. These systems are categorized based on their mechanism types into serial, parallel, hybrid, and collaborative types [26, 27] and based on the mobility of their bases into fixed and mobile types. Mobile classifications include legged, wheeled-legged, and drone types [11, 28]. These advancements are paving the way for artists to explore new dimensions in their work, enabling the creation of more intricate and dynamic pieces with greater ease and precision.

Integrating these advanced robotic systems with RAM technology in artistic design offers unprecedented opportunities to create complex, esthetically appealing artworks. These systems' high precision and flexibility allow artists to achieve intricate designs that seamlessly blend functional and esthetic elements. This intersection of technology and art opens new possibilities for innovative, high-quality artistic creations that were previously unattainable. By leveraging RAM's capabilities, artists can explore new frontiers in design, pushing the boundaries of what is possible by creating visually captivating and structurally sound artworks.

3.3 Topology optimization in artistic design

Topology optimization is crucial for enhancing RAM designs' structural efficiency and esthetic appeal. This process involves systematically reducing the material within a given design space to achieve an optimal distribution of material, leading to lightweight yet strong structures. For artists and designers, topology optimization opens new possibilities for creating intricate and visually striking pieces that maintain structural integrity.

Applications of RAM include the following:

Design efficiency: Using topology optimization, artists can create designs that use materials more efficiently, resulting in lighter sculptures and installations without compromising strength. This is particularly useful in large-scale work, where material cost and weight are significant considerations.

Esthetic complexity: Topology optimization allows for the generation of complex organic shapes that are both functional and beautiful. These shapes often mimic natural forms and add an element of biomimicry to the artwork. Such designs can be achieved more precisely and easily with RAM, which excels at creating intricate geometries.

Integration with artistic vision: The ability to integrate topology optimization within the design process means that artists can start with an initial concept and refine it to balance esthetic appeal with structural requirements. This integration ensures that the final piece looks stunning, is feasible to produce, and is durable over time.

Examples of Studies and Applications include the following:

Researchers have developed methods to convert topology-optimized designs into additive manufacturing-ready models, ensuring optimized structures can be produced using RAM techniques [29].

Studies on the application of artificial intelligence in composite materials highlight how topology optimization can lead to new design paradigms that blend functionality with artistic expression [30].

By leveraging topology optimization, artists and designers can push the boundaries of what is possible with RAM, creating visually captivating, structurally sound, and material-efficient works. The alignment of advanced engineering techniques with artistic creativity represents a significant step forward in artistic design.

4. Aesthetics integration in RAM

4.1 Theoretical framework

The relationship between new productivity and aesthetics has become particularly subtle in modern design. Aesthetics is the science of beauty and esthetic appreciation, focusing on how people perceive and understand beauty and the standards and principles by which beauty is accepted. However, in modern human creative activities, the gradual development of robotic additive manufacturing and intelligent technologies has introduced a third role in creating beautiful objects—robots. Therefore, as a new auxiliary means of artistic creation, RAM needs further linkage and research with "beauty."

Successful design work (generally referring to items with certain practical functions, such as household appliances, automobiles, clothing, jewelry, etc.) is an artwork. From this perspective, design work with a high esthetic value can attract consumers' attention, regardless of its internal functions. People tend to have a favorable impression of beautiful things and are willing to pay more. Therefore, esthetic considerations play an essential role in various designs.

Second, when the design product enters the usage phase, the involvement of aesthetics can enhance the user experience. This requires an excellent design that is visually appealing and comfortable, as the pursuit of comfort is an instinctive human response. From the designer's perspective, esthetic awareness can help them understand how to combine function and form to create a pleasant experience for the customer.

RAM processes have a direct impact on the perceived quality and craftsmanship of a product. Unlike traditional manual manufacturing, RAM offers a novel approach that enables the creation of design works with distinct and innovative esthetic elements. This method allows for a higher degree of precision and control, which is crucial for achieving specific esthetic objectives in design, such as unique textures, forms, and structural details.

As a result, artworks and design pieces created using RAM exhibit unique, recognizable characteristics and expressions that set them apart from those made using traditional manual manufacturing methods. These distinctive features are made possible by the advanced capabilities of RAM, which facilitate the exploration of new forms and structures, thus supporting innovative design practices.

In summary, design esthetic elements closely related to the technical aspects of RAM processes are essential in the following aspects.

4.1.1 Formal aesthetics

Formal aesthetics primarily considers the esthetic characteristics presented by the appearance of design works, including form, material, and process aesthetics. Art is the creation of a symbolic form of human emotions. The purpose of any formative art is to depict visual forms. Although esthetic art forms differ in thousands of ways, the essence of their beauty remains the same. All forms of design art with the property of "expression" show people's physical and mental beauty of sentiment and pleasant beauty [31].

Form aesthetics: The design of an object's external shape, such as the use and combination of elements such as shape, lines, proportions, and colors, as well as the harmony and unity between these elements. Designers often use classic proportionality principles to design products. Robotic additive manufacturing can maximize the precision and accuracy of the design work during the manufacturing process, thereby assisting in accurately realizing design concepts. Designers and artists can operate more efficiently using precise measurement and manufacturing equipment, ensuring that the final effect of the work aligns with the design ideal.

Material aesthetics: 3D printing in product manufacturing can use various materials, including plastic, metals, and ceramics. The choice of material affects the tactile and visual effects of a product. The texture, color, and luster of these materials determine the product's final appearance. Different materials have different textures and provide different esthetic perceptions and style impressions. If more unique visual and tactile sensations are required, machines can also assist artists in creating mixed materials to produce the beauty of material variation. Combining different materials into a single artwork is an essential aspect of machine manufacturing that breaks through traditional craftsmanship.

Process aesthetics: The 3D printing process based on machines, such as the printing speed, layer thickness, and filling density, directly affects the appearance and texture details of the product and presents different esthetic mechanisms. They are generally classified into three types according to the form of the raw materials: powder, solid, and liquid [32]. The processing characteristics of powder materials leave varying degrees of regular matte texture marks (burrs) that are not very smooth. However, when the machine's process planning is reasonable and the accuracy is precise, this matte texture can bring an exceptional rough and rustic beauty to the product's appearance. Solid printing commonly involves heating and extruding filaments, cooling, and solidification. Filament processing left visible layered traces on the product's surface, with a distinctive linear texture being particularly prominent. Photopolymerization technology for liquid material printing generally produces high-density work with a fine texture and smooth appearance, representing delicate beauty [1, 7].

4.1.2 Experiential aesthetics

The aesthetics of design work involve visual appeal and the user's interaction with the product. Modern advanced design consistently emphasizes user experience (UX), requiring designs to be practical and attractive, enhancing overall user satisfaction, and ensuring a pleasant experience with the product. This encompasses the functional aesthetics of the design itself and the empathetic aesthetics arising from the interaction between the design and the user.

Functional aesthetics: Esthetic theory has traditionally occupied itself with fine art in all forms, sometimes with craft and often with notions of beauty and sublimity in art and nature. It has largely ignored the quotidian and familiar objects and experiences that make up our daily lives. Design must be distinguished from art and craft, and the design experience must be defended as having a mainly esthetic nature [33]. Products created by 3D printing or other advanced manufacturing methods must be esthetically pleasing and practical. Many consider it an essential factor that makes designs attractive and engaging to audiences [34]. Designers must consider the functional aspects of a product, such as its structural stability, ease of use, and reliability. These practical functions, which provide consumers convenience and pleasure, align with the human process of appreciating beauty.

Empathetic aesthetics: Beauty can arise from the interaction between a product and the user. The design should make the interaction process intuitive and enjoyable, enhancing the user's esthetic experience through cultural and emotional connections. In modern designs, this can be achieved by utilizing a combination of sensory stimuli—touch, sound, and sight—that are both advanced and optimally aligned to provide direct pleasure to the user [35]. Advanced manufacturing techniques and materials enable designs to incorporate more content, elements, and meaningfulness, which are ultimately reflected in a product and generate richer emotional experiences.

4.1.3 Innovation aesthetics

3D printing technology overcomes the spatial limitations of traditional manual manufacturing, enabling designers to create unprecedented shapes and structures. Innovation is a crucial characteristic of modern aesthetics. Innovation and breakthroughs are primarily reflected in the personalization of design, material innovation, and innovation.

Personalized design: Advanced manufacturing methods enhance the flexibility of product design. Technology and design can collaborate to rapidly prototype and customize production using techniques like modular design and reconfigurable manufacturing systems. This allows quick adaptation to market changes and diverse customer needs. Consequently, designers can create personalized solutions that cater to different consumer preferences, habits, specifications, materials, and colors with lower design times and production costs, making large-scale customized designs feasible.

Material innovation: A close relationship exists between innovative materials and modern intelligent manufacturing and design, which drives manufacturing transformation and design progress. Intelligent materials, self-healing materials, nanomaterials, and composite materials possess unique physical and chemical properties that enable designers and engineers to develop new products that were previously difficult to achieve. These new materials allow for multifunctional integration, such as sensing, adaptability, and deformation capabilities, making products esthetically pleasing, intelligent, and multifunctional.

Form innovation: As a specific symbol, design art has the basic function of formalizing people's internal experiences, or we can say it endows specific forms for expressing human emotions and experiences or various characteristics of their inner life [31]. With the advent of science and technology, people's thoughts should also be expressed through design. RAM manufacturing supports designers in transforming complex geometries and intricate structures into actual design products. Using multi-axis robots, particularly in additive manufacturing and precision machining, makes creating complex and nonlinear shapes possible. This dramatically expands creative freedom and imaginative space for designers. Shapes, structures, and details that could only be imagined before can now be realized, liberating the artistic thinking of designers, and revolutionizing the esthetic experience for consumers and users.

In summary, design encompasses the set of operations organized to fulfill a project's objectives and is no longer just concerned with the above-listed objects but with most society fields [36]. The relationship between advanced manufacturing, robotics, and design aesthetics is inseparable. Design aesthetics can enhance the esthetic appeal of products, improve their practicality and perceived quality, and strengthen multidimensional connections and resonance among users. This, in turn, increases the added value of products, boosts market competitiveness, and promotes the innovative development of modern design aesthetics.

4.2 Case studies

4.2.1 Case study on design flexibility

The vase design is shown in **Figure 2** as an example: the industrial production of traditional container shapes generally employs subtractive manufacturing methods. Whether it is industrial casting and mold-making or handcrafted sculpture, a "removal" process is required. This method is still used in traditional industrial manufacturing involving robots. However, 3D printing employs additive manufacturing, where multi-degree-of-freedom robotic systems use a print head to extrude material, allowing for the direct creation of complex geometric structures and internal designs that are difficult to achieve with traditional manual manufacturing methods.

This vase design in **Figure 2** utilizes 3D printing technology to create dynamic gaps in the vessel walls and flowing curved lines, thereby achieving a dynamic external shape and intricate internal structures. The internal details of vessels are generally overlooked in traditional designs and manufacturing. Thus, the flexibility of 3D printing allows for rapid prototyping and mass customization, significantly reducing development time and costs while offering new esthetic possibilities for modern designs.

3D printing enables rapid prototyping and significantly shortens the product development cycle. It is also suitable for mass customization, avoiding high mold costs, and overcoming the process limitations of traditional manual manufacturing. This achieves flexibility in both design and manufacturing.

4.2.2 Case study on material innovation

The overall material of artworks is generally uniform, such as porcelain, gold, silverware, or jade. Examples of multiple materials in a single work often appear in metal artworks, such as in the craft of inlaid gold and silver, which require multiple processes to complete (see **Figure 3**). This is a bronze beast head from the Warring States period in China decorated with patterns using the inlaid gold and silver technique. This process is essentially a casting process. Grooves were carved into the mold during the production of the bronze object, which was then filled with gold and silver

Figure 2.
Top view of a 3D-printed vase, showcasing the intricate internal and external structures achievable through additive manufacturing.

materials after casting, and the surface was polished to achieve a seamless fit between the materials. However, different materials are separate and prone to detachment.

Modern 3D printing and advanced manufacturing technologies can integrate multiple mixed materials in a single print and metal materials. For example, the delicate component in **Figure 3** was created using a combination of Fused Deposition Modeling (FDM) and Direct Ink Writing (DIW) printing processes. Moreover, with robotic control systems and 3D printing manufacturing methods, it is possible to achieve a mix of different materials in a single piece and a combination of different materials, processes, and colors, allowing a single workflow to achieve multiple effects. This single-step forming technology process reduces the need for multiple processing stages owing to different materials, processes, and colors, thereby improving production efficiency. Additionally, because the 3D printing process involves melting materials, the bonding between different materials and structures is more secure, thereby increasing the sturdiness of the work. Therefore, this technology provides a richer and more flexible space for comprehensive artistic esthetic creation, making it easier to realize more complex ideas.

4.2.3 Case study on form uniqueness

Traditional mechanized porcelain production typically uses molds and rotary machines to shape a porcelain body, as shown in **Figure 4**. This process results in industrial ceramic products that we commonly see, mostly regular circular shapes. Complex and twisted forms are rare because creating such shapes during the body-forming stage involves a complicated process that hinders efficiency. For example, the method of injecting ceramic slurry and demolding after shaping is illustrated in **Figure 4**. The manual second carving of the details after forming the bas are shaped, leading to high labor and time costs for mass production.

Using 3D ceramic printing, we can print complex porcelain shapes in a single step, thereby reducing the need for multiple intricate carving processes and avoiding the limitations of complex forms. The shapes and mechanical structures that traditional ceramic designs cannot achieve can be realized through 3D ceramic printing technology. This technology provides porcelain with plastic-like malleability and allows for a certain degree of mass production, thereby reducing molding costs.

Figure 3.
Horse-head-shaped bronze chariot ornament inlaid with gold and silver, warring states period (403-221 BC), collected in the National Museum of China (left) and curved nano-silver antennas fabricated by FDM and DIW printing techniques (right).

Figure 4.
Industrial machine ceramic body (left), mold casting ceramic body (middle), hand-carved details of the ceramic body (right).

4.3 Benefits and challenges

Advanced robotic manufacturing has brought unprecedented changes to the industrial sector and fields of art and design. For designers and artists, advanced robotic manufacturing presents opportunities and challenges.

The benefits of robotic additive manufacturing are evident when compared with manual craftsmanship and traditional industrial manufacturing. First, in terms of creative and custom design, 3D printing enables artists to transform complex and imaginative designs into actual works. It also allows for high customization with directly and accurately printed materials, significantly reducing the initial molding costs associated with individual customization. Second, 3D printing can significantly reduce the time required for manufacturing, particularly for small-scale production, allowing artists to create prototypes and final works without the need for expensive molds or other manufacturing processes. Additionally, 3D printing is more environmentally friendly than traditional manual manufacturing methods because it uses only the necessary materials and often uses recycled or biodegradable materials. Modern 3D printers have evolved from professional models to more affordable and compact desktop versions. They have become more user-friendly and available in various models that cater to different needs, including educational purposes. These printers can be easily operated, making them accessible for beginners and even young learners, encouraging creativity and early development of design skills. This expansion of advanced manufacturing accessibility and boundaries allows more people to participate, fostering the generation of diverse and rich creative ideas.

Pervasive challenges accompany opportunities.

Quality and resolution challenges: The traces of industrial machine production persist. For example, layered tracks and burrs in 3D printing still present challenges. Although machine-manufactured designs can achieve complex structures, their refinement and detail cannot fully match traditional manual manufacturing methods. This lack of precision poses a challenge for small-scale, high-quality artistic production.

Material limitations: Although 3D printing can use various materials, limitations remain. Some materials suitable for printing may not be appropriate for specific artworks, and the range of available colors and coating methods is limited.

Technical knowledge barriers: In reality, many artists lack technical knowledge. Robotic additive manufacturing (RAM) is an advanced technology that has only recently been applied in arts, making it relatively unfamiliar to many artists. However, specialized training in how to use computers to control robots, models, and design printing paths, as well as the operation of 3D printers, is limited. Most artists or art designers do not have an engineering background and find it challenging to learn and master automation and robotics. Therefore, for artists who are not technically inclined, using advanced technology to support their designs is a significant obstacle.

Intellectual property issues: Because robotic technology lowers production costs and 3D printing increases shaping precision, designs are more easily replicated. This raises concerns about intellectual property among artists and designers. Without proper regulations, advanced robotic manufacturing could facilitate unauthorized production, use, and sale of identical or patented products within a specific timeframe, thus infringing on creators' rights.

Pricing issues: 3D-printed works are priced significantly higher than similar products made by traditional manual manufacturing, limiting their marketability. Consumers are more inclined to purchase a product with a similar appearance and functionality at a lower price to meet their usage and esthetic needs. Compared with price, whether advanced manufacturing technology or traditional manual manufacturing techniques are used is not a primary concern for most consumers.

5. Future directions and research opportunities

5.1 Multi-process integration

RAM is at the forefront of revolutionizing artistic design, enabling the creation of intricate and esthetically appealing structures with high efficiency. However, challenges, such as precision and material performance, persist. To address these issues, multi-process integration combining additive and subtractive methods can be highly effective [19, 37]. This approach allows for the creation of near-net shapes and the removal of surface defects, significantly enhancing artistic works' functional and esthetic qualities.

Multi-process integration involves various manufacturing techniques within a single-machine setup, such as DED and PBF combined with CNC machining. Advanced control systems and software manage the transition between different methods, ensuring seamless operation and high-quality output. Enhanced precision is achieved by switching between processes, allowing for the precise machining of complex features, while improved material properties result from optimizing the final product through combined processes.

By incorporating cycles of additive manufacturing, inspection, and subtractive machining, artists and designers can achieve high-precision final forms, expanding the potential for complex and visually stunning creations. Additionally, this integration reduces the need for multiple setups and transfers, reduces manufacturing time, and increases efficiency. This versatility in design accommodates a broader range of materials and complex geometries, which is beneficial for artistic and custom applications.

In general, multi-process integration in RAM enhances artistic design's esthetic and functional aspects. Artists and designers can create innovative, high-quality, visually captivating, and functionally superior artistic products by leveraging advanced techniques and integrating various processes.

5.2 Digital twin technology

In artistic design, integrating digital twin technology with RAM offers unprecedented opportunities for real-time monitoring and optimization [38]. A Digital

Twin acts as a dynamic, virtual representation of a physical object or process, integrating data from various sources to create a comprehensive model that mirrors the physical counterpart. This enables artists to visualize and modify their designs throughout the manufacturing process. This model allows for real-time interaction and feedback, facilitating adaptive control and optimization of the manufacturing process.

This technology ensures continuous quality control and facilitates fine-tuning esthetic elements, making predicting and preventing defects possible. The Digital Twin combines historical and real-time data to enable direct interaction and control, providing high levels of automation and adaptive process management. Combining digital twins with advanced sensors and AI-driven feedback systems enhances the accuracy and reliability of artistic manufacturing processes. Artists can immediately observe the impact of design changes, allowing for more iterative and refined design processes. Digital twins facilitate high levels of customization and precision, which is essential for creating unique, intricate designs that are visually appealing and structurally sound. Simulating the manufacturing process can help identify and address potential issues before they occur, reducing the risk of defects and failures.

Integrating Digital Twin technology in robotic additive manufacturing provides a robust framework for enhancing artistic design's esthetic and functional aspects. By leveraging real-time data and advanced simulation capabilities, designers can push the boundaries of what is possible in AM, creating innovative and high-quality artistic products.

5.3 Human-machine collaboration

Human-machine collaboration (HMC) in RAM [21] introduces a new paradigm in artistic creation, blending human creativity with the precision and efficiency of robotic systems. This collaborative approach enhances the artistic value of RAM by leveraging human intuition, artistic vision, and robotic precision. Collaborative processes enhance human capabilities through advanced tools and systems that support complex design tasks. Designers can leverage intelligent systems to explore previously unattainable creative possibilities. Artists can directly interact with the manufacturing process by integrating advanced human-machine interfaces, allowing for real-time adjustments and creative inputs. Robotic systems bring precision and repeatability to manufacturing, ensuring that the final products meet high-quality standards and have intricate design specifications. This synergy improves the esthetic quality of the artwork and expands the boundaries of what can be achieved through artistic design. Collaborative systems provide immediate feedback during the design and manufacturing stages, allowing iterative improvements and refinement. This real-time interaction between human designers and machines fosters a dynamic and responsive design process.

The fusion of human creativity and robotic technology has fostered a new era of innovation by creating complex, customized, and visually captivating designs that were previously unattainable. Integrating human input with automated manufacturing allows highly customized and personalized products to be produced and tailored to specific esthetic and functional requirements. This balance is crucial in applications in which visual impact and usability are critical. This approach opens new avenues for artists to experiment with shapes, textures, and materials in ways that are impossible using traditional manual manufacturing methods.

Furthermore, collaboration allows for exploring novel materials and manufacturing techniques, pushing the boundaries of what is possible in design and production.

5.4 Innovative applications

With the advancement of technology, the precision and complexity of collaborative machine manufacturing in design will further improve, enabling designers to realize more intricate and complex ideas and create astonishing works. Simultaneously, the automation and intelligence of machine manufacturing will make the design process more efficient and precise, thereby promoting the development of the design field. Aesthetics plays a significant role in machine manufacturing. Designers can leverage machine manufacturing technology to integrate multidimensional esthetic concepts into product design better, making designs functionally powerful, esthetically pleasing, and reflective of users' spiritual pursuits and identity.

In summary, integrating aesthetics and advanced robotic manufacturing will drive future designs toward more terrific refinement, beauty, individuality, energy efficiency, environmental friendliness, and technological style. This will have a profound and transformative impact on modern design aesthetics and concepts.

5.5 Research opportunities

Research and practice in advanced robotic manufacturing are well-developed in the industrial sector. However, in art and design, technical barriers result in most artists being unable to operate robots and 3D printers or to understand the material selection for 3D printing. Many artists may hire technicians to help realize their creative ideas, but the more intermediaries involved in the creative process, the more the presentation of the artist's vision is affected, making it difficult to integrate technology and art truly. This can add a burden to the artist's creative design process. Therefore, artists are not enthusiastic about using machine methods to replace traditional manual manufacturing techniques as they wish to retain more personal touch and creative freedom, avoiding inconvenience.

The real beginning for artists to adopt 3D printing technology as a means of artistic creation lies in how the advanced robotic manufacturing process can present richer personalized traces of the artist and reduce technical operational barriers. In the future, promoting the integration of aesthetics and RAM will require deeper communication between technology and art.

Research on simplifying equipment operation: The design of the robot control interface, process planning, and 3D printer operation should be more user-friendly, thereby reducing the difficulty of operating professional 3D printing hardware and software. This study aims to broaden the application of advanced robotic manufacturing equipment, making it accessible in a broader range of fields.

Research on material diversification: Robotic additive manufacturing led by 3D printing has significant potential in art and design. However, the range of printable materials and colors supported is insufficient for artistic creation. Artistic creation demands a greater variety of materials and colors than industrial manufacturing. To better integrate industrial additive manufacturing with artistic esthetic production, overcoming material and color limitations and developing more adaptive materials and diverse color options are essential.

Esthetic style exploration in the digital age: Modern design aesthetics is a science that studies the esthetic relationship between people and design. It involves many disciplines, such as ergonomics, materials engineering, environmental science, economics, social markets, culture, and art. The contradiction between mechanical production and esthetic needs is the central issue of modern design aesthetics research, and its mission is to explore the essence of modern design, create new forms with esthetic laws, and promote the continuous development of the modern design movement [39]. 3D printing technology can help artists transform more complex, intricate, and imaginative designs into reality, providing a new dimension and innovative opportunities for "beauty" bestowed by productivity development in the current era. Artists must inherit traditional aesthetics and discover and create beauty in new dimensions as creators of art and beauty. They should seize the opportunities and space brought about by technological advancements and continually explore new aesthetics that bear the imprint of the current era and technological characteristics.

Enhancing the technological literacy of esthetic creators: In recent years, advanced technologies such as robotics, 3D printing, and artificial intelligence have rapidly permeated production and daily life. As practitioners in the art and design industry serve as a sector of esthetic production, it is imperative to consciously focus on and learn theoretical knowledge, technologies, operations, and applications related to advanced production. Keeping pace with the speed of modern development and breaking through personal cognitive boundaries are essential. By embracing and integrating advanced technologies into their practices, artists and designers can ensure they remain at the forefront of innovation in the technological era.

6. Conclusion

Integrating aesthetics and robotic additive manufacturing (RAM) in artistic design represents a significant leap forward in the creative process, merging the precision and efficiency of technology with artistic expression. RAM's ability to produce complex, intricate designs offers artists unprecedented creative freedom rapidly and consistently. This synthesis of art and technology enables the realization of detailed, visually appealing works previously unachievable through traditional manual manufacturing methods.

RAM's advanced robotics, multi-process integration, and real-time monitoring capabilities ensure high precision and quality in artistic creation. These technologies allow artists to explore new forms, materials, and esthetic concepts, enhancing their artistic value and emotional resonance. Furthermore, the potential for human-machine collaboration in RAM introduces a new paradigm in which artists can interact directly with the manufacturing process, incorporating their intuition and creativity into the technological workflow.

However, integrating RAM in artistic design presents challenges, such as the need for advanced technical knowledge and potential intellectual property issues. Addressing these challenges through user-friendly interfaces, diverse material options, and streamlined operational processes will be crucial for broader adoption and innovation in this field.

In summary, RAM offers a transformative approach to artistic creation, enabling complex, customized, and high-quality designs that push the boundaries of traditional art. By embracing RAM, artists can explore new frontiers in design, fostering a dynamic interplay between technology and creativity that shapes the future of artistic expression.

Acknowledgements

This work was supported by the 2023 High-level and Shortage of Talents Scientific Research Start-up Funding Project, Shanghai Technical Institute of Electronics & Information (Grant No. GCC2023018), Shanghai Technical Institute of Electronics and Information High-Level Talent Incentive Program, the China Postdoctoral Science Foundation (Grant No. 2023 M741535), and the Guangdong Provincial Basic and Applied Basic Research Fund Project, Regional Joint Fund-Youth Fund (Grant No. 2023A1515110089).

Appendices and nomenclature

AM	additive manufacturing
BJ	binder jetting
CNC	computer numerical control
DED	directed energy deposition
DIW	direct ink writing
DLP	digital light processing
DMLS	direct metal laser sintering
FDM	fused deposition modeling
HMC	human-machine collaboration
MEX	material extrusion
PBF	powder bed fusion
RAM	robotic additive manufacturing
SL	sheet lamination
SLA	stereolithography
SLS	selective laser sintering
UX	user experience
VP	vat photopolymerization

Author details

Shuying Wang[1] and Donghua Zhao[2]*

1 Design and Art College, Shanghai Technical Institute of Electronics and Information, Shanghai, China

2 School of System Design and Intelligent Manufacturing, Southern University of Science and Technology, Shenzhen, China

*Address all correspondence to: dongdong5212a@163.com

References

[1] Gibson I, Rosen D, Stucker B. Additive Manufacturing Technologies: 3D Printing, Rapid Prototyping, and Direct Digital Manufacturing. 2nd ed. New York, NY: Springer New York; 2015. 1-498 pp. Available from: http://link.springer.com/book/10.1007/978-1-4939-2113-3

[2] Zhao D, Guo W. Shape and performance controlled advanced Design for Additive Manufacturing: A review of slicing and path planning. Journal of Manufacturing Science and Engineering, Transactions of the ASME. 2020;**142**(1):1-23

[3] Ding D, Pan Z, Cuiuri D, Li H, Larkin N, Van Duin S. Automatic multi-direction slicing algorithms for wire based additive manufacturing. Robotics and Computer-Integrated Manufacturing. 2016;**37**:130-150

[4] Zhao D, Guo W. Mixed-layer adaptive slicing for robotic additive manufacturing (AM) based on decomposing and regrouping. Journal of Intelligent Manufacturing. 2020;**31**(4):985-1002

[5] Bhatt PM, Kabir AM, Peralta M, Bruck HA, Gupta SK. A robotic cell for performing sheet lamination-based additive manufacturing. Additive Manufacturing. 2019;**27**:278-289

[6] Attaran M. The rise of 3-D printing: The advantages of additive manufacturing over traditional manufacturing. Business Horizons. 2017;**60**(5):677-688

[7] Gibson I, Rosen D, Stucker B, Khorasani M. Development of Additive Manufacturing Technology. In: Additive Manufacturing Technologies. Cham: Springer; 2021. DOI: 10.1007/978-3-030-56127-7_2

[8] Bourell DL. Perspectives on additive manufacturing. Annual Review of Materials Research. 2016;**46**(1):1-18

[9] Pires JN, Azar AS, Nogueira F, Zhu CY, Branco R, Tankova T. The role of robotics in additive manufacturing: Review of the AM processes and introduction of an intelligent system. Industrial Robot: The International Journal of Robotics Research and Application. 2021;**49**(2):311-331

[10] Riccio A, Saputo S, Sellitto A, Russo A, Di Caprio F, Di Palma L. An insight on the crashworthiness behavior of a full-scale composite fuselage section at different impact angles. Aerospace. 2019;**6**(6):72

[11] Wang Z, Xie F, Xie Z, Liu X. Technical status and prospect of measurement and machining integration technology for robotized equipment. ROBOT. 2024;**46**(1):118-128

[12] Weissenböck R. Robotic Design-Fabrication. In: Real Time-Proceedings of the 33rd eCAADe Conference. eCAADe (Education and Research in Computer Aided Architectural Design in Europe) and Faculty of Architecture and Regional Planning. Vol. 2. TU Wien. 2015. pp. 309-318

[13] Schwartz M, Hwang S, Lee Y, Won J, Kim S, Park J. Aesthetic design and development of humanoid legged robot. In: 2014 IEEE-RAS International Conference on Humanoid Robots. Madrid, Spain: IEEE; 2014. pp. 13-19. Available from: http://ieeexplore.ieee.org/document/7041311/

[14] Galimberti G, Doubrovski EL, Guagliano M, Previtali B, Verlinden JC. Investigating the links between the process parameters and their influence on the aesthetic evaluation of selective laser melted parts. In: Proceedings of the 27th Annual International Solid Freeform Fabrication Symposium. Austin, Texas, USA: The University of Texas at Austin; 2016. pp. 2367-2387

[15] Liu Z, Liu Q, Xu W, Wang L, Zhou Z. Robot learning towards smart robotic manufacturing: A review. Robotics and Computer-Integrated Manufacturing. 2022;**77**:102360

[16] Urhal P, Weightman A, Diver C, Bartolo P. Robot assisted additive manufacturing: A review. Robotics and Computer-Integrated Manufacturing. 2019;**59**:335-345

[17] Lu L, Yuan S, Yao X, Li Y, Zhu J, Zhang W. In-situ process evaluation for continuous fiber composite additive manufacturing using multisensing and correlation analysis. Additive Manufacturing. 2023;**74**:103721

[18] Lu L, Hou J, Yuan S, Yao X, Li Y, Zhu J. Deep learning-assisted real-time defect detection and closed-loop adjustment for additive manufacturing of continuous fiber-reinforced polymer composites. Robotics and Computer-Integrated Manufacturing. 2022;**79**(August):102431

[19] Flynn JM, Shokrani A, Newman ST, Dhokia V. Hybrid additive and subtractive machine tools—Research and industrial developments. International Journal of Machine Tools and Manufacture. 2016;**101**:79-101

[20] Wang L, Lu B. Development of additive manufacturing technology and industry in China. Strategic Study of CAE. 2022;**24**(4):202-211

[21] Xiong Y, Tang Y, Kim S, Rosen DW. Human-machine collaborative additive manufacturing. Journal of Manufacturing Systems. 2022;**2023**(66):82-91

[22] Pérez-Castillo JL, Cuan-Urquizo E, Roman-Flores A, Olvera-Silva O, Romero-Muñoz V, Gómez-Espinosa A, et al. Curved layered fused filament fabrication: An overview. Additive Manufacturing. 2021;**47**(September):102354

[23] Ding Y, Dwivedi R, Kovacevic R. Process planning for 8-axis robotized laser-based direct metal deposition system: A case on building revolved part. Robotics and Computer-Integrated Manufacturing. 2017;**44**:67-76

[24] Gibson BT, Mhatre P, Borish MC, Atkins CE, Potter JT, Vaughan JE, et al. Controls and process planning strategies for 5-axis laser directed energy deposition of Ti-6Al-4V using an 8-axis industrial robot and rotary motion. Additive Manufacturing. 2022;**58**:103048

[25] Jiang J, Newman ST, Zhong RY. A review of multiple degrees of freedom for additive manufacturing machines. International Journal of Computer Integrated Manufacturing. 2021;**34**(2):195-211

[26] Wu Y, Wang H, Li Z. Quotient KinematicsMachines: Concept, analysis, and synthesis. Journal of Mechanisms and Robotics. 2011;**3**(4):041004

[27] Zhao D, Guo W, Han Y. General function sets theory-based type synthesis of the collaborative 3D printer for planar and non-planar printing. Proceedings of the Institution of Mechanical Engineers, Part C: Journal of Mechanical Engineering Science. 2022;**236**(16):8965-8979

[28] Zhang K, Chermprayong P, Xiao F, Tzoumanikas D, Dams B, Kay S, et al. Aerial additive manufacturing with multiple autonomous robots. Nature. 2022;**609**(7928):709-717

[29] Li Y, Zhang W, Zhang Q. A method for converting topology optimization designs into additively manufacturable structures. Engineering. 2018;**4**(2):179-189

[30] Zhang J, Yang W, Li Y. Application of artificial intelligence in composite materials. Advances in Mechanics. 2021;**51**(4):865-900

[31] Guo F. On the Component of Design Beauty in the Design Aesthetics Principles. In: 2015 2nd International Conference on Education, Language, Art and Intercultural Communication (ICELAIC-15). Paris, France: Atlantis Press; 2015. pp. 478-480

[32] ASTM International. Standard Terminology for Additive Manufacturing Technologies (ASTM F2792-12). 2012. Available from: https://www.astm.org/f2792-12.html

[33] Forsey J. The Aesthetics of Design. USA: OUP; 2013

[34] Kostelnick C. The art of visual design: The rhetoric of aesthetics in technical communication. Technical Communication. 2020;**67**(4):6-27

[35] Hekkert P. Design aesthetics: Principles of pleasure in design. Psychology Science. 2006;**48**(2):157

[36] Findeli A. Ethics, aesthetics, and design. Design Issues. 1994;**10**(2):49-68

[37] Kanishka K, Acherjee B. A systematic review of additive manufacturing-based remanufacturing techniques for component repair and restoration. Journal of Manufacturing Processes. 2023;**89**:220-283

[38] Xiong Y, Tang Y, Zhou Q, Ma Y, Rosen DW. Intelligent additive manufacturing and design state of the art and future perspectives. Additive Manufacturing. 2022;**59**(PA):103139

[39] Huang Q. On the value characteristics and development process of modern design aesthetics. Decoration. 2004;**9**:24-25

Chapter 4

A Review of Solid-State Additive Manufacturing: Method, Microstructural Evolution, Mechanical Properties, Applications and Challenges

Ying Li, Xiwu Li, Changshu He, Yanan Li, Kai Wen, Lizhen Yan, Wei Xiao, Yongan Zhang and Baiqing Xiong

Abstract

Solid-state additive manufacturing (SSAM) is an emerging technology that offers a unique approach by avoiding the melting and solidification of metals during the fabricating process. As a result, it eliminates common solidification related defects like porosity and cracking. This method is characterized by low residual stress, low distortion, and the formation of fine equiaxed grain structures, making it particularly advantageous for fabricating large-scale components in sectors like aerospace, aviation, and rail transportation. This review explores the fundamental principle and recent development of solid-state additive techniques, namely, friction stir additive manufacturing (FSAM), additive friction stir deposition (AFSD), friction and rolling-based additive manufacturing (FRAM), and wire-based friction stir additive manufacturing (W-FSAM). Furthermore, the advantages, challenges, and future trends of these SSAM techniques are reviewed. A comprehensive review of the characteristics and current research trends in solid-state additive manufacturing methods is crucial for driving further advancements in this technology.

Keywords: solid-state additive manufacturing, working principle, developments, microstructure, mechanical property

1. Introduction

Additive manufacturing (AM), also known as 3D printing, uses computer-aided design (CAD) and a high-energy beam (such as a laser or electron beam) as a heat source to fabricate components layer by layer [1]. This technology originated in the late 1960s, and with the development of computer technology and joining/welding techniques in the past 30 years, AM technology has rapidly developed in the field of intelligent manufacturing [2].

In 2012, Guan Qiao from the Beijing Aeronautical Manufacturing Engineering Research Institute introduced the concept of "Generalized Additive Manufacturing" (**Figure 1**), which expanded the range of available heat sources. In addition to using high-energy beams (electron beams, laser beams) as welding heat sources, other types of welding heat sources (arc, plasma, friction heat, etc.) are also utilized for additive manufacturing. Furthermore, it has expanded the manufacturing modes, such as the rapid prototyping manufacturing of block assembly welding, breaking through the layer-by-layer accumulation manufacturing mode of 3D printing [3].

With the increasing demand for lightweight materials in aviation, aerospace, weapons and equipment, and rail transit industries, the development of high-performance, high-reliability, and lightweight integral structural parts has become an important trend. These integral component structures are often complex and large-scale. The traditional manufacturing mode with "casting + forging + machining" is challenging in terms of the requirements for efficient and low-cost production processes. The metal additive manufacturing (MAM) is an effective way to fabricate large-scale integral components with high efficiency and low cost [4–7]. Currently, MAM technologies that employ high-intensity energy sources like lasers [8–10], electron beams [11, 12], and electric arcs [13, 14] have been extensively researched over the past three decades. These methods provide notable advantages, such as no need for molds, offering high production flexibility, and delivering enhanced efficiency [15]. However, there are some limitations to such fusion-based AM. Such as:

1. After melting and rapid solidification, forming metals are prone to severe deformation and high residual stress in the components [16].

2. During the fusion-based AM process, defects such as pores, cracks, and inadequate fusion are prone to occur in formed components. Additionally, alloys with a higher sensitivity for hot cracking are difficult to manufacture with high-energy beam AM [17–19].

3. The grain size of the formed components is large, and the segregation is easy to occur, resulting in an inhomogeneous structure [20, 21].

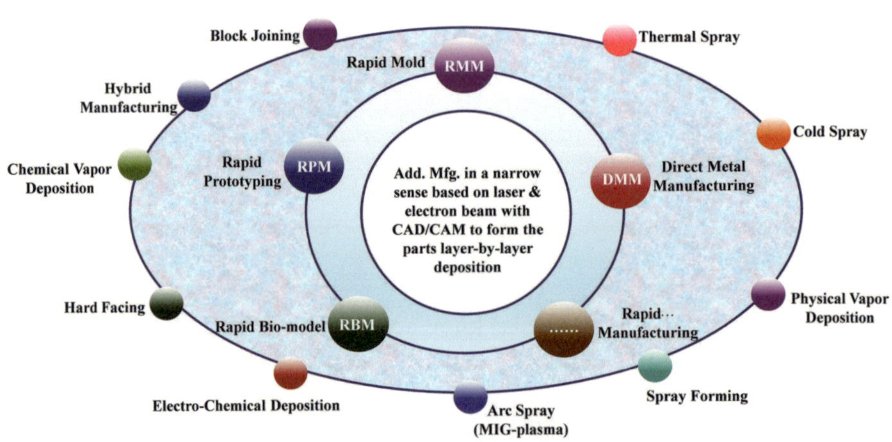

Figure 1.
The connotation and classification of generalized additive manufacturing technology [3].

To solve the aforementioned issues and satisfy the requirements for components with small deformation, low residual stress, and free from defects such as porosity and cracks, the solid-state additive manufacturing (SSAM) technology [22], which is based on the principle of friction stirring, has attracted the increasing attention of researchers. White [23], in year 2002, filed a patent for a process of consolidating metal powders/materials using friction joining. In 2005, Thomas et al. [24] introduced a novel application of friction stir welding-based processes for additive manufacturing. The following year, in 2006, Airbus utilized friction-based additive manufacturing (FBAM) technology to fabricate 2050 Al-Li wing ribs, demonstrating the potential of FBAM for producing low-cost, lightweight components [25, 26]. Boeing, in year 2012, believed that FBAM could be used for the pre-forming of structural components, achieving rapid manufacturing of these components and significantly improving the material utilization rate [27]. Based on the differences in working principles, FBAM technology is primarily divided into rotary friction welding (RFW), friction deposition (FD) [28], friction surfacing (FS) [29], linear friction welding (LFW) [30, 31], friction stir additive manufacturing (FSAM) [32–35], additive friction stir deposition (AFSD), friction and rolling-based additive manufacturing (FRAM), and wire friction stir additive manufacturing (W-FSAM) (shown in **Figure 2**). Among these technologies, FSAM, AFSD, FRAM, and W-FSAM are the most noteworthy in recent years. This paper provides a comprehensive review of the fundamental principles, current research on microstructures and mechanical properties, advantages, limitations, and future trends of FBAM technologies. It offers the latest research progress, serving as a valuable resource for individuals or companies seeking an in-depth understanding of FBAM technology.

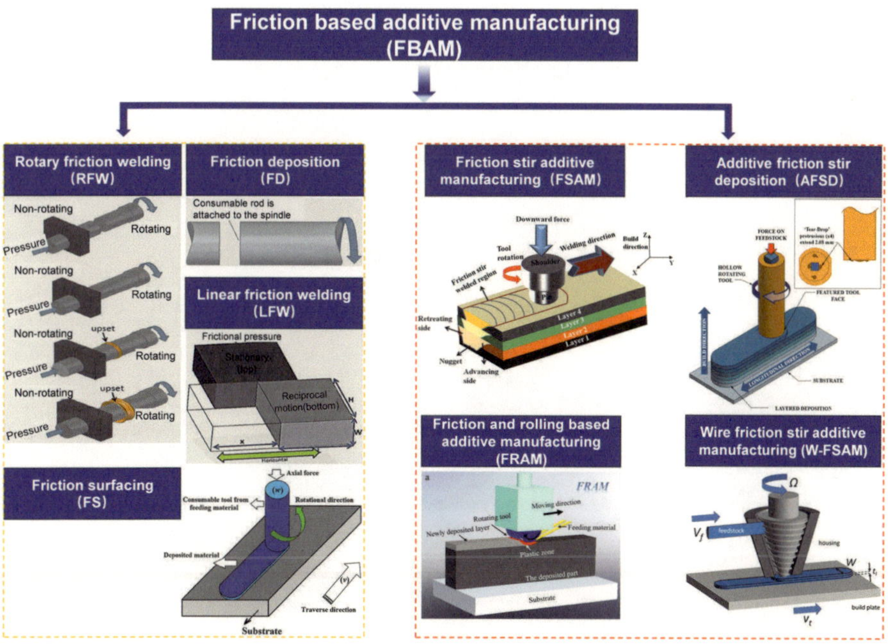

Figure 2.
Classification of FBAM technologies [22, 36–40].

2. Friction stir additive manufacturing (FSAM)

2.1 Working principle of FSAM

The basic working principle of the FSAM is similar to that of friction stir lap welding (FSLW). As illustrated in **Figure 3** [22], a non-consumable rotating tool is pressed into a set of overlapping plates under axial pressure, and the subsequent FSLW is carried out along predetermined paths. The materials around the pin undergo severe plastic deformation with the heat and stirring effects; thus, two plates were connected together [41]. If the build height achieves the desired height after the first run, the process for adding subsequent layers will be paused. However, if the required build height has not been reached, the top surface of the build is smoothed using suitable methods to eliminate any flash that developed during the FSLW process. Once this is done, a new plate or sheet is positioned on top of the existing one. These steps are continued until the desired build height is reached [42].

2.2 Current research

As the friction stir welding (FSW) and friction stir processing (FSP) technologies become increasingly mature in the field of joining or fabricating lightweight alloys such as aluminum and magnesium, FSAM technology is currently mainly applied to the additive manufacturing of aluminum and magnesium alloy components [43]. Palanivel et al. [36] in year 2015, it found that the FSAM technique could effectively produce WE43 magnesium alloy components with high performance. The strength of the component could reach 400 MPa, and the elongation achieved 17%. As shown in **Figure 4**, this remarkable performance is primarily attributed to the fine grain structure and favorable precipitate characteristics that develop as a result of the severe plastic deformation and elevated temperatures experienced during the FSAM process. This research has attracted widespread attention from scholars, and subsequently, a large number of related studies have emerged [33, 44].

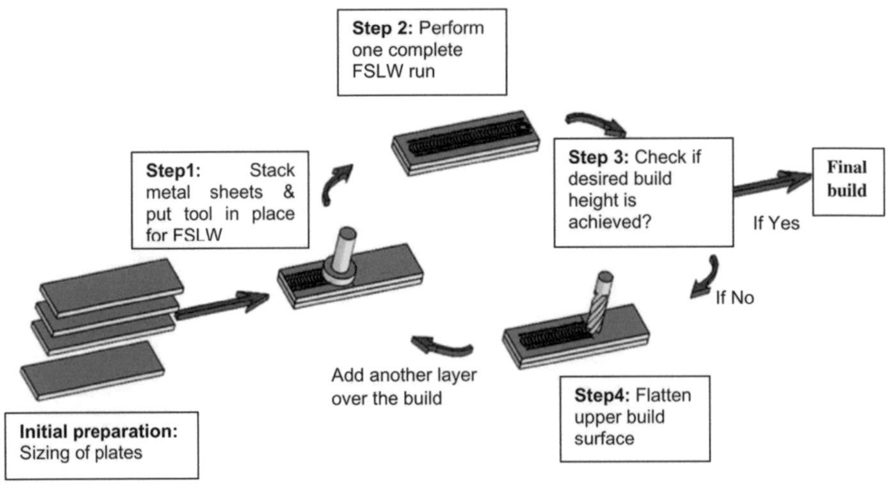

Figure 3.
Steps utilized in FSAM [22].

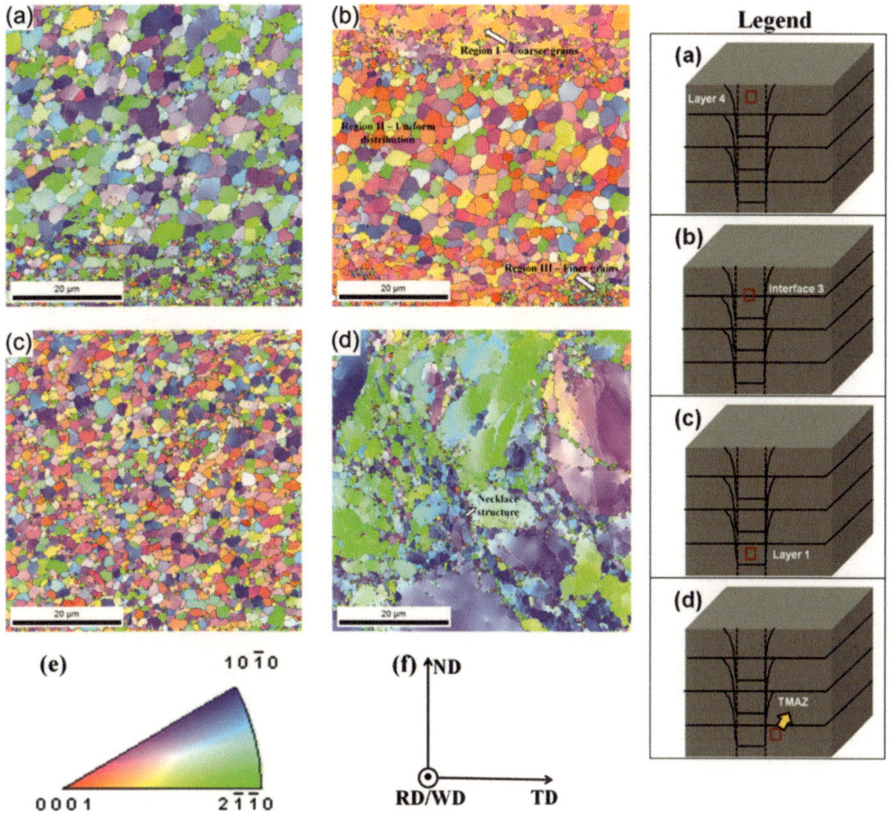

Figure 4.
EBSD maps in different regions of the build: (a) top region, (b) interface of layer 3–4, (c) bottom region, and (d) thermo-mechanically affected zone. Color correspondence scale and the sample coordinate system have been shown in (e) and (f) [36].

The hot topics mainly focus on the optimization of the formation, microstructure, and mechanical performance of the build. As in-depth and systematic research on FSAM continues, some challenges have been identified. Defects related to single-layered FSLW, such as hook or kissing bonds, are also present in the builds fabricated by the FSAM (shown in **Figure 5**) [45]. These defects are generally influenced by the shape of the tool, the rotation speed, and the travel speed, all of which can negatively impact the mechanical properties of the components produced by FSAM [42]. Zhao [46] proposed that the back-and-forth processing path can significantly reduce the hook defects that extend into the stir zone.

Apart from the various defects, the inhomogeneous microstructures and properties produced along the building direction are another important issue, particularly in heat-treatable alloys. During the FSAM process, fine equiaxed grains are formed as a result of severe plastic deformation and high temperatures. Simultaneously, the nano-scale particles dissolved into the matrix during the heating process, potentially re-precipitating during the cooling stage. Moreover, the layer-by-layer additive manufacturing process results in varying thermal exposure across different layers of the fabricated structure. Layers closer to the bottom of the build experience more thermal cycles, which would lead to grain coarsening and precipitate growth, ultimately

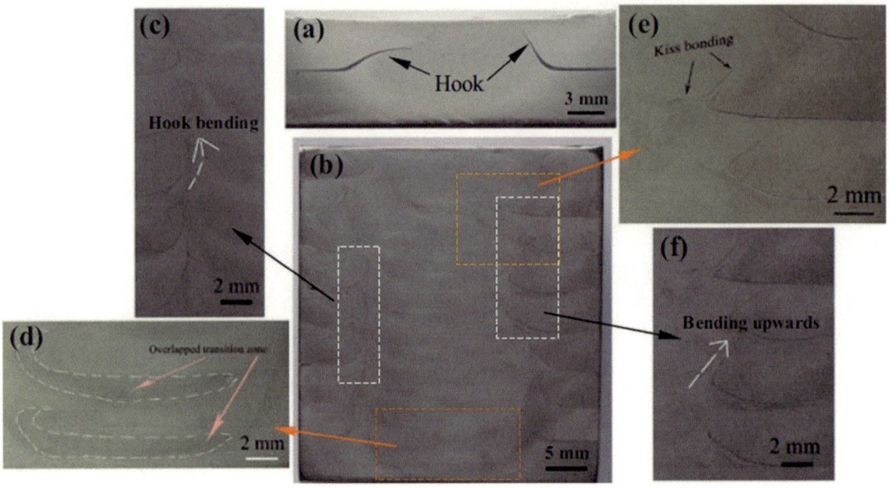

Figure 5.
Typical defects in the build: (a) hook defect in the single-layered build, (b) multi-layered build, and (c–f) local magnifications in (b) [45].

deducing the mechanical properties of the final build [47]. The macro-softening phenomenon in 7xxx aluminum alloy build, where the mechanical properties decreased from the top to the bottom region, was observed by Mao [45] and He [42].

In-process water cooling during FSAM has been shown to be an effective method for solving the aforementioned macro-softening issue. As shown in **Figure 6**, relatively uniform microstructures and mechanical properties were achieved along the building direction of the 7 N01 aluminum alloy build fabricated by the underwater

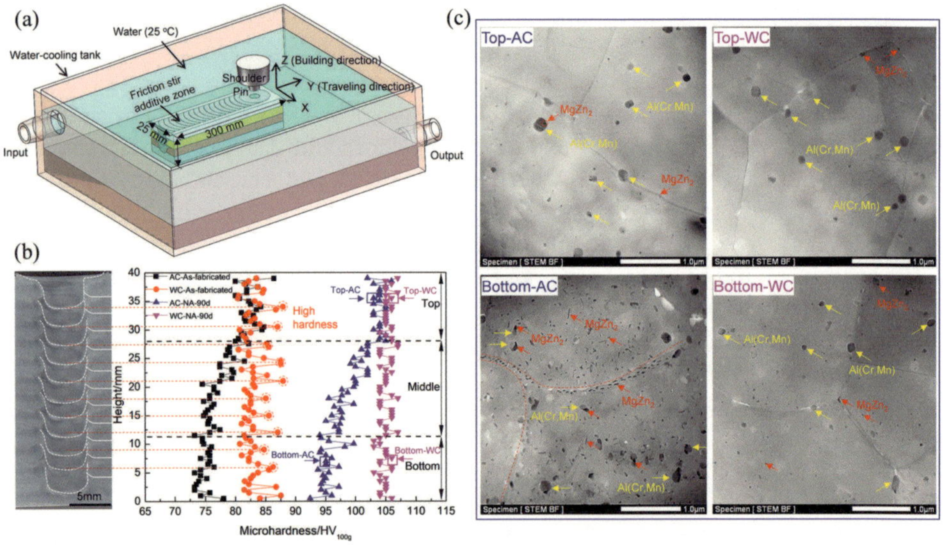

Figure 6.
(a) Experimental setup of the underwater FSAM process, (b) microhardness distributions on the cross-section of the air-cooled (AC) and water-cooled (WC) builds along the building direction, and (c) precipitated images in the top and bottom regions of the air-cooled and water-cooled build [47].

FSAM [47]. Underwater FSAM suppresses the impact of thermal cycles generated by new additive processes on the previously formed layers. This reduces the coarsening of the precipitates and grains, effectively mitigating or eliminating the macro-softening in the build and improving the mechanical properties of the metals in the lower part of the build.

The inhomogeneous characteristics observed in the FSAMed build are apparent not only on a macro scale but also on a micro-scale level. The periodic nature of the FSAM process induces variations in both the microstructure and the mechanical properties of the build along the building direction. In the FSAMed build fabricated using 5083-O aluminum alloy, increased microhardness was observed at the bottom region of each stir zone. This may be related to the fine equiaxed grains in this region, leading to fine grain strengthening [48]. However, Lu et al. [44] found a reduction in microhardness at the bottom region of each stir zone in the 2050 aluminum alloy build. A similar phenomenon was noted in the 7A04 aluminum alloy build (as shown in **Figure 7**), where a low-hardness region periodically appeared at the bottom of each overlapping zone along the build direction. This phenomenon was present in the as-fabricated build and became more visible after artificial aging [43].

To clarify the mechanism behind the abovementioned phenomenon, Li et al. [41] investigated the effects of restirring and reheating on the microstructural evolution during the underwater FSAM. As shown in **Figure 8**, restirring reduced the grain size at the bottom of the overlapping region from 1.97 to 0.87 μm, decreased the recrystallization degree from 74.0 to 29.8%, and transformed the initial random texture into a strong shear texture with the C {110}<110 > orientation. After reheating, static recrystallization occurred near the newly additive zones, increasing both grain size and recrystallization degree. The main factor of the decrease in mechanical properties at the bottom of the overlapping region was the high density of $Mg(ZnAlCu)_2$ phase precipitates, driven by the fine grain structure and the high density of subgrains and dislocations in this region.

Li et al. [49] further discovered that the high density of grain boundaries, subgrain boundaries, and dislocations facilitated the precipitation and coarsening of nano-scale phases during conventional aging at 120°C for 24 hours (as shown in **Figure 9**). This resulted in over-aging in the low-hardness zone (LHZ) of the build, further

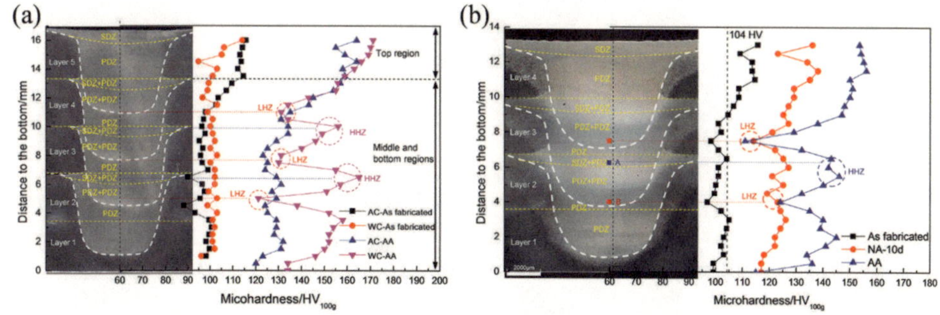

Figure 7.
(a) Microhardness distributions of the air- and water-cooled builds in the as-fabricated and artificial aging states; (b) microhardness distributions of the water-cooled build in the as-fabricated, natural aging for 10 days, and artificial aging states [43].

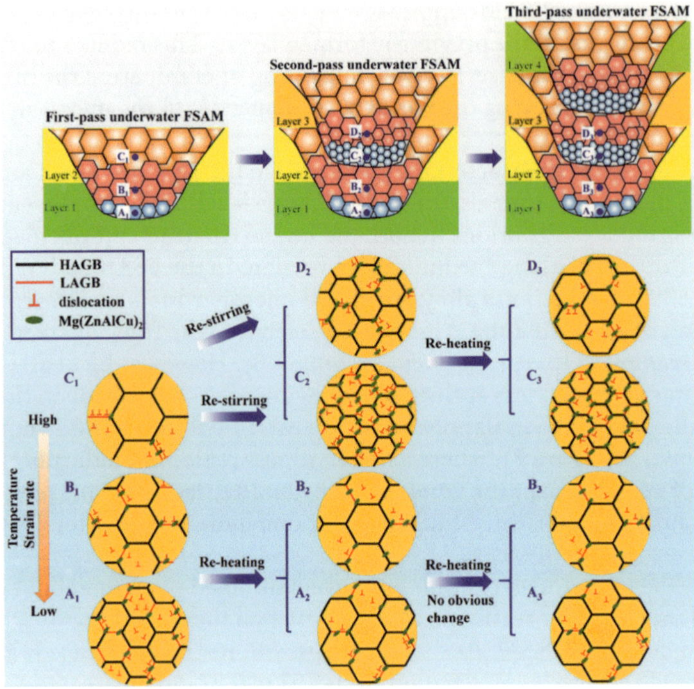

Figure 8.
Schematic of microstructural evolution during underwater FSAM. The specified positions of the microstructures are marked with blue dots and letters (A1, A2, A3, B1, B2, B3, C1, C2, C3, D2, D3) [41].

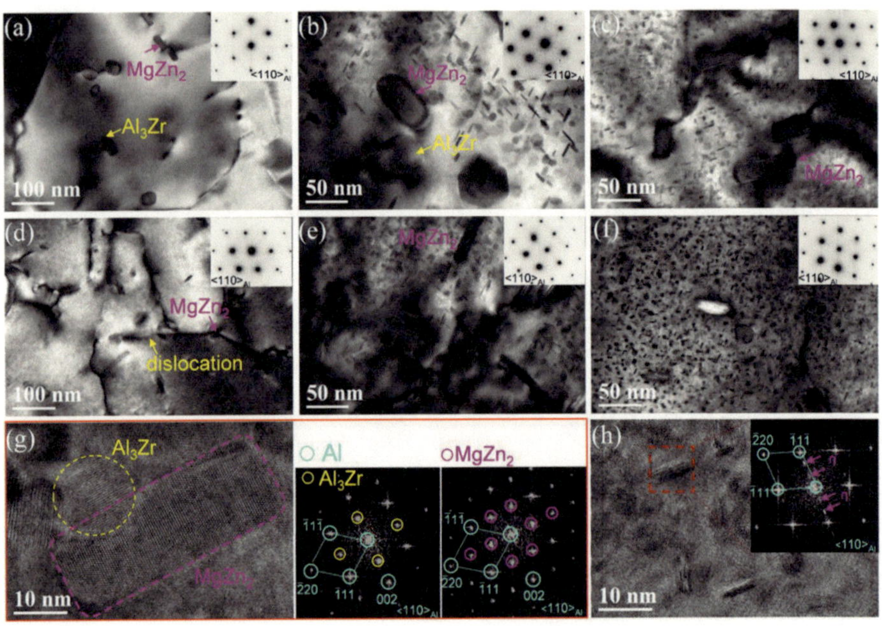

Figure 9.
TEM images of the precipitates in the low-hardness zone (LHZ) and high-hardness zone (HHZ) of the builds in the different aging states: (a, d) LHZ and HHZ after natural aging for 7 days; (b, e) LHZ and HHZ in the after aging at 120°C for 24 hours; (c, f) LHZ and HHZ after aging at 100°C for 48 hours; (g) HRTEM and FFTs of Al_3Zr and $MgZn_2$ phases; and (h) HRTEM and FFT of η phase [49].

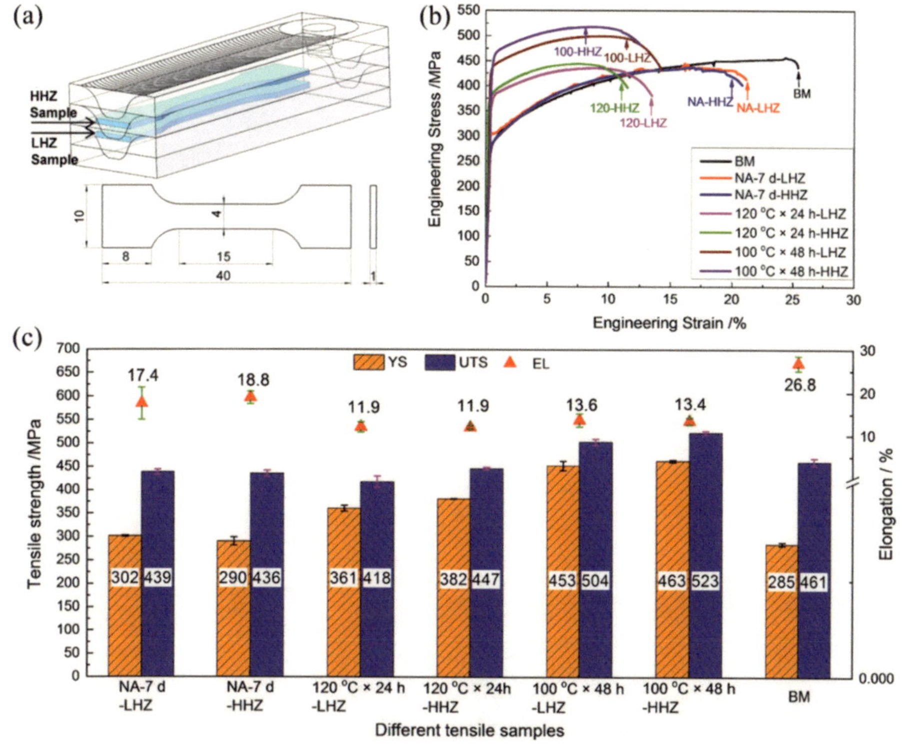

Figure 10.
Tensile properties of the samples extracted from LHZ and HHZ of the builds: (a) schematic of the extraction position and dimension of the tensile samples; (b) tensile curves of the typical tensile samples; and (c) tensile strength and elongation [49].

aggravating the micro-scale softening in the structure. Low-temperature aging treatments proved effective in mitigating the over-aging problem. As shown in **Figure 10**, after aging at 100°C for 48 hours, the yield strength in the LHZ and high-hardness zone (HHZ) reached 453 and 463 MPa, respectively, with the ultimate tensile strength increasing to 504 MPa in the LHZ and 523 MPa in the HHZ. Compared to the aging condition at 120°C for 24 hours, the micro-scale softening was significantly reduced.

Previous research primarily focused on multi-layer and single-pass FSAM processes. However, Liu et al. [50] employed a multi-layer, multi-pass FSAM technique (as shown in **Figure 11**) with rapid water cooling to fabricate a three-dimensional, large-scale bulk pure copper structure. This work demonstrated that large-scale bulk UFG copper can be successfully fabricated using FSAM, achieving uniform microstructure and high mechanical performance.

In recent years, the FSAM technologies were not only used to fabricate Al and Mg alloy builds. It is also possible to achieve the customization of microstructure and properties of the component fabricated by FSAM. Based on this, FSAM technology has been applied to the fabrication of hybrid structures [51] and gradient materials [51–56]. Derazkola et al. [57] fed the Fe_2O_3 nanoparticles into the interface of the base materials by FSAM (shown in **Figure 12**). The result indicated that the *in situ* feeding mechanism improves the physical and chemical properties of the printed structure.

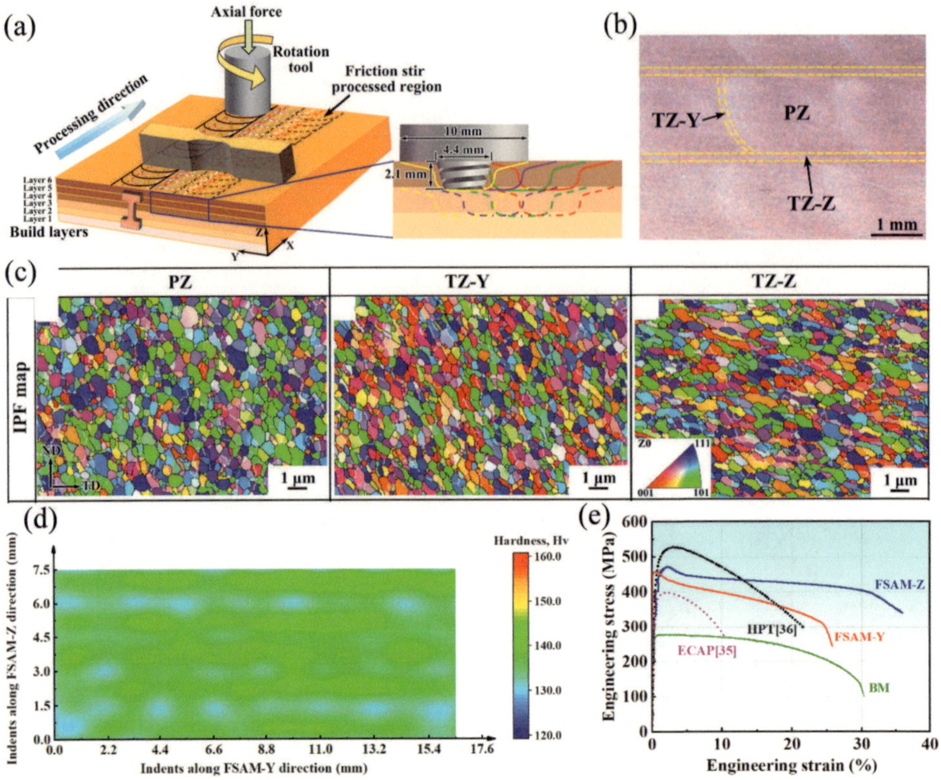

Figure 11.
(a) schematic diagram of multi-layer and multi-pass FSAM process, (b) local magnification of the FSAM Cu sample, (c) EBSD microstructures of different regions in FSAM Cu sample, (d) cross-sectional hardness distribution feature and (e) engineering tensile stress-strain curves of UFG Cu processed by different SPD methods [50].

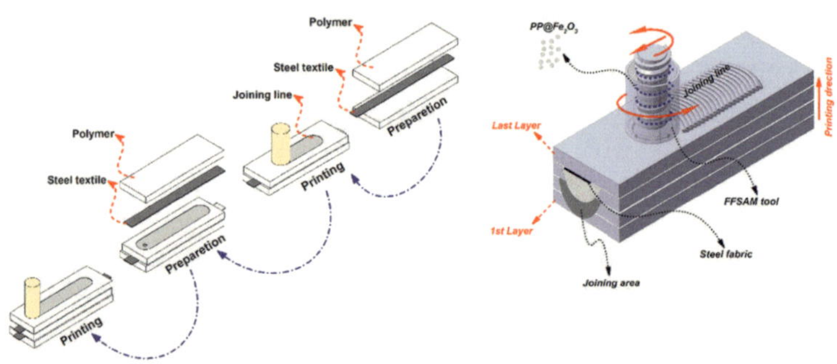

Figure 12.
Schematic view of nanocomposite structure printed process [57].

3. Additive friction stir deposition (AFSD)

3.1 Working principle of AFSD

The first commercialized AFSD technology was developed and patented by Aeroprobe Corporation in 2016 [58]. The schematic process of the AFSD process is illustrated in **Figure 13(a)** [59]. In this process, a non-consumable hollow, rotating

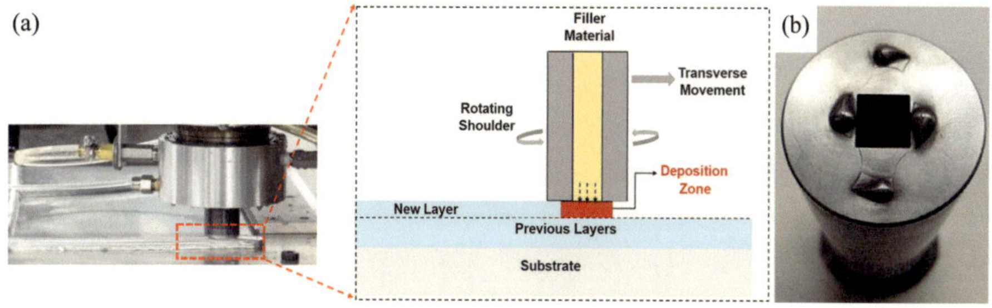

Figure 13.
(a) Actual image of the AFSD process and schematic diagrams of the process [59] and (b) Image of the teardrop featured tool [61].

tool with a shoulder is utilized. Powders and metal rods are used as feed material, which is continuously delivered to the processing area through the hollow section of the rotating tool. The shoulder rapidly rotates and generates heat through dynamic contact friction between the tool shoulder, feed materials, and material substrate [60]. To promote the flow of materials during the AFSD process, a rotation tool with four "teardrop" protrusions on the tool face (**Figure 13(b)**) [61].

3.2 Current research

The American Aeroprobe Corporation manufactured the first AFSD equipment in 2014 [62], and in recent years, in collaboration with the University of Alabama and Virginia Tech, they have carried out AFSD for 2219 [63], 6061 [38, 61, 64–66], 7075 [67, 68], 7050 [69], and 2050 [70] aluminum alloys. It provides a new path for developing fully dense, near-net shape components with a refined-equiaxed grain morphology.

The effects of the process parameters [71–73], deposition path [66], material flow features, thermal behavior during the AFSD process [74], and the effect of post-deposition heat treatment on the mechanical behavior of the build [75, 76] were explored. Such investigations were useful for tailoring the desired microstructures to reach the desired mechanical properties [65]. Complex thermo-mechanical behavior appeared during the AFSD process. To efficiently achieve the desired microstructure by AFSD parameter optimization, Han et al. [77] introduced a deep learning framework to quantitatively analyze the microstructural variations in builds produced by AFSD under various processing conditions (as shown in **Figure 14**). This framework allows for the quantitative analysis of trends and the prediction of microstructures, providing a pathway toward addressing several key materials science challenges. These include predicting microstructural evolution, modeling the processing-structure relationship, and optimizing the design of heterogeneous materials.

As AFSD is still in its early stages of development, a comprehensive understanding of the thermal process fundamentals—such as temperature evolution and heat generation mechanisms—has yet to be achieved. To bridge this gap, Garcia et al. [64] investigated these aspects during the AFSD process by employing *in situ* monitoring techniques. They used complementary infrared imaging, thermocouple measurements, and optical imaging (as illustrated in **Figure 15**) to gain insights into the thermal field and material flow behavior.

For further obtained high-performance components, Beck et al. [38] presents for the first time the influence of a standard post-deposition heat treatments (PDHT) on

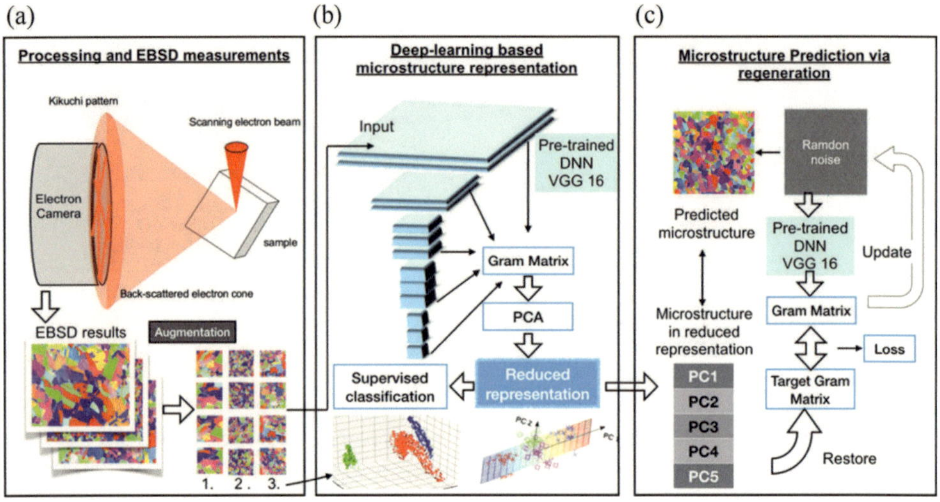

Figure 14.
Experimental workflow: (a) microstructures analyzed using EBSD, (b) deep learning-based microstructure representation, and (c) microstructure prediction via *regeneration [77].*

the nanostructure, microstructure, and mechanical properties of AA6061 processed *via* AFSD. While abnormal grain growth (AGG) was observed after solutionized/quenched (SQ), and solutionized/quenched/artificially aged (SQA). Despite the AGG, PDHT increases the as-deposited (AD) strength and hardness in the longitudinal direction, producing peak aged mechanical behavior. Particulary, an 8% increase in yield and 6% in ultimate strength were observed compared to the wrought control (shown in **Figure 16**).

In addition to the aluminum alloys, AFSD has also been applied to fabricate the Inconel [62], titanium, and steel [78, 79] alloys components. Patil et al. [80] found

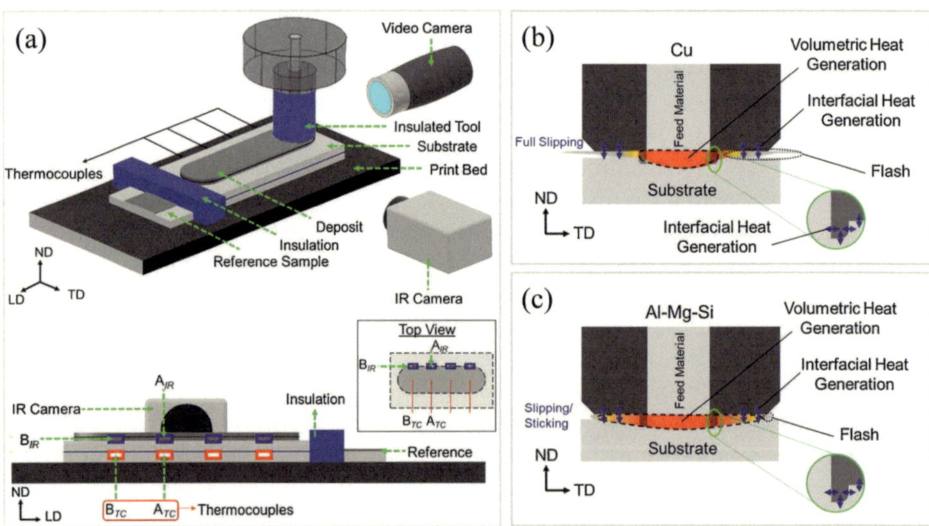

Figure 15.
(a) An illustration of the monitoring setup during the AFSD process. A diagram showing the heat generation mechanisms for (b) Cu and (c) Al-Mg-Si [64].

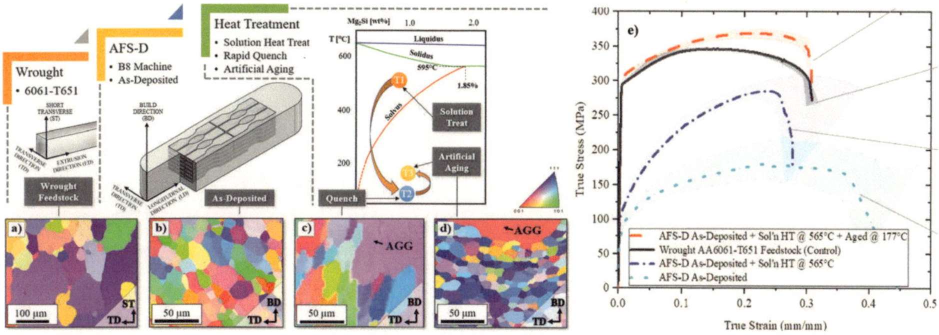

Figure 16.
Inverse pole figure maps for (a) extruded wrought AA6061-T651, (b) AFSD AD, (c) AFSD SQ, and (d) AFSD SQA. (e) Experimental true stress-strain tensile test results of wrought AA6061-T651, AFSD AD, AFSD SQ, and AFSD SQA [38].

that the fine equiaxed grains with size of ~6 μm in the AFSD IN625, along with bands of very fine grains of size ~ 1.2 μm within the interface regions, were formed. The grain size reduction led to a substantial improvement in the yield strength of the build (reaching 753 MPa), which is notably higher than that of the feed material (485 MPa), the as-cast condition (350 MPa), and other AM builds (shown in **Figure 17**).

The AFSD of Ti6Al4V alloy was successfully performed over a wide processing window. High performance was achieved, with ductility values reaching approximately 20% and maximum yield and ultimate tensile strengths of 1010 and 1233 MPa, respectively [81]. However, in the SS304 build fabricated by AFSD, strain-induced

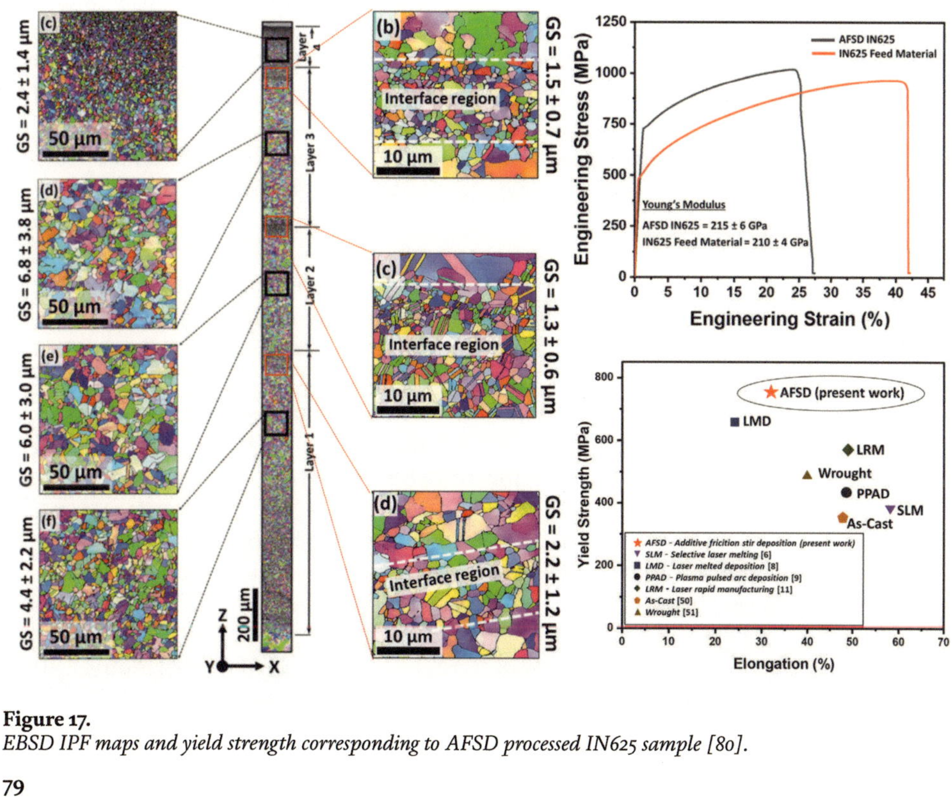

Figure 17.
EBSD IPF maps and yield strength corresponding to AFSD processed IN625 sample [80].

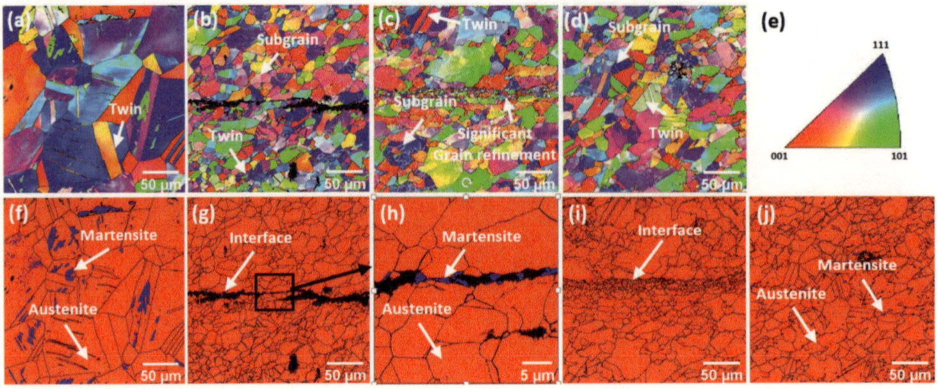

Figure 18.
IPF map of (a) base material, (b) interface region, (c) interface region, (d) inner layer region, and (e) IPF color code. Phase map of (f) base material, (g) interface region, (h) high-resolution phase map of region marked with square in (g), (i) interface region, and (j) inner layer region [82].

martensitic transformation needs to be noticed at the interface of the build (shown in **Figure 18**), which decreases the mechanical properties [82].

4. Friction and rolling-based additive manufacturing (FRAM)

4.1 Working principle of FRAM

To address the limitations of continuous material addition during the FSAM and AFSD processes, Xie et al. [83] developed a novel solid-state additive manufacturing (AM) process called friction and rolling-based additive manufacturing (FRAM) for the first time [83]. The FRAM process involves the following steps, as illustrated in **Figure 1**:

1. *Tool head rotation:* Before deposition, the tool head is rapidly rotated and pressed against the top of the substrate to generate heat through contact friction, as shown in **Figure 19(a)**.

2. *Material feeding:* Two feeding wires are continuously delivered into the contact interface from the side, where they are stirred into the interaction region, as illustrated in **Figure 19(b)**.

3. *Material mixing:* As the tool head moves forward, the mixed materials—comprising the feeding wires and the substrate material—are stirred and migrated to the rear side of the tool head, as depicted in **Figure 19(c)**.

4. *Layer completion:* Once a layer is completed, the tool head is raised to the next layer height, and the above process is repeated to construct the component, as shown in **Figure 19(d)**.

The feedstocks used in the FRAM process can be in the form of strips (**Figure 19(e)**) or wires (**Figure 19(f)**). **Figure 19(g)** presents a schematic of the tool head.

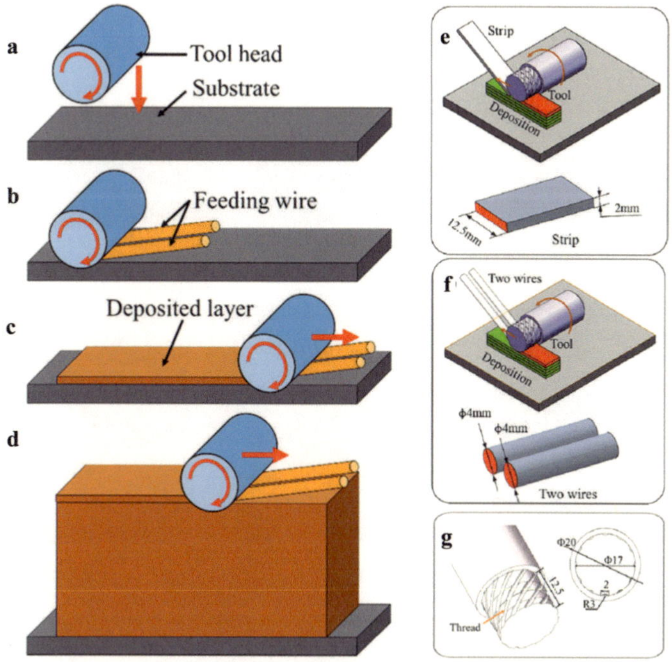

Figure 19.
Schematic of the FRAM process [83, 84]: (a) the tool head was pushed onto the upper surface of the substrate, (b) feeding wires were steadily supplied to the contact interface, (c) a new layer produced as the tool moved forward, (d) the tool head lifted to the next layer's height and repeated to fabricate the component, (e) the feedstock of strip, (f) the feedstock of wires, and (g) image of the tool head.

4.2 Current research

FRAM is a new solid-state additive manufacturing technology that has emerged in the past 5 years. Xie et al.'s team epitomizes the forefront of the FRAM technology. In recent years, the FRAM technique has been used to fabricate 6061 [83], 5A90 [85], 2319 [39], and 2195 [86] aluminum alloy components. A High-strength and high-elongation component was fabricated by FRAM using commercial ER2319 welding wires [39]. **Figure 20** displays a uniform and fine-grained structure with approximately equiaxed grains that range from 4 to 6 μm in size. The ultimate tensile strength and yield strength of the deposited build exhibited minimal variation in all directions. Moreover, the mechanical properties of the build were higher than those of the builds fabricated by the fusion-based AM like wire arc additive manufacturing and electron beam fabrication.

Furthermore, Xie et al. [87] successfully revealed the mechanisms for good bonding interlayers between the AA2319 and AA6061 hybrid aluminum alloys produced by FRAM. As shown in **Figure 21**, the good bonding between adjacent layers in FRAM can be attributed to several factors. Firstly, simultaneous material plastic flow and macroscopic migration resulted in a macroscopic mechanically interlocking non-planar interface. Additionally, the convex thread on the tool head enhances material flow, promoting better mixing of materials in the adjacent layer. Secondly, direct contact with no obvious oxides resulted in complete metallurgical interface bonding and eliminated potential defects.

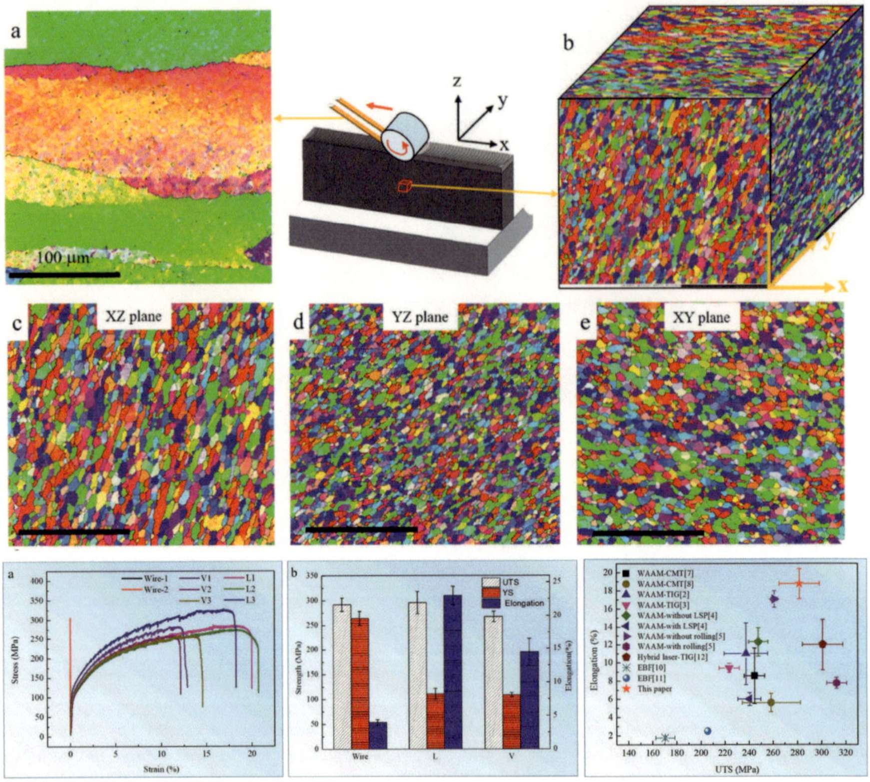

Figure 20.
Microstructure and mechanical properties of the 2319 aluminum alloy build fabricated by the FRAM [39].

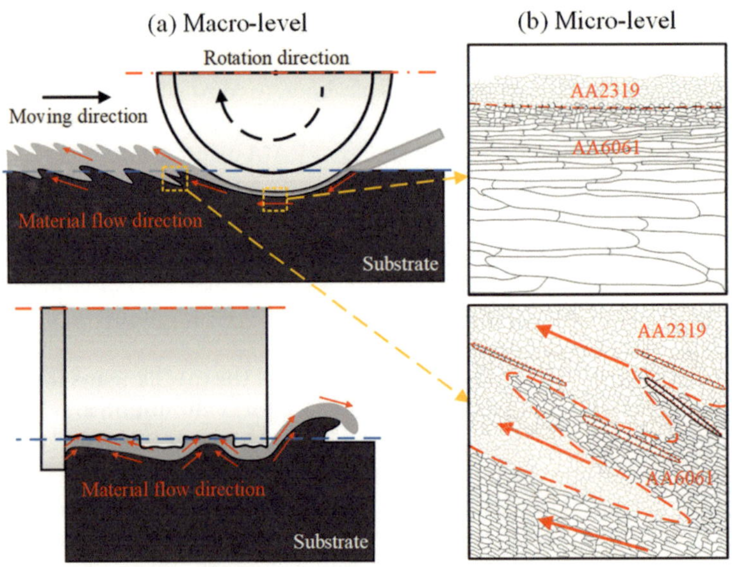

Figure 21.
Interlayer bonding mechanisms of FRAM [87]: (a) mechanically interlocked, non-planar interface formed between adjacent layers at the macro level and (b) dynamic recrystallization occurring at the interface, resulting in complete metallurgical bonding at the micro-level.

Heat accumulation is a prevalent issue across all additive manufacturing methods as the number of layers increases, leading to numerous thermal cycles in the deposited material. Excessive peak temperatures and extended periods of high-temperature result in grain coarsening and softening of the deposited material, significantly compromising forming accuracy and microstructural uniformity. To address this challenge, a heat-balance-control method known as plasma-arc preheating and instant water cooling (PC-FRAM) was developed. This approach creates a temperature field characterized by rapid increases and decreases in temperature around the tool head. The combination of preheating and instant cooling improves the deposition efficiency of FRAM and weakens the thermal-softening effect, thus increasing the tensile strength and elongation of the deposited samples [86].

5. Wire friction stir additive manufacturing (W-FSAM)

5.1 Working principle of W-FSAM

Wire-based friction stir additive manufacturing (W-FSAM) was introduced by Chen as a method to achieve solid-state manufacturing of large metallic structures through the continuous feeding of wire materials [88]. **Figure 22** illustrates the schematic of the W-FSAM experimental setup and process. As shown in **Figure 22(a)**, the W-FSAM system consists of a stationary storage chamber equipped with a wire feeding port, a screwed transport mechanism, and three stirring probes. The screwed transport structure is designed to continuously convey and extrude the feeding wires. The stirring probes enhance the dynamic fluidity of the thermo-plasticized materials, improving the metallurgical bonding between adjacent layers. During the dwelling stage of the W-FSAM process (illustrated in **Figure 22(b)**), the wire material is sheared into metal particles by the screwed transport mechanism and continuously fed downward. In this stage, the metal particles undergo continuous compaction and thermo-plasticization, allowing them to maintain a thermoplastic state within the storage chamber for deposition. As the feeding wire is continuously sheared and thermo-plasticized by the transport structure, the materials are ultimately deposited onto the base plate or previously deposited layers. The stirring probes facilitate the dynamic fluidity of the therms-plasticized materials, enhancing the bonding between

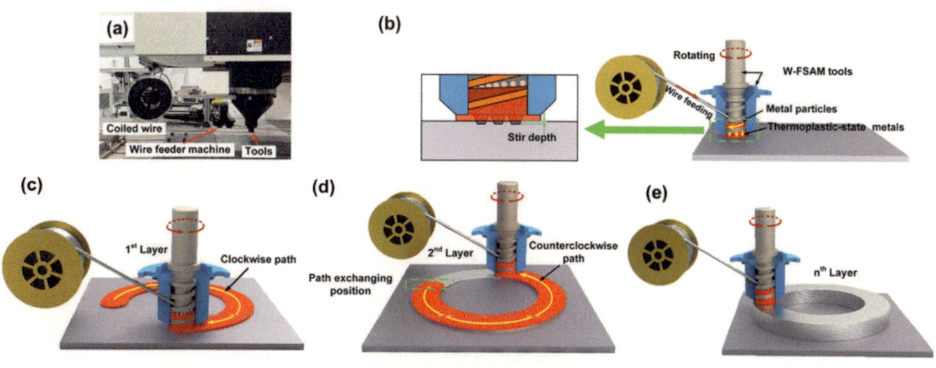

Figure 22.
(a) W-FSAM experimental setup, (b) dwell stage of the W-FSAM process, (c) the first layer produced, (d) the second layer produced, and (e) the repeated process.

layers. Throughout the W-FSAM process, alternating clockwise and counterclockwise cyclic paths are executed (as shown in **Figure 22(c)** and **(d)**). This alternating procedure is repeated until the desired build height for the flange ring structure is achieved (as depicted in **Figure 22(e)**).

5.2 Current research

W-FSAM technology is a recent addition to the field of additive manufacturing, with limited reports available on this topic. Huang's team reported that W-FSAM can be used to fabricate large structures without kissing bond defects. And the uniform microstructures with fine and equiaxed grains were obtained. The ultimate tensile strength of the build was evaluated to achieve 111% of wire base metal [88]. Bor et al. [40] successfully fabricated builds comprising about 50 layers of AA6060 Al alloy. The quality of the builds was primarily influenced by the feed ratio, which is defined as the volume of feedstock material supplied per unit of time relative to the volume of material needed for a layer. Builds fabricated with feed ratios of 1.7 and 1.3 achieved commendable mechanical properties in the build direction, resulting in ultimate tensile strength values of around 100 MPa and homogeneous elongation exceeding 12.5%. These measurements were taken from tensile test specimens extracted from the interior of the builds.

6. Advantages, challenges, and applications in FBAM techniques

The key feature of these FBAM techniques is that they are carried out in the solid state without melting or re-solidification during the additive manufacturing process. Owing to this feature, the FSAM, AFSD, FRAM, and w-FSAM possess distinct advantages over fusion-based additive manufacturing. **Table 1** shows a detailed comparison of these FBAM techniques. All these FBAM techniques require lower consumed energy and reduce the residual stress and distortion in the components;

S. No.	Technology	Feedstock	Strength	Challenges	Potential applications
1	FSAM	Plate	• Low consumed energy. • Low residual stress and distortion. • Fine microstructures with equiaxed grains. • Higher production rates. • No solidification related defects. • Ability to produce a variety of materials including difficult by fusion welding	• Material clamping is needed. • Tool wear rates higher. • Difficult to fabricate intricate structures. • Need for additional subtractive manufacturing. • Microstructure inhomogeneity. • Kissing bond defect is the key issue affecting the performance	• Stiffer assemblies such as ribs, fuselage stringers, and I beams in the aerospace sectors • Granted structure components

S. No.	Technology	Feedstock	Strength	Challenges	Potential applications
2	AFSD	Powder, rob	• No solidification related defects. • Fine microstructures with equiaxed grains. • Almost isotropic properties obtained. • Ability to fabricate large components. • Ability to produce a variety of materials including difficult by fusion welding. • Ability to forming relatively complex structures	• Special purpose equipment required. • The interlayer region is prone to becoming a weak region in the formed parts after heat treatment	• Suited for the deposition of large-scale components • Structural repair of functional components
3	FRAM	Wire, strip	• Continuously deliver materials. • No solidification related defects. • Fine microstructures with equiaxed grains. • Ability to produce a variety of materials including difficult by fusion welding	• Difficult to fabricate complex geometries	• Suited for the deposition and repair of large-scale high-strength components
4	W-FSAM	Wire	• Achieved complex large-sized parts without mold supporting. • No solidification related defects. • Fine microstructures with equiaxed grains. • Ability to produce a variety of materials including difficult by fusion welding	• Lower production rates	• Fully dense and near-net-shaped typical components, such as variable-diameter, hemispheroid, and flange ring structures

Source: Refs. [22, 83, 85, 88–90].

Table 1.
Summary of FBAM techniques.

furthermore, fine microstructures with equiaxed grains and no solidification-related defects are obtained. Despite the aforementioned advantages, several limitations still hinder the widespread adoption of these emerging technologies in industrial manufacturing processes. For instance, fabricating highly intricate structures remains a challenge. Additionally, there is often a need for supplementary subtractive manufacturing processes, such as machining or grinding, to achieve the final shape of the components. Moreover, the uniform microstructure and high-performance components need further exploration.

In the future, components made *via* these FBAM techniques may find applications for stiffer assemblies such as ribs, fuselage stringers, I-beams in launch vehicles aerospace, tailored graded structures, and wear-resistant coats.

7. Conclusions

This review presents a comprehensive introduction to the processing characteristics and current research progress of FSAM, AFSD, FRAM, and W-FSAM technologies. It also highlights the advantages, challenges, and potential applications of these solid-state additive manufacturing methods. Despite these technologies having gained significant attention due to their unique forming benefits, challenges related to achieving uniform microstructures and optimal performance components continue to hinder their widespread adoption. Additionally, most studies have focused primarily on mechanical properties at room temperature, leaving a gap in understanding how these materials perform in specialized conditions, such as extreme temperatures, fatigue, corrosion, and residual stress. To advance the industrial application of SSAM technologies, it is essential to expand the available data on performance across a broader range of operating environments.

Acknowledgements

This work was financially supported by the Youth Fund Project of GRINM and the National Key R&D Program of China (2023YFB3710400).

Conflict of interest

The authors declare no conflict of interest.

Author details

Ying Li[1,2,3*], Xiwu Li[1,2,3], Changshu He[4], Yanan Li[1,2,3], Kai Wen[1,2,3], Lizhen Yan[1,2,3], Wei Xiao[1,2,3], Yongan Zhang[1,2,3] and Baiqing Xiong[1,3]

1 State Key Laboratory of Nonferrous Metals and Processes, GRINM Group Co., LTD., Beijing, China

2 GRIMAT Engineering Institute Co., LTD, Beijing, China

3 General Research Institute for Nonferrous Metals, Beijing, China

4 School of Materials Science & Engineering, Northeastern University, Shenyang, China

*Address all correspondence to: liying@grinm.com

References

[1] Lehmann T, Rose D, Ranjbar E, et al. Large-scale metal additive manufacturing: A holistic review of the state of the art and challenges. International Materials Reviews. 2021;**67**(4):410-459. DOI: 10.1080/09506608

[2] Karayel E, Bozkurt Y. Additive manufacturing method and different welding applications. Journal of Materials Research and Technology. 2020;**9**(5):11424-11438. DOI: 10.1016/j. jmrt

[3] Guan Q. Generallized additive manufacturing based on welding/joining technologies. The Paton Welding Journal. 2013;**10-11**:33-38

[4] Wong KV, Hernandez A. A review of additive manufacturing. ISRN Mechanical Engineering. 2012;**2012**:1-10. DOI: 10.5402/2012/208760

[5] Guo N, Leu MC. Additive manufacturing: Technology, applications and research needs. Frontiers of Mechanical Engineering. 2013;**8**(3):215-243. DOI: 10.1007/s11465-013-0248-8

[6] Tapia G, Elwany A. A review on process monitoring and control in metal-based additive manufacturing. Journal of Manufacturing Science and Engineering. 2014;**136**(6):060801. DOI: 10.1115/1.4028540

[7] Long J, Nand A, Ray S. Application of spectroscopy in additive manufacturing. Materials (Basel). 2021;**14**(1):203. DOI: 10.3390/ma14010203

[8] He P, Webster RF, Yakubov V, et al. Fatigue and dynamic aging behavior of a high strength al-5024 alloy fabricated by laser powder bed fusion additive manufacturing. Acta Materialia. 2021;**220**:117312. DOI: 10.1016/j.actamat

[9] Santos Macías JG, Douillard T, Zhao L, et al. Influence on microstructure, strength and ductility of build platform temperature during laser powder bed fusion of alsi10mg. Acta Materialia. 2020;**201**:231-243. DOI: 10.1016/j.actamat.2020.10.001

[10] Martin JH, Yahata BD, Hundley JM, et al. 3d printing of high-strength aluminium alloys. Nature. 2017;**549**(7672):365-369. DOI: 10.1038/nature23894

[11] Dadkhah M, Mosallanejad MH, Iuliano L, et al. A comprehensive overview on the latest progress in the additive manufacturing of metal matrix composites: Potential, challenges, and feasible solutions. Acta Metallurgica Sinica (English Letters). 2021;**34**(9):1173-1200. DOI: 10.1007/s40195-021-01249-7

[12] Brice C, Shenoy R, Kral M, et al. Precipitation behavior of aluminum alloy 2139 fabricated using additive manufacturing. Materials Science and Engineering: A. 2015;**648**:9-14. DOI: 10.1016/j.msea.2015.08.088

[13] Tawfik MM, Nemat-Alla MM, Dewidar MM. Enhancing the properties of aluminum alloys fabricated using wire + arc additive manufacturing technique - A review. Journal of Materials Research and Technology. 2021;**13**:754-768. DOI: 10.1016/j. jmrt.2021.04.076

[14] Pattanayak S, Sahoo SK. Gas metal arc welding based additive manufacturing—A review. CIRP Journal of Manufacturing Science and Technology. 2021;**33**:398-442. DOI: 10.1016/j.cirpj.2021.04.010

[15] Sames WJ, List FA, Pannala S, et al. The metallurgy and processing science of metal additive manufacturing. International Materials Reviews. 2016;**61**(5):315-360. DOI: 10.1080/09506608.2015.1116649

[16] Madhuri N, Jayakumar V, Sathishkumar M. Recent developments and challenges accompanying with wire arc additive manufacturing of mg alloys: A review. Materials Today: Proceedings. 2021;**46**:8573-8577. DOI: 10.1016/j.matpr.2021.03.548

[17] Hauser T, Reisch RT, Breese PP, et al. Porosity in wire arc additive manufacturing of aluminium alloys. Additive Manufacturing. 2021;**41**:101993. DOI: 10.1016/j.addma.2021.101993

[18] Zhang C, Gao M, Zeng X. Workpiece vibration augmented wire arc additive manufacturing of high strength aluminum alloy. Journal of Materials Processing Technology. 2019;**271**:85-92. DOI: 10.1016/j.jmatprotec.2019.03.028

[19] Altıparmak SC, Yardley VA, Shi Z, et al. Challenges in additive manufacturing of high-strength aluminium alloys and current developments in hybrid additive manufacturing. International Journal of Lightweight Materials and Manufacture. 2021;**4**(2):246-261. DOI: 10.1016/j.ijlmm.2020.12.004

[20] Dong B, Cai X, Xia Y, et al. Effects of interlayer temperature on the microstructures of wire arc additive manufactured al-zn-mg-cu alloy: Insights into texture responses and dynamic precipitation behaviors. Additive Manufacturing. 2021;**48**:102453. DOI: 10.1016/j.addma.2021.102453

[21] Dong B, Cai X, Lin S, et al. Wire arc additive manufacturing of al-zn-mg-cu alloy: Microstructures and mechanical properties. Additive Manufacturing. 2020;**36**:101447. DOI: 10.1016/j.addma.2020.101447

[22] Srivastava M, Rathee S, Maheshwari S, et al. A review on recent progress in solid state friction based metal additive manufacturing: Friction stir additive techniques. Critical Reviews in Solid State and Materials Sciences. 2018;**44**(5):345-377. DOI: 10.1080/10408436.2018.1490250

[23] White D, Arbor A. Object Consolidation Employing Friction Joining: United States. Patent US6457629B1. 2002

[24] Thomas WM, Norris IM, Staines DG, et al. Friction stir welding–process developments and variant techniques. In: The SME Summit Oconomowoc. Milwaukee, USA: TWI Ltd; 2005. pp. 1-22

[25] Lequeu P, Smith KP, Daniélou A. Aluminum-copper-lithium alloy 2050 developed for medium to thick plate. Journal of Materials Engineering and Performance. 2009;**19**(6):841-847. DOI: 10.1007/s11665-009-9554-z

[26] Palanivel S, Mishra RS. Building without melting: A short review of friction-based additive manufacturing techniques. International Journal of Additive and Subtractive Materials Manufacturing. 2017;**1**(1):82-103. DOI: 10.1504/IJASMM.2017.10003956

[27] Baumann JA. Production of energy efficient preform structures (PEEPS). Office of Scientific and Technical Information Technical Reports. 2012. DOI: 10.2172/1042703

[28] Dilip JJS, Ram GDJ, Stucker BE. Additive manufacturing with friction welding and friction deposition processes. International Journal of Rapid Manufacturing. 2012;**3**(1):56. DOI: 10.1504/ijrapidm.2012.046574

[29] Dilip JJS, Babu S, Rajan SV, et al. Use of friction surfacing for additive manufacturing. Materials and Manufacturing Processes. 2013;**28**(2):189-194. DOI: 10.1080/10426914.2012.677912

[30] Dalgaard E, Wanjara P, Gholipour J, et al. Linear friction welding of a near-β titanium alloy. Acta Materialia. 2012;**60**(2):770-780. DOI: 10.1016/j.actamat.2011.04.037

[31] Chamanfar A, Jahazi M, Cormier J. A review on inertia and linear friction welding of ni-based superalloys. Metallurgical and Materials Transactions A. 2015;**46**(4):1639-1669. DOI: 10.1007/s11661-015-2752-4

[32] Puleo SM. Additive Friction Stir Manufacturing of 7055 Aluminum Alloy. New Orleans: University of New Orleans; 2016

[33] Shen Z, Chen S, Cui L, et al. Local microstructure evolution and mechanical performance of friction stir additive manufactured 2195 al-li alloy. Materials Characterization. 2022;**186**:111818. DOI: 10.1016/j.matchar.2022.111818

[34] Singh Rajawat M, Pagrut S, Dwivedi S, et al. Microstructural characterization of friction stir assisted laminated lap welding of aa6063 sheets. Materials Today: Proceedings. 2022;**56**:949-953. DOI: 10.1016/j.matpr.2022.02.633

[35] Tan Z, Zhang Z. Band gap characteristics of friction stir additive manufactured phononic crystals. Physica Scripta. 2022;**97**(2):025702. DOI: 10.1088/1402-4896/ac48aa

[36] Palanivel S, Nelaturu P, Glass B, et al. Friction stir additive manufacturing for high structural performance through microstructural control in an mg based we43 alloy. Materials and Design. 2015;**65**:934-952. DOI: 10.1016/j.matdes.2014.09.082

[37] Khodabakhshi F, Gerlich AP. Potentials and strategies of solid-state additive friction-stir manufacturing technology: A critical review. Journal of Manufacturing Processes. 2018;**36**:77-92. DOI: 10.1016/j.jmapro.2018.09.030

[38] Beck SC, Rutherford BA, Avery DZ, et al. The effect of solutionizing and artificial aging on the microstructure and mechanical properties in solid-state additive manufacturing of precipitation hardened Al-Mg-Si alloy. Materials Science and Engineering: A. 2021;**819**:141351. DOI: 10.1016/j.msea.2021.141351

[39] Xie R, Chen X, Shi Y, et al. Printing high-strength high-elongation aluminum alloy using commercial er2319 welding wires through deformation-based additive manufacturing. Materials Science and Engineering: A. 2023;**868**:144773. DOI: 10.1016/j.msea.2023.144773

[40] Bor T, Leede MD, Deunk F, et al. Friction screw extrusion additive manufacturing of an al-mg-si alloy. Additive Manufacturing. 2023;**72**:103621. DOI: 10.1016/j.addma.2023.103621

[41] Li Y, He C, Wei J, et al. Restirring and reheating effects on microstructural evolution of Al-Zn-Mg-Cu alloy during underwater friction stir additive manufacturing. Materials. 2022;**15**(11):3804. DOI: 10.3390/ma15113804

[42] He C, Li Y, Zhang Z, et al. Investigation on microstructural evolution and property variation along building direction in friction stir additive manufactured Al-Zn-Mg alloy. Materials Science and Engineering:

A. 2020;**777**:139035. DOI: 10.1016/j.msea.2020.139035

[43] Li Y, He C, Wei J, et al. Correlation of local microstructures and mechanical properties of Al-Zn-Mg-Cu build fabricated via underwater friction stir additive manufacturing. Materials Science and Engineering: A. 2021;**805**:140590. DOI: 10.1016/j.msea.2020.140590

[44] Lu IK, Reynolds AP. Innovative friction stir additive manufacturing of cast 2050 Al-Cu-Li aluminum alloy. Progress in Additive Manufacturing. 2021;**6**(3):471-477. DOI: 10.1007/s40964-021-00175-5

[45] Yuqing M, Liming K, Chunping H, et al. Formation characteristic, microstructure, and mechanical performances of aluminum-based components by friction stir additive manufacturing. The International Journal of Advanced Manufacturing Technology. 2015;**83**(9-12):1637-1647. DOI: 10.1007/s00170-015-7695-9

[46] Zhao Z, Yang X, Li S, et al. Interfacial bonding features of friction stir additive manufactured build for 2195-t8 aluminum-lithium alloy. Journal of Manufacturing Processes. 2019;**38**:396-410. DOI: 10.1016/j.jmapro.2019.01.042

[47] He C, Li Y, Wei J, et al. Enhancing the mechanical performance of Al-Zn-Mg alloy builds fabricated via underwater friction stir additive manufacturing and post-processing aging. Journal of Materials Science and Technology. 2022;**108**:26-36. DOI: 10.1016/j.jmst.2021.08.050

[48] Palanivel S, Sidhar H, Mishra RS. Friction stir additive manufacturing: Route to high structural performance. Journal of Metals. 2015;**67**(3):616-621. DOI: 10.1007/s11837-014-1271-x

[49] Li Y, He C, Wei J, et al. Effect of post-fabricated aging on microstructure and mechanical properties in underwater friction stir additive manufacturing of Al–Zn–Mg–Cu alloy. Materials. 2022;**15**:3368. DOI: 10.3390/ma15093368

[50] Liu M, Wang B, An XH, et al. Friction stir additive manufacturing enabling scale-up of ultrafine-grained pure copper with superior mechanical properties. Materials Science and Engineering: A. 2022;**857**:144088. DOI: 10.1016/j.msea.2022.144088

[51] Ardalanniya A, Nourouzi S, Jamshidi AH. Fabrication of the laminated Al-Zn cup/Al-Zn composite using friction stir additive manufacturing. Materials Today Communications. 2021;**27**:102268. DOI: 10.1016/j.mtcomm.2021.102268

[52] Sharma A, Bandari V, Ito K, et al. A new process for design and manufacture of tailor-made functionally graded composites through friction stir additive manufacturing. Journal of Manufacturing Processes. 2017;**26**:122-130. DOI: 10.1016/j.jmapro.2017.02.007

[53] Derazkola HA, Khodabakhshi F, Simchi A. Evaluation of a polymer-steel laminated sheet composite structure produced by friction stir additive manufacturing (FSAM) technology. Polymer Testing. 2020;**90**:106690. DOI: 10.1016/j.polymertesting.2020.106690

[54] Wang W, Han P, Wang Y, et al. High-performance bulk pure al prepared through cold spray-friction stir processing composite additive manufacturing. Journal of Materials Research and Technology. 2020;**9**(4):9073-9079. DOI: 10.1016/j.jmrt.2020.06.034

[55] Srivastava M, Rathee S. Microstructural and microhardness study

on fabrication of al 5059/sic composite component via a novel route of friction stir additive manufacturing. Materials Today: Proceedings. 2021;**39**:1775-1780. DOI: 10.1016/j.matpr.2020.07.137

[56] Kumar Jha K, Kesharwani R, Imam M. Microstructural and micro-hardness study on the fabricated al 5083-o/6061-t6/7075-t6 gradient composite component via a novel route of friction stir additive manufacturing. Materials Today: Proceedings. 2022;**56**:819-825. DOI: 10.1016/j.matpr.2022.02.262

[57] Derazkola HA, Mohammadiabokheili R, Kordani N, et al. Evaluation of nanocomposite structure printed by solid-state additive manufacturing. CIRP Journal of Manufacturing Science and Technology. 2022;**37**:174-184. DOI: 10.1016/j.cirpj.2022.01.006

[58] Kandasamy K. Solid state joining using additive friction stir processing. United States Patent. 2016;**2**(12)

[59] Yu HZ, Jones ME, Brady GW, et al. Non-beam-based metal additive manufacturing enabled by additive friction stir deposition. Scripta Materialia. 2018;**153**:122-130. DOI: 10.1016/j.scriptamat.2018.03.025

[60] Rathee S, Srivastava M, Pandey PM, et al. Metal additive manufacturing using friction stir engineering: A review on microstructural evolution, tooling and design strategies. CIRP Journal of Manufacturing Science and Technology. 2021;**35**:560-588. DOI: 10.1016/j.cirpj.2021.08.003

[61] Phillips BJ, Avery DZ, Liu T, et al. Microstructure-deformation relationship of additive friction stir-deposition Al-Mg-Si. Materialia. 2019;**7**:100387. DOI: 10.1016/j.mtla.2019.100387

[62] Rivera OG, Allison PG, Jordon JB, et al. Microstructures and mechanical behavior of inconel 625 fabricated by solid-state additive manufacturing. Materials Science and Engineering: A. 2017;**694**:1-9. DOI: 10.1016/j.msea.2017.03.105

[63] Rivera OG, Allison PG, Brewer LN, et al. Influence of texture and grain refinement on the mechanical behavior of aa2219 fabricated by high shear solid state material deposition. Materials Science and Engineering: A. 2018;**724**:547-558. DOI: 10.1016/j.msea.2018.03.088

[64] Garcia D, Hartley WD, Rauch HA, et al. In situ investigation into temperature evolution and heat generation during additive friction stir deposition: A comparative study of cu and Al-Mg-Si. Additive Manufacturing. 2020;**34**:101386. DOI: 10.1016/j.addma.2020.101386

[65] Griffiths RJ, Garcia D, Song J, et al. Solid-state additive manufacturing of aluminum and copper using additive friction stir deposition: Process-microstructure linkages. Materialia. 2021;**15**:100967. DOI: 10.1016/j.mtla.2020.100967

[66] Phillips BJ, Mason CJT, Beck SC, et al. Effect of parallel deposition path and interface material flow on resulting microstructure and tensile behavior of Al-Mg-Si alloy fabricated by additive friction stir deposition. Journal of Materials Processing Technology. 2021;**295**:117169. DOI: 10.1016/j.jmatprotec.2021.117169

[67] Griffiths RJ, Petersen DT, Garcia D, et al. Additive friction stir-enabled solid-state additive manufacturing for the repair of 7075 aluminum alloy. Applied Sciences. 2019;**9**(17):3486. DOI: 10.3390/app9173486

[68] Avery DZ, Phillips BJ, Mason CJT, et al. Influence of grain refinement and microstructure on fatigue behavior for solid-state additively manufactured Al-Zn-Mg-Cu alloy. Metallurgical and Materials Transactions A. 2020;**51**(6):2778-2795. DOI: 10.1007/s11661-020-05746-9

[69] Mason CJT, Rodriguez RI, Avery DZ, et al. Process-structure-property relations for as-deposited solid-state additively manufactured high-strength aluminum alloy. Additive Manufacturing. 2021;**40**:101879. DOI: 10.1016/j.addma.2021.101879

[70] Ghadimi H, Ding H, Emanet S, et al. Hardness distribution of al2050 parts fabricated using additive friction stir deposition. Materials. 2023;**16**(3):1-278. DOI: 10.3390/ma16031278

[71] Jin Y, Wang T, Liu T, et al. Gradient process parameter optimization in additive friction stir deposition of aluminum alloys. International Journal of Machine Tools and Manufacture. 2024;**195**:104113. DOI: 10.1016/j.ijmachtools.2023.104113

[72] Ghadimi H, Talachian M, Ding H, et al. The effects of layer thickness on the mechanical properties of additive friction stir deposition-fabricated aluminum alloy 6061 parts. Metals. 2024;**14**(1). DOI: 10.3390/met14010101

[73] Cahalan LP, Williams MB, Brewer LN, et al. Parametric investigation of parallel deposition passes on the microstructure and mechanical properties of 7075 aluminum alloy processed with additive friction stir deposition. Applied Sciences. 2024;**14**(1):457. DOI: 10.3390/app14010457

[74] Hu F, Chen G, Lin Y, et al. Numerical and experimental study on the thermal process during additive friction stir deposition. CIRP Journal of Manufacturing Science and Technology. 2024;**48**:55-66. DOI: 10.1016/j.cirpj.2023.12.002

[75] Rutherford BA, Avery DZ, Phillips BJ, et al. Effect of post-deposition heat treatment on the mechanical behavior and deformation mechanisms of a solid-state additively manufactured Al-Mg-Si alloy. Journal of Engineering Materials and Technology. 2024;**146**(2):021005. DOI: 10.1115/1.4064179

[76] Chen L, Zhu L, Lu L, et al. The effect of heat treatment on the microstructure and electrochemical corrosion behavior of multilayer aa6061 alloy fabricated by additive friction stir deposition. Applied Surface Science. 2024;**650**:159167. DOI: 10.1016/j.apsusc.2023.159167

[77] Han Y, Griffiths RJ, Yu HZ, et al. Quantitative microstructure analysis for solid-state metal additive manufacturing via deep learning. Journal of Materials Research. 2020;**35**(15):1936-1948. DOI: 10.1557/jmr.2020.120

[78] Agrawal P, Haridas RS, Yadav S, et al. Additive friction stir deposition of ss316: Effect of process parameters on microstructure evolution. Materials Characterization. 2023;**195**:112470. DOI: 10.1016/j.matchar.2022.112470

[79] Agrawal P, Shiau C-H, Sharma A, et al. Ion irradiation and examination of additive friction stir deposited 316 stainless steel. Materials and Design. 2024;**238**:112730. DOI: 10.1016/j.matdes.2024.112730

[80] Patil SM, Krishna KVM, Sharma S, et al. Thermo-mechanical process variables driven microstructure evolution during additive friction stir deposition of in625. Additive Manufacturing.

2024;**80**:103958. DOI: 10.1016/j.addma.2024.103958

[81] Farabi E, Babaniaris S, Barnett MR, et al. Microstructure and mechanical properties of ti6al4v alloys fabricated by additive friction stir deposition. Additive Manufacturing Letters. 2022;**2**:100034. DOI: 10.1016/j.addlet.2022.100034

[82] Gotawala N, Kumar Mishra N, Shrivastava A. Solid-state depositions of multilayer ss304 by friction stir metal deposition. Materials Letters. 2022;**314**:131881. DOI: 10.1016/j.matlet.2022.131881

[83] Xie R, Shi Y, Liu H, et al. A novel friction and rolling based solid-state additive manufacturing method: Microstructure and mechanical properties evaluation. Materials Today Communications. 2021;**29**:103005. DOI: 10.1016/j.mtcomm.2021.103005

[84] Xie R, Chen P, Shi Y, et al. Effect of feeding material shape on microstructures and mechanical properties in friction rolling additive manufacturing. Materials and Design. 2024;**241**:112952. DOI: 10.1016/j.matdes.2024.112952

[85] Liu H, Sun Y, Xie R, et al. Characterization of microstructure and local mechanical properties of friction rolling additive manufactured al li alloy under repeated thermal-mechanical cycles. Materials Characterization. 2023;**204**:113169. DOI: 10.1016/j.matchar.2023.113169

[86] Sun Y, Liu H, Xie R, et al. Heat-balance control of friction rolling additive manufacturing based on combination of plasma preheating and instant water cooling. Journal of Materials Science and Technology. 2025;**205**:168-181. DOI: 10.1016/j.jmst.2024.03.054

[87] Xie R, Liang T, Shi Y, et al. Revealing the bonding mechanisms between deposit and substrate of the friction rolling additive manufactured hybrid aluminum alloys. Additive Manufacturing. 2022;**56**:102942. DOI: 10.1016/j.addma.2022.102942

[88] Chen H, Meng X, Chen J, et al. Wire-based friction stir additive manufacturing. Additive Manufacturing. 2023;**70**:103557. DOI: 10.1016/j.addma.2023.103557

[89] Shao J, Samaei A, Xue T, et al. Additive friction stir deposition of metallic materials: Process, structure and properties. Materials and Design. 2023;**234**:112356. DOI: 10.1016/j.matdes.2023.112356

[90] Korgancı M, Bozkurt Y. Recent developments in additive friction stir deposition (AFSD). Journal of Materials Research and Technology. 2024;**30**:4572-4583. DOI: 10.1016/j.jmrt.2024.04.179

Chapter 5

Topology Optimization for Metal Additive Manufacturing: Effect of Build Orientation and Overhang Constraint

Babin Dangal, Harshith Boppana and Sangjin Jung

Abstract

Additive manufacturing (AM) is an advanced manufacturing technology that offers significant advantages for producing complex geometries, but it also presents unique challenges, particularly in managing overhangs and selecting appropriate build orientations. Overhangs, which are unsupported features that extend beyond the main structure, can lead to defects such as warping, sagging, and incomplete fusion during fabrication. To address these challenges, this chapter explores the integration of overhang constraints and build orientation as critical factors in the topology optimization process for AM. The study demonstrates that incorporating overhang constraints into the optimization algorithm can significantly reduce the need for support structures, thereby minimizing material usage, post-processing requirements, and overall manufacturing costs. Additionally, this chapter examines the effect of build orientation on the optimized geometry, highlighting how changes in orientation can dramatically affect the volume of support structures and the manufacturability of the final part. We also show that fixing the build orientation for AM in a particular orientation and then performing topology optimization often fails to produce a converged, feasible design solution. The findings underscore the need for a holistic approach to design optimization for metal additive manufacturing, where these AM design factors are simultaneously optimized to achieve practical and efficient designs.

Keywords: additive manufacturing, topology optimization, build orientation, support structure, overhang constraint

1. Introduction

Additive Manufacturing (AM) encompasses a variety of advanced manufacturing technologies that can deposit materials layer-by-layer to create objects from 3D digital models [1, 2]. A wide range of materials [3], including polymers [4], ceramics [5], and metals [6], can be utilized in AM. In particular, metal AM processes such as Powder Bed Fusion (PBF) [7, 8], Directed Energy Deposition (DED) [9], and Binder Jetting (BJ) [10, 11] are effective for producing various shapes of metal components. Metal AM has changed the manufacturing landscape by enabling the production of intricate geometries that are difficult or impossible to achieve with conventional manufacturing methods. This unique capability of AM has opened new avenues for design innovation, especially when combined with topology optimization.

Topology optimization is a mathematical approach used to determine the optimal material distribution within a given design space to achieve a specific performance goal, such as minimizing compliance or weight [12–15]. The process involves defining an objective function, constraints, and using optimization algorithms to iteratively adjust the material layout. **Figure 1** shows an example of the topology optimization process, including the initial design domain, boundary and loading conditions, structural analysis, and optimization results. Common methods include density-based approaches, such as the Solid Isotropic Material with Penalization (SIMP) method [16, 17], and level-set methods [18, 19], which use implicit surfaces to represent the material boundaries. The field of topology optimization has its roots in the late twentieth century, with early work focusing on structural optimization for mechanical components [12–15]. Over the years, advancements in computational power and algorithms have expanded its application to various engineering disciplines [20–22].

The topology optimization problem is generally formulated as:

$$\begin{aligned} &\min_{\mathbf{x}} \; f(\mathbf{x}) \\ &\text{subject to}: \frac{V(\mathbf{x})}{V_0} - V_{\max} \le 0 \\ &\qquad G_j(\mathbf{x}) \le 0, \quad j = 1,\ldots,n_G \end{aligned} \tag{1}$$

where **x** is the design variables, $f(x)$ is the objective function, $G_j(x)$ is the jth constraint, V is the material volume, V_0 is the volume of the initial design domain, and V_{max} is the allowable volume fraction. In Eq. (1), the objective function typically aims to minimize the compliance (maximize stiffness) of the structure.

$$f(\mathbf{x}) = \mathbf{U}^{T}\mathbf{K}\mathbf{U} \tag{2}$$

where **U** is the displacement vector obtained by solving the system of linear eqs. **KU** = **F**, where **K** is the stiffness matrix. While compliance minimization is widely

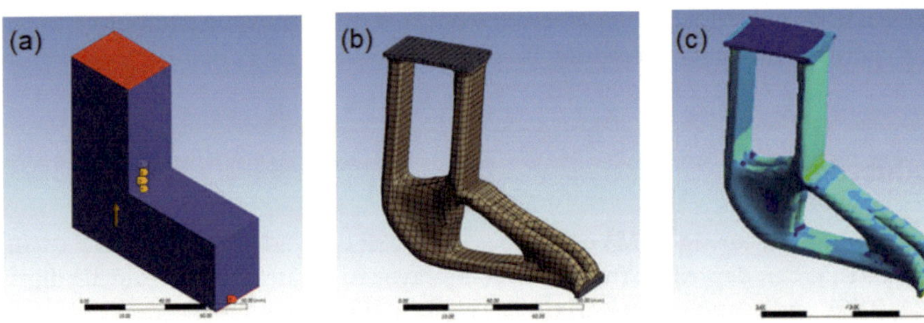

Figure 1.
Topology optimization: (a) initial design domain, (b) optimal geometry, and (c) structural analysis result.

used, there are other formulations that can be appropriate depending on the specific goals of the optimization (e.g., stress minimization) [23, 24].

One of the most popular methods for topology optimization is the density-based approach [16, 17]. The density-based approach uses constant material properties in each element of the discretized design domain. This approach depends on the Young's modulus E_e and the relative density x_e of each element as follows:

$$E_e(x_e) = (x_e)^p E \qquad 0 \le x_e \le 1 \tag{3}$$

where p is the penalization factor greater than 1, and E is the Young's modulus in the base material.

When assigning the Young's modulus to a void material, the modified SIMP method uses non-zero values to prevent singularities in the finite element stiffness matrix [17] as follows:

$$E_e(x_e) = E_{\min} + (x_e)^p (E - E_{\min}) \qquad 0 \le x_e \le 1 \tag{4}$$

where $E_{\min}$ is a very small value of the stiffness.

In the literature, optimization algorithms such as the Method of Moving Asymptotes (MMA) and optimality criteria methods have been employed to solve nonlinear programming problems for minimizing compliance, stress, volume, or other mechanical performance [13, 25].

The integration of topology optimization with AM technologies allows for the creation of components with optimized geometries, resulting in improved performance, reduced weight, and enhanced functionality [26–31]. Recently, it has also been possible to optimize porous infills such as lattice structures, grid patterns, and honeycomb structures in AM parts [32–35], which can significantly reduce the material consumption of AM parts. A review of topologically optimized AM parts in the aerospace industry found that AM parts can reduce material usage by up to 65% compared to traditionally manufactured counterparts [36, 37]. This not only reduces material costs but also directly reduces manufacturing costs, as less material leads to faster build times for AM.

However, the layer-by-layer nature of metal AM introduces challenges such as anisotropic mechanical properties, residual stresses, and the need for support structures to prevent deformation during the build process. Among the various challenges associated with AM, managing overhangs and selecting the appropriate build orientation are critical for ensuring the structural integrity and manufacturability of the final AM part [38–44]. Overhangs pose significant challenges, particularly in metal AM processes such as DED and PBF, where high thermal gradients and rapid solidification rates can lead to defects like sagging. These challenges necessitate advanced optimization strategies, such as topology optimization, to design parts that are both structurally sound and efficient to manufacture [38–40].

This chapter focuses on the role of overhang constraints and build orientation in the topology optimization process for AM. By integrating these factors into the optimization algorithm, it is possible to minimize the need for support structures, reduce manufacturing costs, and improve the overall performance of the final part. The paper presents a detailed analysis of how overhang constraints and build orientation affect the optimized geometry and discusses the complexities involved in achieving an optimal design that balances structural performance with manufacturability.

2. Overhang constraints in topology optimization for AM

Overhangs are the portions of a part that extend beyond the support structure, posing challenges during fabrication due to the lack of support. These areas are prone to defects such as warping, sagging, and incomplete fusion. Managing overhangs is critical for ensuring the structural integrity and dimensional accuracy of the final part. Overhangs are particularly challenging in metal AM due to the high thermal gradients and rapid solidification rates. Without adequate support, overhanging features can deform under their own weight or due to residual stresses, leading to poor surface quality and dimensional inaccuracies. In extreme cases, unsupported overhangs can result in building failures. To address these issues, various strategies have been developed, such as optimizing part orientation, designing support structures, and incorporating overhang constraints into topology optimization algorithms [38–40].

Support structures are commonly used to stabilize overhanging features during the build process. However, they require additional material and post-processing to remove, increasing the overall manufacturing cost and time. Therefore, designing parts with minimal support requirements is highly desirable. One approach is to orient the part in a way that minimizes the number of overhangs. For example, orienting a part to align its flat surfaces with the build platform can reduce the need for support structures. Another approach involves modifying the part geometry to include self-supporting features, such as fillets and chamfers, which can minimize overhangs and eliminate the need for support structures. In the case of metal AM such as PBF and DED, when the down-skin angle α is 45 degrees or less, support structures are required to prevent the part from collapsing or warping, as shown in **Figure 2**.

Several strategies have been developed to mitigate overhang issues, including the use of support structures, re-orienting the build direction, and modifying the part geometry. In topology optimization, overhang constraints can be incorporated to design self-supporting structures. Techniques such as support structure generation, build direction optimization, and explicit overhang constraint formulation are employed to address these challenges. For example, Gaynor and Guest [40] integrated overhang constraints into the topology optimization process, resulting in designs that reduced the need for support structures while maintaining structural integrity. The overhang constraints can be applied by defining the maximum allowable overhang angle, typically set at 45 degrees, for a particular part with a designated build direction. This constraint ensures that overhanging features within the part do not exceed the defined angle, thereby maintaining structural integrity during the manufacturing process.

Figure 3(a) shows a connecting rod example for topology optimization, including the initial design domain, boundary and loading conditions, and build orientation. The boundary conditions are applied based on the fundamental operating

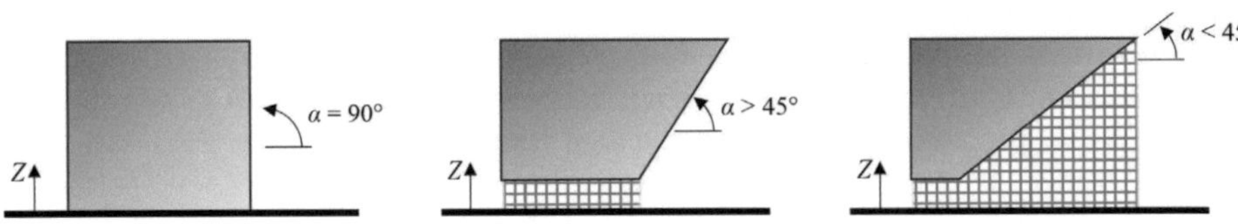

Figure 2.
Down-skin angles and support structures.

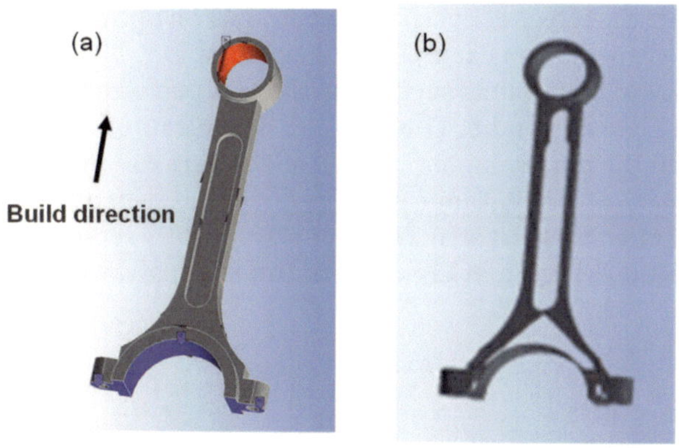

Figure 3.
Topology optimization of a connecting rod: (a) Initial design domain, boundary and loading conditions, and build orientation, and (b) optimized geometry.

characteristics of the connecting rod. As seen in the initial design domain, the frictionless support is applied to the connecting rod bearing, and the two bolt hubs have the fixed support. In addition, a load of 100 N is applied to the rod bushing of the rear bearing hole.

Figure 3(b) illustrates the optimized connecting rod geometry after topology optimization with the overhang constraint. The topology optimization results were obtained using ANSYS [45]. The maximum allowable overhang angle was set to 45 degrees in the optimization formulation. Due to the build orientation shown in **Figure 3(a)**, the down-skin angles of the optimal geometry are greater than 45 degrees, as shown in **Figure 3(b)**. The brake pedal design in **Figure 4** yields similar results to the connecting rod example. **Figure 4(a)** shows the initial design domain of the brake pedal and the build direction for metal AM, and the build orientation is along the +Y axis direction. As seen in **Figure 4(b)**, the topology optimization result was obtained using the overhang constraint, and the optimized geometry ensures that the down-skin angles for this orientation are greater than 45 degrees.

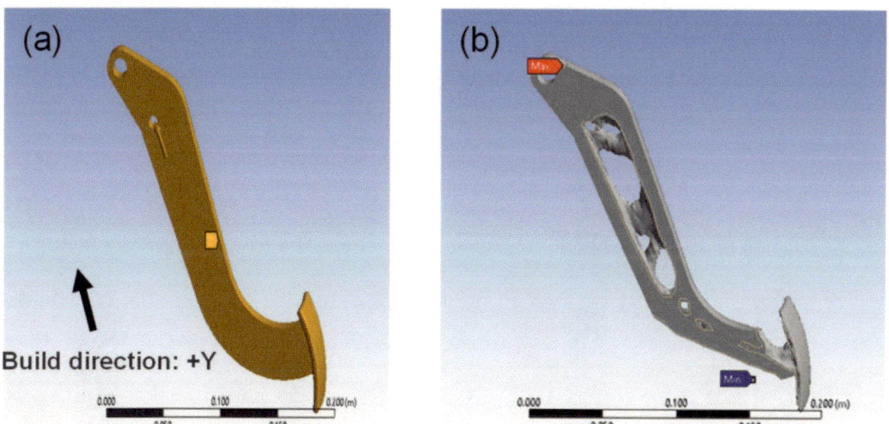

Figure 4.
Topology optimization of a brake pedal: (a) Initial design domain, boundary and loading conditions, and build orientation, and (b) optimized geometry.

To demonstrate the effect of overhang constraints on optimal part geometries, the cantilever and L-shaped beam examples are also presented. Topology optimization with and without AM constraints such as overhang is performed to minimize compliance as shown in **Figures 5** and **6**. The topology optimization result with AM constraints shows that the total volume (i.e., optimized part volume + support structure volume) is smaller than that without AM constraints, as seen in **Figure 6(a)**. This is because the required support structure was significantly reduced when using the overhang constraint, as shown in **Figure 6(b)**. Interestingly, when AM constraints

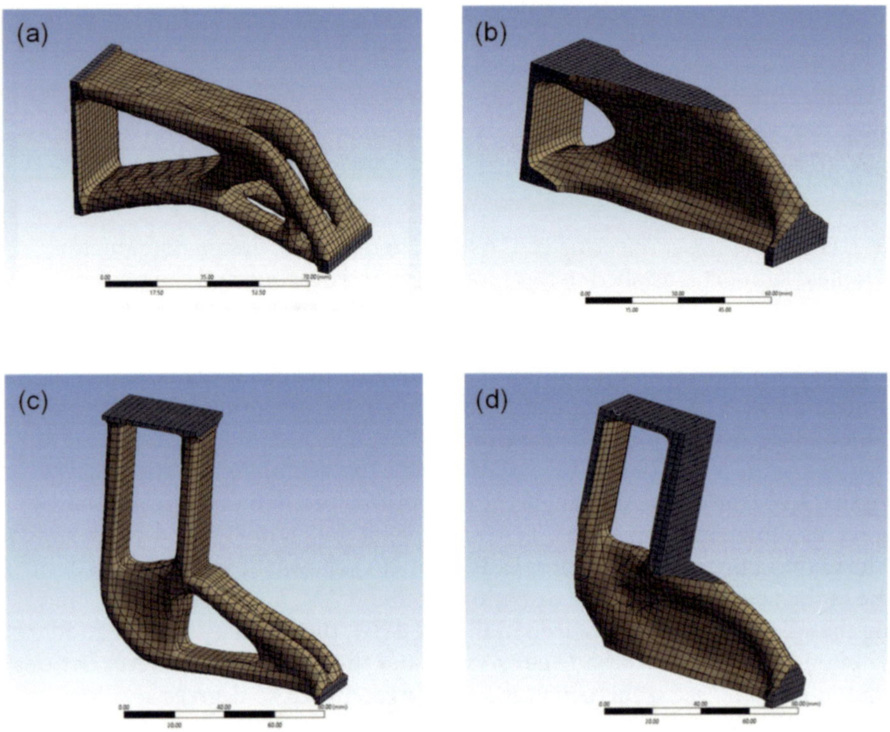

Figure 5.
Topology optimization results: (a) optimized cantilever beam without overhang constraints; (b) optimized cantilever beam with overhang constraints; (c) optimized L-shaped beam without overhang constraints; and (d) optimized L-shaped beam with overhang constraints.

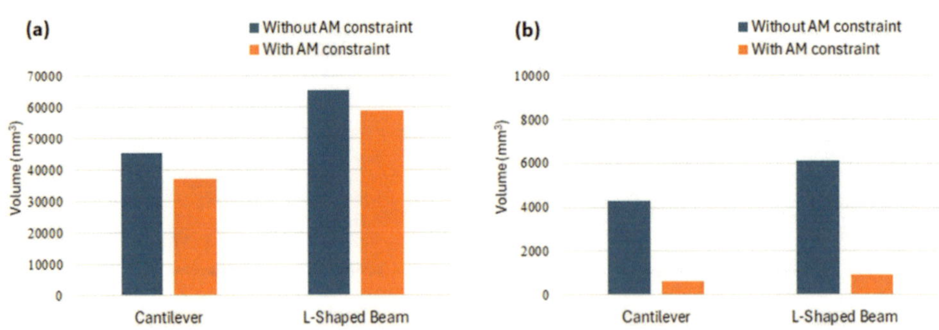

Figure 6.
Comparison of the volume: (a) total volume (i.e., part volume + support volume); and (b) support volume.

are added, the optimizer tends to converge toward overly simplified shapes in order to minimize support structures, as seen in **Figure 5**. Topology optimization using AM constraints does not always yield these simple geometry results because it varies depending on the design domain, boundary and loading conditions, and build orientation. However, these results show that adding overhang constraints may limit the ability to obtain more varied and unique designs and restrict the potential benefits of AM.

In addition to overhang constraints, selecting the appropriate build orientation is critical and depends on the specific part geometries. The results in **Figures 3–6** are all obtained when the build orientation is fixed in a specific direction. However, changing the build orientation can lead to very different optimization results, and improper build orientation may result in suboptimal designs, non-symmetric geometries, and difficulty in achieving convergence during topology optimization. It is important to carefully consider the build orientation to ensure that topology optimization is effectively applied in AM processes. The next section presents more details on the effect of build orientation on optimal part design for AM.

3. Impact of build orientation on topology optimization for AM

Build orientation is the direction in which a part is fabricated relative to the build platform. It affects various aspects of the manufacturing process, including support structure requirements, build time, surface finish, and mechanical properties. Proper selection of the build orientation is critical to minimizing support structures and achieving desired mechanical properties.

The build orientation plays a role in determining the anisotropic mechanical properties of the manufactured part. Due to the layer-by-layer manufacturing process, the material properties can vary significantly in different directions. For example, parts built in a vertical orientation may have different tensile strength and fatigue resistance than those built in a horizontal orientation. Understanding these variations is essential to designing parts that meet specific performance criteria.

In the context of topology optimization for AM, build orientation is a critical factor that must be considered during the design phase [41–44]. Build orientation affects the material properties of the part, which in turn affects the optimal geometry. For example, a part oriented to minimize support structures may have different mechanical performance compared to one oriented for optimal mechanical performance. This interdependency requires a holistic approach to topology optimization, where build orientation is integrated into the optimization algorithms to ensure that the final design is both manufacturable and performs as intended.

Integrating build orientation into topology optimization involves defining the build direction as a design variable. Existing optimization techniques such as density-based methods, level-set methods, and gradient-based optimization can be commonly used for topology optimization, but advanced optimization formulation and algorithms are also required to simultaneously optimize the material layout and the build orientation to achieve the minimized support structures and the best possible performance while ensuring manufacturability. This integration requires the development of new computational methods and mathematical formulations for topology optimization that can effectively converge to the optimal design solution for AM.

Figure 7 shows the results of topology optimization with overhang constraints for six different build orientations. The figure indicates that the optimization algorithm

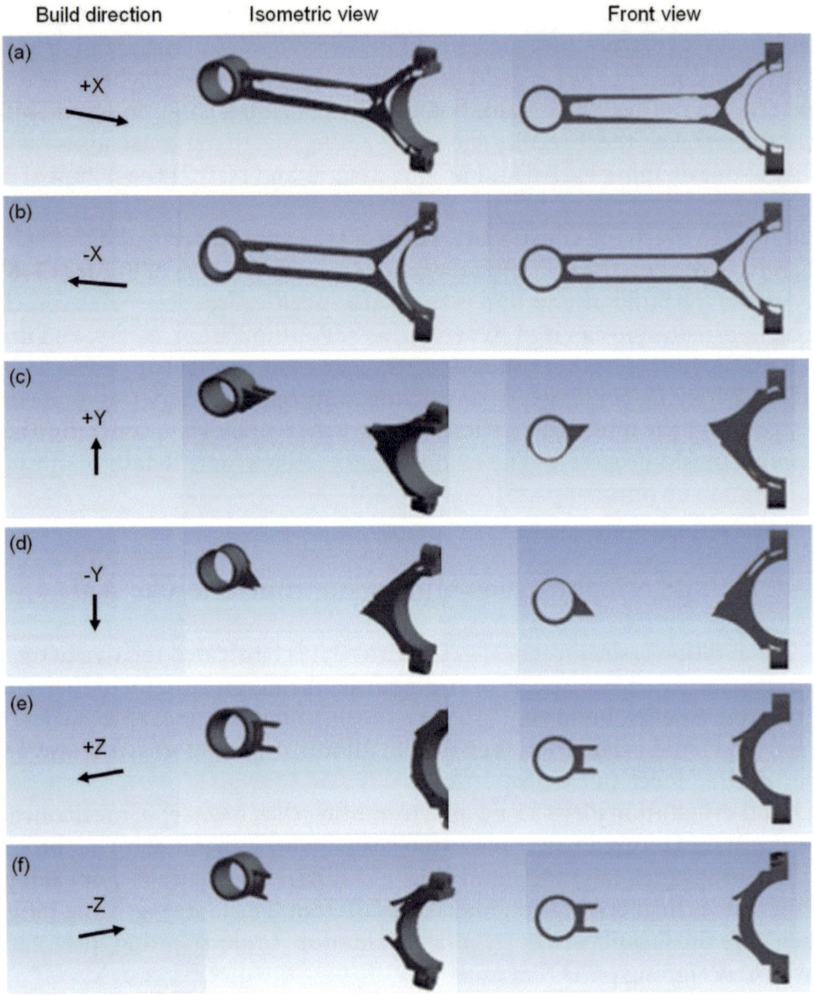

Figure 7.
Optimized connecting rod in six different build orientations.

successfully converged for the +X and -X build orientations. However, it did not converge for the +Y, -Y, +Z, and -Z orientations. This failure to converge in the +Y, -Y, +Z, and -Z directions is due to the inability to reduce the part volume to the desired fraction while ensuring that the down-skin angles exceed 45 degrees. Specifically, the thinness of the connecting rod makes it challenging to reshape the part so that the down-skin angles on the thin surfaces are greater than 45 degrees for the +Z or -Z orientations. Thus, these results show that fixing the build orientation for AM in a particular orientation and then performing topology optimization often fails to produce a converged, feasible design solution. In a given build direction, the topology optimization algorithm only converged to the optimal design when the initial design domain had sufficient volume to adjust the down-skin angles and satisfy the overhang constraint.

Figure 8 presents similar topology optimization results for a brake pedal. For the +Y and -Y build orientations, which correspond to the longitudinal direction of the brake pedal, the optimization algorithm successfully converged to an optimized

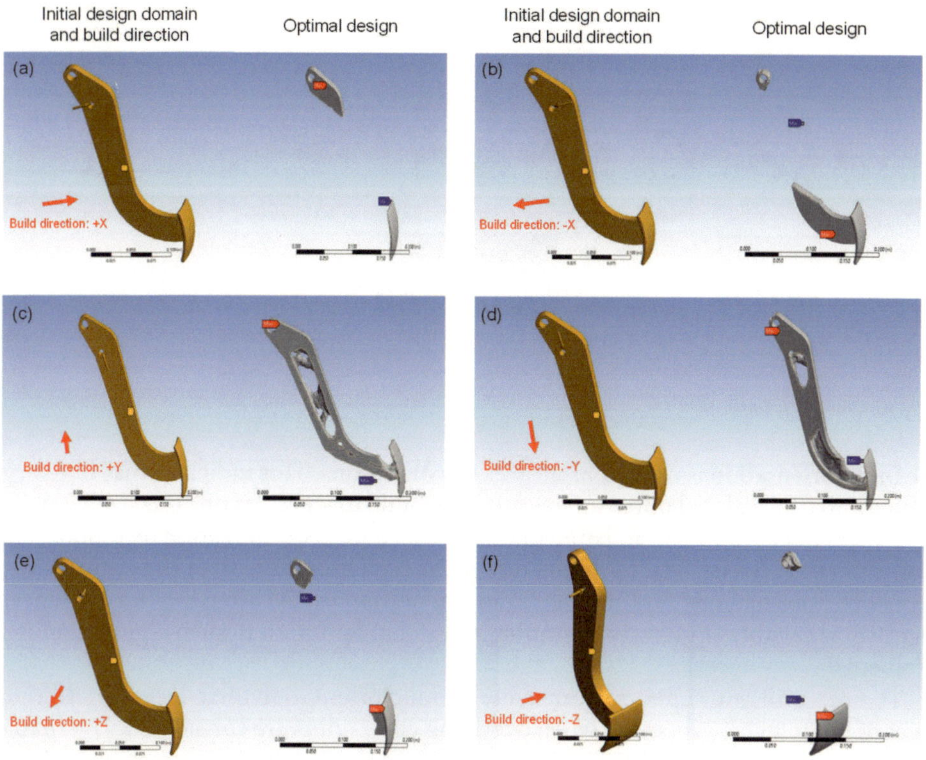

Figure 8.
Optimized brake pedal in six different build orientations.

design. The down-skin angles exceeded 45 degrees, resulting in a self-supporting shape that did not require support structures in these directions. However, as with the connecting rod, the optimization did not converge for the +X, −X, +Z, and -Z orientations. The optimization process stopped at a disconnected geometry without further optimization due to the thinness of the part. Again, it was challenging to maintain down-skin angles greater than 45 degrees for the +Z or -Z orientations, as shown in **Figures 8(e)** and **8(f)**.

These results demonstrate the critical importance of selecting the appropriate build orientation, which depends on the specific part geometry. Improper build orientation can lead to suboptimal designs or difficulties in achieving convergence during topology optimization. Therefore, careful consideration of build orientation is essential to effectively applying topology optimization in AM processes.

In the topology optimization process for AM, if the build orientation is defined as a design variable to be optimized, the build orientation can be a critical variable that can significantly impact the final AM part geometry. As the build orientation continuously changes during the optimization process, the volume of support structures required to manufacture the part can vary dramatically, as shown in **Figure 9**. When the build orientation is adjusted by rotating the part in the θ or φ direction, the support volume changes substantially depending on the angles θ and φ. In the figure, the x, y, and z axes in each plot represent θ, φ, and the volume of the support structure, respectively. Although the three AM parts do not have particularly complex shapes, the volume of the support structure changes significantly when

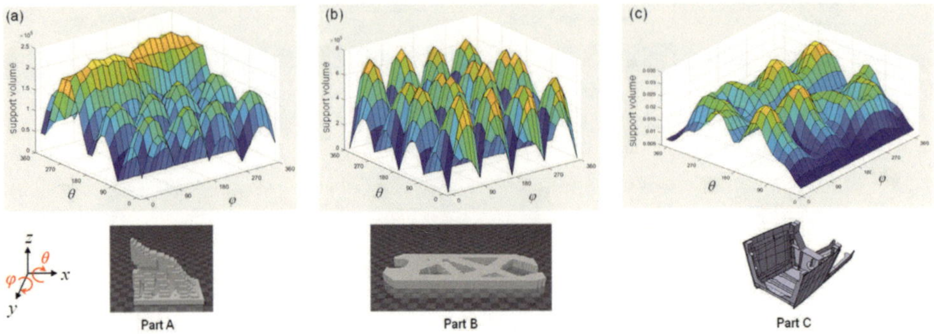

Figure 9.
Support structure volume.

the build orientation is rotated in the θ and φ directions. This indicates that it is very difficult to find a global optimum because the build orientation optimization is a highly nonlinear problem. Existing gradient-based local optimization techniques may easily converge to the local optimum due to the high nonlinearity of the optimization problem. Consequently, optimizing build orientation in AM often requires advanced algorithms capable of effectively finding the global optimum in such highly nonlinear problems.

The variability in support volume due to build orientation introduces an additional layer of complexity, especially when overhang constraints are considered. Overhangs in the design necessitate additional support material to prevent deformation and ensure the part is built accurately. As a result, optimizing the part geometry and build orientation simultaneously becomes a highly complex design problem. The interdependent nature of these AM design factors means that a change in build orientation affects not only structural performance but also the manufacturability and material usage. Therefore, advanced algorithms and computational techniques are essential to navigate the trade-offs between minimizing support structures, satisfying AM constraints, and maximizing the structural integrity of the final product. This advanced optimization for AM ensures that the design is not only feasible and efficient but also practical for real-world AM applications.

4. Closing remarks

This chapter has examined the critical role of overhang constraints and build orientation in the topology optimization process for AM. Overhang constraints are essential for minimizing the need for support structures, which in turn reduces material usage, post-processing time, and overall manufacturing costs. Build orientation is another critical factor that directly affects the manufacturability and performance of AM parts. The results presented in this chapter show that improper build orientation can lead to local optimums or challenges in achieving convergence during topology optimization.

In topology optimization for AM, when the build orientation is treated as a design variable, it significantly affects the part geometry and support structure volume due to its high nonlinearity. Existing gradient-based methods for solving this optimization problem may converge to local optima, making it difficult to find a global optimum. Therefore, an advanced integrated approach to optimizing both topology and build orientation with AM constraints is needed to ensure that the final design is not only globally optimal but also practical for AM.

Future work should focus on developing topology optimization algorithms that can more effectively handle the complexities of integrating overhang constraints and build orientation into the topology optimization process. In addition, machine learning techniques and multi-objective optimization algorithms should be incorporated to explore more complex design spaces and simultaneously balance structural performance, material usage, production cost, and manufacturability constraints. Experimental validation of the optimized designs will also be essential to confirm their manufacturability and performance in real-world applications.

Conflict of interest

The authors declare no conflict of interest.

Author details

Babin Dangal, Harshith Boppana and Sangjin Jung*
School of Mechanical, Aerospace, and Materials Engineering, Southern Illinois University, Carbondale, IL, USA

*Address all correspondence to: sangjin.jung@siu.edu

References

[1] Gibson I, Rosen DW, Stucker B, Khorasani M, Rosen D, Stucker B, et al. Additive Manufacturing Technologies. 3rd ed. Switzerland: Springer Cham; 2021. DOI: 10.1007/978-3-030-56127-7

[2] Bikas H, Stavropoulos P, Chryssolouris G. Additive manufacturing methods and modelling approaches: A critical review. The International Journal of Advanced Manufacturing Technology. 2016;**83**:389-405. DOI: 10.1007/s00170-015-7576-2

[3] Bourell D, Kruth JP, Leu M, Levy G, Rosen D, Beese AM, et al. Materials for additive manufacturing. CIRP Annals. 2017;**66**(2):659-681. DOI: 10.1016/j.cirp.2017.05.009

[4] Tan LJ, Zhu W, Zhou K. Recent progress on polymer materials for additive manufacturing. Advanced Functional Materials. 2020;**30**(43):2003062. DOI: 10.1002/adfm.202003062

[5] Travitzky N, Bonet A, Dermeik B, Fey T, Filbert-Demut I, Schlier L, et al. Additive manufacturing of ceramic-based materials. Advanced Engineering Materials. 2014;**16**(6):729-754. DOI: 10.1002/adem.201400097

[6] Frazier WE. Metal additive manufacturing: A review. Journal of Materials Engineering and Performance. 2014;**23**:1917-1928. DOI: 10.1007/s11665-014-0958-z

[7] King WE, Anderson AT, Ferencz RM, Hodge NE, Kamath C, Khairallah SA, et al. Laser powder bed fusion additive manufacturing of metals; physics, computational, and materials challenges. Applied Physics Reviews. 2015;**2**:041304. DOI: 10.1063/1.4937809

[8] Singh R, Gupta A, Tripathi O, Srivastava S, Singh B, Awasthi A, et al. Powder bed fusion process in additive manufacturing: An overview. Materials Today: Proceedings. 2020;**26**:3058-3070. DOI: 10.1016/j.matpr.2020.02.635

[9] Svetlizky D, Das M, Zheng B, Vyatskikh AL, Bose S, Bandyopadhyay A, et al. Directed energy deposition (DED) additive manufacturing: Physical characteristics, defects, challenges and applications. Materials Today. 2021;**49**:271-295. DOI: 10.1016/j.mattod.2021.03.020

[10] Li M, Du W, Elwany A, Pei Z, Ma C. Metal binder jetting additive manufacturing: A literature review. Journal of Manufacturing Science and Engineering. 2020;**142**(9):090801. DOI: 10.1115/1.4047430

[11] Ziaee M, Crane NB. Binder jetting: A review of process, materials, and methods. Additive Manufacturing. 2019;**28**:781-801. DOI: 10.1016/j.addma.2019.05.031

[12] Sigmund O, Maute K. Topology optimization approaches: A comparative review. Structural and Multidisciplinary Optimization. 2013;**48**(6):1031-1055. DOI: 10.1007/s00158-013-0978-6

[13] Sigmund O. A 99 line topology optimization code written in Matlab. Structural and Multidisciplinary Optimization. 2001;**21**:120-127. DOI: 10.1007/s001580050176

[14] Bendsøe MP, Kikuchi N. Generating optimal topologies in structural design using a homogenization method. Computer Methods in Applied Mechanics and Engineering. 1988;**71**(2):197-224. DOI: 10.1016/0045-7825(88)90086-2

[15] Wu J, Sigmund O, Groen JP. Topology optimization of multi-scale structures: A review. Structural and Multidisciplinary Optimization. 2021;**63**:1455-1480. DOI: 10.1007/s00158-021-02881-8

[16] Bendsøe MP. Optimal shape design as a material distribution problem. Structural Optimization. 1989;**1**:193-202. DOI: 10.1007/BF01650949

[17] Rozvany G. The SIMP method in topology optimization-theoretical background, advantages and new applications. In: Proceedings of the 8th Symposium on Multidisciplinary Analysis and Optimization; 6-8 September 2000. Long Beach, CA: AIAA; 2000. DOI: 10.2514/6.2000-4738

[18] Wang MY, Wang X, Guo D. A level set method for structural topology optimization. Computer Methods in Applied Mechanics and Engineering. 2003;**192**(1-2):227-246. DOI: 10.1016/S0045-7825(02)00559-5

[19] Van Dijk NP, Maute K, Langelaar M, Van Keulen F. Level-set methods for structural topology optimization: A review. Structural and Multidisciplinary Optimization. 2013;**48**:437-472. DOI: 10.1007/s00158-013-0912-y

[20] Yang RJ, Chahande AI. Automotive applications of topology optimization. Structural Optimization. 1995;**9**:245-249. DOI: 10.1007/BF01743977

[21] Cavazzuti M, Baldini A, Bertocchi E, Costi D, Torricelli E, Moruzzi P. High performance automotive chassis design: A topology optimization based approach. Structural and Multidisciplinary Optimization. 2011;**44**:45-56. DOI: 10.1007/s00158-010-0578-7

[22] Zhu JH, Zhang WH, Xia L. Topology optimization in aircraft and aerospace structures design. Archives of Computational Methods in Engineering. 2016;**23**:595-622. DOI: 10.1007/s11831-015-9151-2

[23] Yang RJ, Chen CJ. Stress-based topology optimization. Structural Optimization. 1996;**12**:98-105. DOI: 10.1007/BF01196941

[24] Cheng G, Jiang Z. Study on topology optimization with stress constraints. Engineering Optimization. 1992;**20**(2):129-148. DOI: 10.1080/03052159208941276

[25] Svanberg K. The method of moving asymptotes—A new method for structural optimization. International Journal for Numerical Methods in Engineering. 1987;**24**(2):359-373. DOI: 10.1002/nme.1620240207

[26] Brackett D, Ashcroft I, Hague R. Topology optimization for additive manufacturing. In: Proceedings of the 2011 International Solid Freeform Fabrication Symposium; 8-10 August 2011. Austin, TX: The University of Texas at Austin; 2011. DOI: 10.26153/tsw/15300

[27] Liu J, Gaynor AT, Chen S, Kang Z, Suresh K, Takezawa A, et al. Current and future trends in topology optimization for additive manufacturing. Structural and Multidisciplinary Optimization. 2018;**57**(6):2457-2483. DOI: 10.1007/s00158-018-1994-3

[28] Meng L, Zhang W, Quan D, Shi G, Tang L, Hou Y, et al. From topology optimization design to additive manufacturing: Today's success and tomorrow's roadmap. Archives of Computational Methods in Engineering. 2020;**27**:805-830. DOI: 10.1007/s11831-019-09331-1

[29] Zhu J, Zhou H, Wang C, Zhou L, Yuan S, Zhang W. A review of topology optimization for additive manufacturing:

Status and challenges. Chinese Journal of Aeronautics. 2021;**34**(1):91-110. DOI: 10.1016/j.cja.2020.09.020

[30] Bruggi M, Duysinx P. Topology optimization for minimum weight with compliance and stress constraints. Structural and Multidisciplinary Optimization. 2012;**46**(3):369-384. DOI: 10.1007/s00158-012-0759-7

[31] Zegard T, Paulino GH. Bridging topology optimization and additive manufacturing. Structural and Multidisciplinary Optimization. 2016;**53**(1):175-192. DOI: 10.1007/s00158-015-1274-4

[32] Lu L, Sharf A, Zhao H, Wei Y, Fan Q, Chen X, et al. Build-to-last: Strength to weight 3D printed objects. ACM Transactions on Graphics (ToG). 2014;**33**(4):1-10. DOI: 10.1145/2601097.2601168

[33] Lee J, Lee K. Block-based inner support structure generation algorithm for 3D printing using fused deposition modeling. The International Journal of Advanced Manufacturing Technology. 2017;**89**:2151-2163. DOI: 10.1007/s00170-016-9239-3

[34] Zhang P, Liu J, To AC. Role of anisotropic properties on topology optimization of additive manufactured load bearing structures. Scripta Materialia. 2017;**135**:148-152. DOI: 10.1016/j.scriptamat.2016.10.021

[35] Wu J, Wang CC, Zhang X, Westermann R. Self-supporting rhombic infill structures for additive manufacturing. Computer-Aided Design. 2016;**80**:32-42. DOI: 10.1016/j.cad.2016.07.006

[36] Huang R, Riddle M, Graziano D, Warren J, Das S, Nimbalkar S, et al. Energy and emissions saving potential of additive manufacturing: The case of lightweight aircraft components. Journal of Cleaner Production. 2016;**135**:1559-1570. DOI: 10.1016/j.jclepro.2015.04.109

[37] Tomlin M, Meyer J. Topology optimization of an additive layer manufactured (ALM) aerospace part. In: Proceedings of the 7th Altair CAE Technology Conference; 10 May 2011. Gaydon, Warwickshire, UK: Altair Engineering; 2011

[38] Mirzendehdel AM, Suresh K. Support structure constrained topology optimization for additive manufacturing. Computer-Aided Design. 2016;**81**:1-13. DOI: 10.1016/j.cad.2016.08.006

[39] Langelaar M. Topology optimization of 3D self-supporting structures for additive manufacturing. Additive Manufacturing. 2016;**12**:60-70. DOI: 10.1016/j.addma.2016.06.010

[40] Gaynor AT, Guest JK. Topology optimization considering overhang constraints: Eliminating sacrificial support material in additive manufacturing through design. Structural and Multidisciplinary Optimization. 2016;**54**(5):1157-1172. DOI: 10.1007/s00158-016-1551-x

[41] Das P, Chandran R, Samant R, Anand S. Optimum part build orientation in additive manufacturing for minimizing part errors and support structures. Procedia Manufacturing. 2015;**1**:343-354. DOI: 10.1016/j.promfg.2015.09.041

[42] Zhang Y, Bernard A, Harik R, Karunakaran KP. Build orientation optimization for multi-part production in additive manufacturing. Journal of Intelligent Manufacturing. 2017;**28**:1393-1407. DOI: 10.1007/s10845-015-1057-1

[43] Wang C, Qian X. Simultaneous optimization of build orientation and topology for additive manufacturing. Additive Manufacturing. 2020;**34**:101246. DOI: 10.1016/j.addma.2020.101246

[44] Olsen J, Kim IY. Design for additive manufacturing: 3D simultaneous topology and build orientation optimization. Structural and Multidisciplinary Optimization. 2020;**62**(4):1989-2009. DOI: 10.1007/s00158-020-02590-8

[45] ANSYS. Ansys Mechanical Finite Element Analysis (FEA) Software for Structural Engineering. 2024. Available from: https://www.ansys.com/products/structures/ansys-mechanical

Chapter 6

4D Printing-3D Printing of Smart Materials

Rajesh Pungampalayam Kannappan and Manikandan Nachimuthu

Abstract

The advent of additive manufacturing (AM) or 3-dimensional (3D) printing in the late 1980s has revolutionized the manufacturing sector and made it accessible even to hobbyists. AM involves producing parts by sequentially layering material, one layer at a time. Despite ongoing advancements, several challenges persist in the adoption of AM, including the speed of printing, the surface finish of the final product, and the development of fully functional materials. Recently, a novel manufacturing technique known as 4-dimensional (4D) printing has emerged. Four-dimensional printing involves the use of smart materials that, once 3D printed, can change shape over time. Key research areas in 4D printing include the responsive materials used, the stimuli triggering the change, and the duration of the transformation process. To ensure the future viability of this technology, it is crucial to address issues like the standardization of the printing process and the interaction mechanisms between materials and processes. This chapter provides an overview of 4D printing, its applications, and the latest advancements, focusing on the materials and methods employed in fabricating 4D-printed structures.

Keywords: 3D printing, 4D printing, smart materials, shape change, interaction mechanism

1. Introduction

The foundation of 4-dimensional (4D) printing is deeply rooted in the advancements and innovations of 3-dimensional (3D) printing technology. Therefore, understanding the history of 3DP is crucial, as it paved the way for the development of 4DP. A fundamental knowledge of the history, processes, and techniques of 3D printing will enhance the reader's comprehension of 4D printing and its advancements. Charles Hull's patented stereolithographic process marked a significant breakthrough in the design and creation of three-dimensional structures [1]. Unlike traditional subtractive manufacturing processes, stereolithography (SLA) is an additive manufacturing (AM) technique that builds parts layer by layer. Over the years, the 3D printing process has evolved, incorporating innovative materials and processing methods. As new advanced materials are developed, the applications for 3D printing continue to expand. A key factor contributing to the technology's increased accessibility was the expiration of early patents, which led to the development and proliferation of innovative 3D printing devices, reducing costs, and broadening applications to laboratories, schools, homes, and libraries.

The widespread adoption of 3D printing methods offers numerous advantages, including the precise manufacture of complex geometries, significant material savings, design flexibility, and product customization. Understanding the technology's limitations and ongoing research efforts will help uncover more benefits. One notable advantage of 3D printing is its capacity for mass customization, allowing for the production of personalized goods tailored to individual consumers while maintaining cost efficiency. Additionally, 3D printing enables the creation of complex geometries, such as lattice structures, which would require extensive tooling time using traditional methods like casting and molding. However, to fully capitalize on these benefits, improvements in fabrication speed through better design of 3D printing equipment are necessary. To address some of the challenges associated with 3D printing, a new method referred to as 4D printing was introduced in 2013. The following sections summarize in detail the process of 4D printing, its applications, materials, and methods used to achieve the 4D-printed structures.

2. 4D printing

Four-dimensional printing gained popularity when Skylar Tibbits demonstrated how a static 3D-printed object could alter its shape over time during a TED Talk in 2013 [2]. When exposed to external stimuli like temperature, pressure, light, magnetic fields, water, or a combination of these, 4D-printed structures can gradually change their structure or function. This shape or function change is attributed to the use of Shape Memory Materials (SMMs) or Stimuli Responsive Materials (SRMs). These materials are ideal for smart systems where the material senses stimuli and generates a pre-programmed movement by acting as a machine by itself. **Figure 1** illustrates the key difference between 3D and 4D printing processes.

2.1 Key elements of the 4D printing process

The five fundamental components of a 4D printing process are:

3D printing method: The process begins with the printing of a 3D structure using a single or a combination of smart materials in appropriate proportions. Shape-shifting

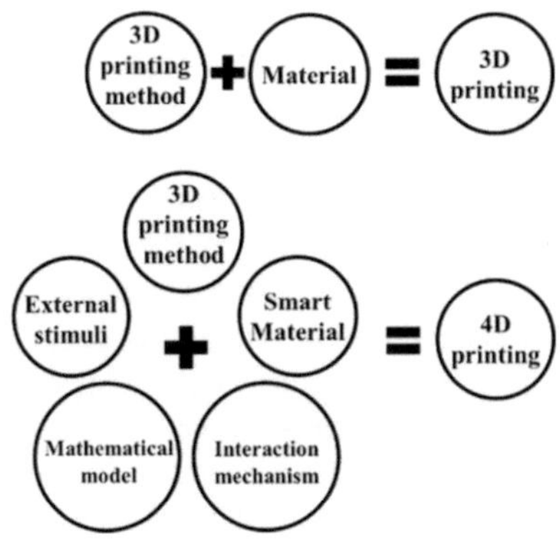

Figure 1.
Illustration of differences between 3D and 4D printing.

behavior is induced by changes in the physical or chemical parameters of the structure. Thus, the selection of an appropriate 3D printing process is crucial for the printing of these multi-material structures.

Stimulus: Common stimuli used to induce shape change in 4D printing include light, heat, water, pH, and combinations of these factors. The type of stimuli depends on the application-specific requirements.

Stimuli-responsive SMM: The selection of suitable stimuli-responsive SMM is critical for the success of 4D printing. These materials are characterized by properties, such as sensing, responsiveness, self-healing, multifunctionality, and shape recovery.

Interaction mechanism: The stimuli must be applied in a specific sequence and for a defined duration. The most commonly used interaction mechanism is constrained thermo-mechanics, where heat triggers the SMM to exhibit shape recovery.

Mathematical model: Mathematical modeling is essential in 4D printing to develop the necessary material properties that enable desired shape recovery. Theoretical and numerical models must be established to predict and relate the interactions among the process, material, stimuli, and interaction mechanisms, ensuring the achievement of the intended shape or functionality changes.

2.2 Advances in 4D printing technology and applications

Using features like size modification, the use of new materials with unique properties, multi-material printing, self-repairing, self-healing, and self-assembling capabilities. 4D printing overcomes the drawbacks of 3D printing, including the need for support structures, post-processing, and functional requirements. Even though materials, concepts, and printers are developing quickly, 4D printing is still in its infancy and still has applications in robotics, electronics, aerospace, and medicine.

2.2.1 Aerospace

The International Space Station, where intelligent acrylonitrile butadiene styrene (ABS) structures were successfully produced in microgravity, is a prime example of the promise of 4D printing in aerospace. Because of their self-assembly capabilities, short production times, and sustainability, these printed structures have potential for use in space missions. Furthermore, because of its great temperature stability and stress tolerance, polyetheretherketone (PEEK), a thermoplastic that is 80% lighter than nitinol, is being used in space applications.

2.2.2 Flexible electronics and sensors

Since 4D-printed structures are composed of different materials and can respond to various stimuli, such as heat, light, electric fields, and magnetic fields, they are finding growing applications in advanced sensor technologies. These sensors are lightweight, accurate, responsive, and reasonably priced. Living quality can be improved by 4D printing in appliances and smart houses. As an illustration, an intelligent photovoltaic system that can modify its configuration to maximize solar energy harvesting based on the amount of sunlight available was demonstrated [3].

2.2.3 Medical

Customizable noninvasive medical gadgets with 4D printing's ability to use the advantages of 3D printing and form morphing offer great potential. Biocompatible printing materials that can adapt their structure in physiological situations are essential for this technology. Tissue engineering, medication administration, and organ regeneration are examples of potential uses. A recent development uses a near-infrared-sensitive material-based 4D-printed brain model to investigate stem cell behavior [4]. Four-dimensional-printed structures are used as biological sensors, actuators, and biorobots in nano-biomedicine to track physiological changes. With its ability to self-heal and exhibit deformation features, orthopedic structures made via 4D printing can treat anomalies in bone structure and change their properties in response to external stimuli.

2.2.4 Soft robotics

The flexibility and adaptability of traditional robotics materials, like metals, stiff polymers, and high-temperature ceramics, are constrained. The goal of the developing field of soft robotics is to build robots that possess human-like dexterity and adaptability in order to overcome these constraints. Soft robotics can benefit greatly from the flexibility, deformability, and environmental adaptability of 4D printed structures. They make it possible to design self-assembling structures and soft robot actuators. In order to accomplish this self-assembly, active elements are embedded in stiff structures, which, when stimulated by outside factors, deform and assemble.

The potential uses of 4D printing are predicted to grow substantially in the future, including a multitude of new sectors. 4D printing is an interdisciplinary field with applications in material research, healthcare, nanoscience, and other areas. In conclusion, the creation of a new field centered on intelligent, controlled, and integrated structure printing is one of the promises of the future of 4D printing, in addition to developments in shape morphing. This will encourage more research into cutting-edge materials and 3D printing methods for a variety of uses. The review of new materials and related 3D printing methods that were investigated through a number of studies showcasing the potential of 4D printing can be found in the following section.

3. Shape memory materials for 4D printing

Stimuli such as heat, stress/pressure, electrical current/voltage, magnetic field, pH-change/solvent/moisture, and light, change the shape and alter the properties of certain groups of materials referred to as SRMs. SMMs are one category of SRMs wherein the materials exhibit shape recovery to their original shape when deformed and exposed to any stimuli mentioned above. Various research carried out based on the material categories as shown in Ref. [5] is illustrated in **Figure 2**.

3.1 Shape memory alloys and 4D printing

Shape memory alloys (SMAs) have been developed to exhibit shape memory effect (SME) through transitions between the martensite and austenite phases. This section summarizes key findings on the 3D printing of SMAs published to date. SMAs

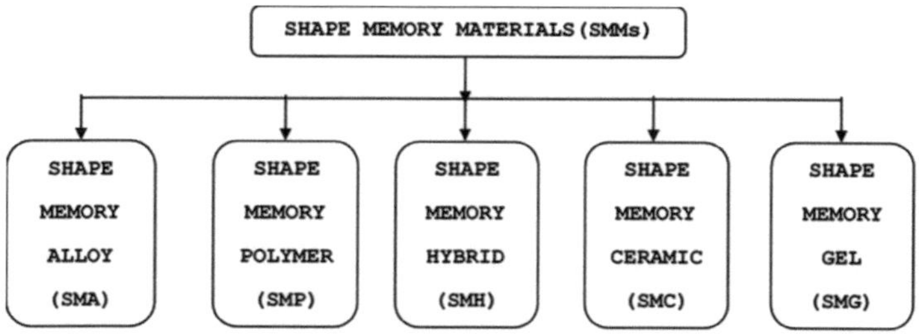

Figure 2.
Categories of SMMs used in 4D printing.

are generally responsive to thermal or magnetic stimuli, with thermo-responsive SMAs being the most commercially advanced [6, 7]. Among the various types of SMAs, copper (Cu), and nickel-titanium (NiTi) based alloys are particularly suitable for engineering applications. NiTi alloys are notable for their shape memory effect, superelasticity, low stiffness, biocompatibility, damping properties, and corrosion resistance, making them valuable in medical and engineering fields [8].

For additive manufacturing of NiTi, selecting the appropriate powder material is crucial. Techniques for producing SMA powders include mechanical attrition, hydriding, water, and gas atomization. Various combinations of nickel and titanium powders have been tested to determine the optimal composition for achieving the shape memory effect. Medium-sized particle fractions between 25 and 75 μm are particularly suitable for 3D printing due to their balance between size, flowability, and shape [9–14]. Selective laser melting (SLM) is an effective method for manufacturing dense NiTi parts, with low power and low speed laser parameters yielding good shape memory effects. **Figure 3** illustrates parts created using this method.

Three-dimensional printed NiTi structures can also exhibit shape memory behavior following pseudoelasticity. However, impurities can significantly degrade the structural and shape memory properties of NiTi alloys, and impurity levels should be kept below 500 ppm as per ASTM F2063–05 standards [15, 16]. Controlling laser parameters, such as power, scanning speed, and hatch distance during the powder production or SLM process can help manage impurity levels [17–19].

Several flow-based deposition processes are also used to 3D print NiTi alloys, including:

Figure 3.
Various structures printed from NiTi powder using the SLM process.

- Directed Energy Deposition (DED) [20]
- Laser Engineered Net Shaping (LENS) [21]
- Laser Rapid Manufacturing (LRM) [22]
- Direct Metal Deposition (DMD) [23]
- Laser Solid Forming (LSF) [24]

These processes vary mainly in the type of laser used and the NiTi alloy composition. Although copper-based SMAs are considered an alternative to NiTi alloys, their shape recovery strain is comparatively lower. Nevertheless, Cu-based SMAs like Cu-Al-Ni are preferred for their ambient transformation temperatures, thermal stability, and nontoxicity [25]. For Cu-based SMAs, direct metal laser fabrication (DMLF) and direct laser deposition (DLD) are reported as viable 3D fabrication methods. The SLM process has successfully produced high-density martensitic samples of 81.95Cu-11.85Al-3.2Ni-3Mn (wt.%) [26]. SLM processing significantly impacts grain size, packing density, and morphology. For the additive manufacturing of magnetic Ni-Mn-Ga SMA powders, powder jet binder jetting is recognized for its ability to control complex shapes and part porosity [27]. The research reviewed shows that selective laser melting (SLM) and powder jet binder jetting processes are suitable for the 4D printing of Ni-based and Cu-based SMA structures.

3.2 Shape memory polymers and 4D printing

Shape memory polymers (SMPs) have the unique capability of remembering a permanent shape while forming a temporary shape that can be fixed and later reverted to the original shape through an external stimulus such as heat, light, electrochemical changes, pH, mechanical force, electrical fields, or biological factors [28]. SMPs offer several advantages over SMAs, including lower density, lower cost, higher strain recovery, biocompatibility, and biodegradability [29–31]. However, they also have drawbacks, such as low modulus and low operating temperatures.

The 4D printing process for SMPs involves several programming steps, including heating, loading, cooling, load removal, and reheating. Early advancements in 4D printing of SMPs involved embedding glass polymer fibers in an elastomeric matrix, achieving a shape fixity ratio of 80% [32]. Similar work was carried out in printers where active hinges with different shape memory behaviors were printed, resulting in digital SMPs—a combination of two base materials, VeroWhite and TangoBlack. Sequential folding mechanisms were demonstrated by printing a clear inkjet shrink film on a pre-strained polymer sheet, which transformed into complex 3D shapes when exposed to light. A more direct method of 3D printing SMPs integrated the programming steps into the printing process, where a strip was printed with an elastomer and SMP [33]. The pre-strain of the elastomer was controlled through parameters like material composition, curing light energy, and time, resulting in shape change above the glass transition temperature (T_g).

Stereolithography (SLA) and digital light projection (DLP) have also been explored for 3D printing SMPs. These methods require special preparation of photocurable resins. Using a commercial Picoplus39, Asiga SLA printer, various shapes such as a cardiovascular stent, an Eiffel Tower, and an eagle were printed using

Polycaprolactone (PCLA) material [34]. Depositing a layer of Carbon Nanotubes (CNTs) over the 3D-printed SMPs enabled them to function as electric heaters. PCLA has also been used for medical and fashion applications, including the creation of a shape memory shoe designed to fit different users' foot profiles. The customization and reduction of injury during stent deployment are significant advantages of 4D-printed medical stents. Shape memory rings, flowers, and shoe heel attachments have also been created, demonstrating the SME using PCLA [35] as the base material as shown in **Figure 4**.

Recent research has shown multiple shape memory cycles, using commercially available Gray 60 resin to 4D print a basic square shape programmed for a triple shape memory effect [36]. Customized photocurable resins such as tert-butyl acrylate-co-di (ethylene glycol) diacrylate, soybean oil epoxidized acrylate, and epoxy-acrylate hybrid have also been used to create 3D structures through SLA printing [37]. The potential use of liquid crystal elastomers (LCE) was demonstrated by Yuan et al. LCEs were prefabricated using TangoBlack, and silver electrodes were sintered and embedded.

Fused Filament Fabrication (FFF) and Digital Ink Writing (DIW) are other techniques used in 4D printing SMPs [38, 39]. DIW is more compatible with a broader range of materials than FFF. Controlling the ink's rheological properties is crucial for effective DIW, ensuring shape retention once the ink exits the nozzle. Diels-Alder reversible thermoset (DART) resins and UV cross-linkable inks (PLA, dichloromethane, and benzophenone) have been explored for DIW, showing good shape memory behavior. Remote actuation is achieved using Fe_3O_4 particles. SMPs can be actuated by heat, solvent, moisture, pH change, light, and other stimuli, offering contactless actuation through induction heating inside the human body [40–44]. It is observed that SMPs can be effectively 4D printed using SLA, DLP, and DIW. The fabrication of thermoset polymers, however, poses challenges for FFF, limiting its use.

3.3 Hydrogels for 4D printing

Hydrogels, known for their ability to swell with the addition of water, are three-dimensional, cross-linked, smart materials. Due to their biocompatibility and storage capacity, hydrogels have become a preferred category of materials for fabricating smart devices [45]. Various 3D bioprinting techniques for hydrogels have been

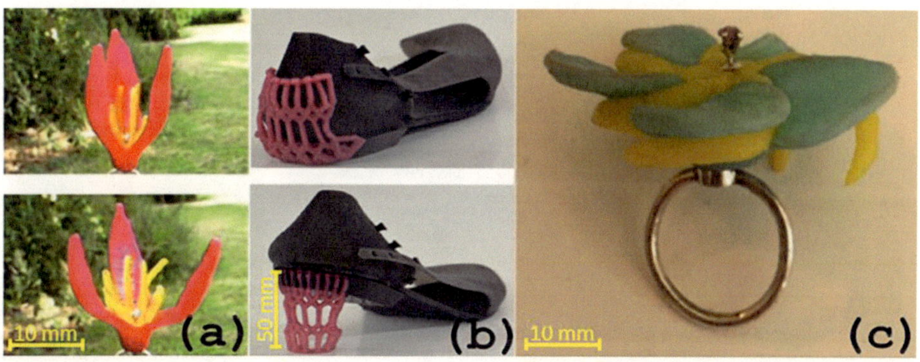

Figure 4.
Four-dimensional printed PCLA structures (a) a flower, (b) a shoe heel, and (c) a ring.

explored, as highlighted in a comprehensive review of their biocompatibility. The primary 3D bioprinting methods for hydrogels include:

- Extrusion-based bioprinting: Gelatin-methacrylate (GelMA) material is 3D printed using Stereolithography (SLA) and digital light processing (DLP) methods [46, 47].

- Inkjet-based bioprinting [48].

- Laser-based techniques: GelMA scaffolds are fabricated using dynamic SLA methods [49].

Initial advancements in hydrogel applications came from Tibbits' group, which demonstrated primitive shape-changing behaviors, such as linear stretching, ring stretching, and folding [50]. These behaviors were achieved by printing hydrogels on rigid plastic bases using a commercially available Stratasys Connex 500 printer. The printed hydrogels exhibited a volume expansion of 200% when exposed to water, thereby showcasing their shape-changing properties. M. Jamal et al. reviewed the fundamentals of hydrogel 3D fabrication, highlighting the use of conventional photolithography to create Polyethylene Glycol (PEG)-based bi-layer structures. Differential swelling of the hydrogel allowed for predictable and controllable folding of the constructs. A custom-built dual extrusion printer capable of printing Poly (NIPAM) and Poly (HEMA) at varying extrusion rates was also explored [51]. This setup included ink deposition from the dual extruder, controlled by linear actuators mounted on a computer numerical control (CNC) milling stage as shown in **Figure 5**.

Hydrogels are responsive to various external stimuli, including temperature, pH, electric current, magnetic field, and light. A notable example is a smart temperature-responsive hydrogel, designed using alginate/poly (N-isopropylacrylamide)

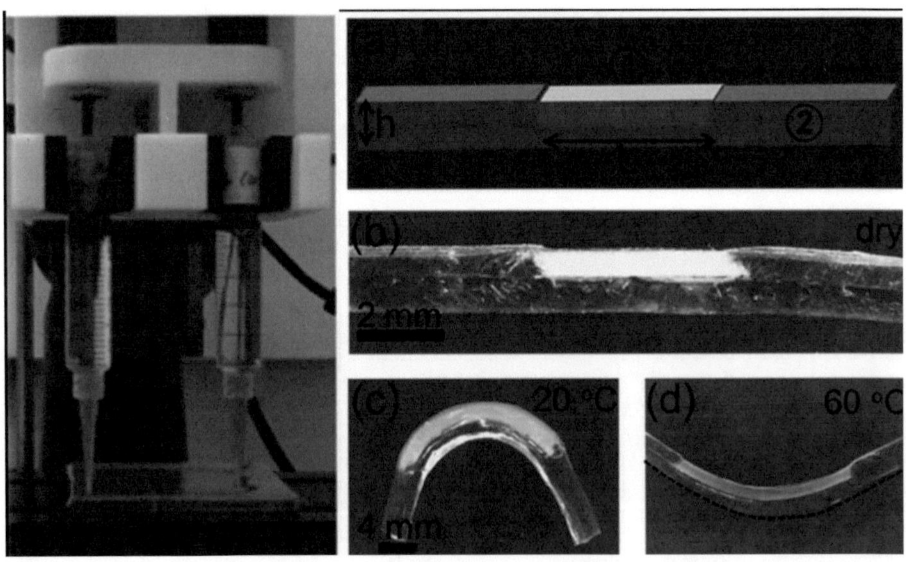

Figure 5.
Custom hydrogel ink extrusion printer. Left: Extrusion printer; Right: Poly (NIPAM)- and Poly (HEMA)-printed structure.

(PNIPAAm) inks to produce a smart valve [52]. The flow rate through this valve could be reduced by closing the smart valve. Additionally, chemo-responsive and pH-responsive hydrogels have also been extensively studied. The key experiments reviewed indicate that hydrogels' flexibility and their ability to be 3D printed using extrusion, ink-based, and laser-based methods make them highly adaptable. When stimulated by external factors, such as temperature, pH, electric current, magnetic field, and light, hydrogels demonstrate significant flexibility [53–56].

3.4 Shape memory hybrids and composite for 4D printing

Shape memory hybrids (SMHs) are advanced materials created by combining SMPs, SMAs, and hydrogels. The synthesis of SMAs and SMPs for specific applications demands significant expertise in chemical and polymer sciences. SMHs, however, can be developed from simple materials with well-known properties, even if the individual materials lack shape memory effect on their own. SMHs consist of a matrix material and an inclusion material, which can be metallic, organic, or inorganic. The inclusion material imparts the shape memory properties to the SMH.

SMHs are categorized based on their response to stimuli: pressure-responsive, thermo-responsive, and multi-stimuli responsive materials. For instance, a combination of silicone and wax was used to demonstrate a basic SMH property where a pre-bent shape reverted to its original form when immersed in water at 25°C, taking about 15 seconds for full recovery as shown in **Figure 6** [57]. Similarly, another study utilized water as the inclusion material in a silicone matrix, achieving sharp shape recovery due to the phase change from ice to water at approximately 0°C. SMHs responsive to cooling have also been developed, such as a plastic sponge filled with Poloxamer (P407) gel that exhibited shape recovery when immersed in 0°C water [58]. Pressure-responsive SMHs can be created by incorporating special fillers into thermo-responsive SMHs or compressing the inclusion material, resulting in notable shape recovery upon pressure release. Additionally, SMHs with self-healing capabilities have been explored. For example, a silicone/melting glue SMH combined with a metal spring demonstrated self-healing when the metal spring was joule-heated [59].

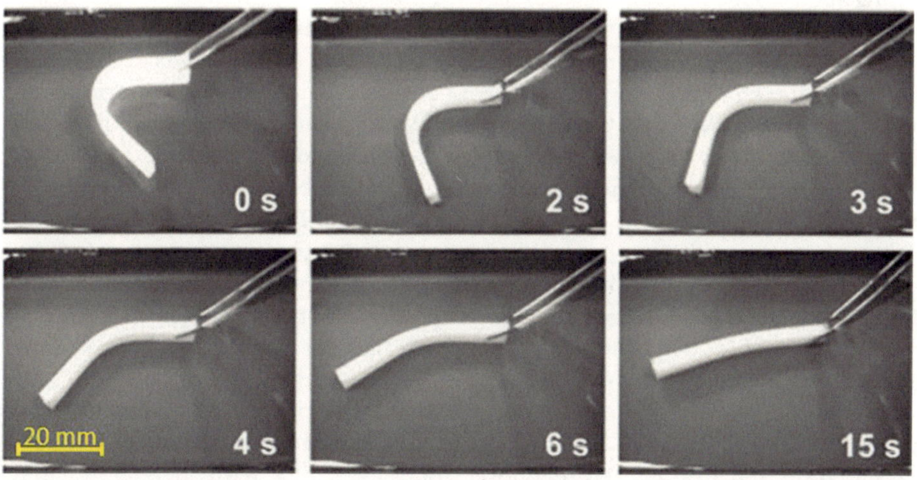

Figure 6.
A silicone/wax SMH shape recovery in 15 s.

Smart composites, which combine SMPs with conventional materials, are also significant. Research has investigated the combination of SMPs with inactive stiff plastic plates and polymer fibers [60]. Advances in 4D printing have led to the development of smart nanocomposites, such as a Barium Titanate/Polyethylene Glycol Diacrylate (PEGDA) nanocomposite fabricated using Digital Projection Printing (DPP), achieving high resolution and reproducibility. Another study created a three-layer composite material from TangoBlack+ and two layers of SMP fibers, demonstrating controlled bending deformation based on fiber volume fraction and alignment using an Objet260 Connex 3D printer [61].

Nanocomposites like carbon nanotube (CNT) combined with poly (D,L-lactide-co-trimethylene carbonate) showed rapid shape-changing behavior when actuated with 24 V and were used to fabricate a sensor for detecting solvent leaks. Another composite made of PLA and Fe_3O_4 particles exhibited promising applications in healthcare and biomedical fields due to its magnetic actuation properties. Furthermore, a CNT/PLA-based 4D printed structure demonstrated high shape recovery percentages up to 97.3% [62].

From the reviewed studies, it is evident that SMHs and smart composites offer a superior alternative for developing application-specific smart materials, circumventing the need for deep polymer synthesis knowledge. These materials leverage the diverse 3D printing techniques to achieve desired properties for various applications.

4. Summary and future of 4D printing

This chapter investigated 4D printing, emphasizing the smart materials used and 3D printing methods employed. Since the advent of 4D printing in 2013, coupled with the use of SMMs, traditional design and manufacturing processes have undergone significant transformation. The evolution of 4D printing has introduced the concept that "the material is the machine," representing a fusion of material, manufacturing method, external stimuli, and their interactions. These interactions facilitate changes in shape, functionality, or properties of the printed smart material over time. SMAs, SMPs, and shape memory composites (SMCs) used in 4D printing have opened up exciting engineering applications and addressed issues encountered with conventional materials. The potential of SMHs in 4D printing is vast, promising new research avenues.

The review identifies the most prevalent 4D printing techniques as SLA, FFF, SLM, DLP, and DIW. Among these, SLM is considered highly suitable for printing SMAs, particularly NiTi and Cu-based alloys, with thermal actuation being the primary stimulus. SMPs have gained favor over SMAs due to advantages like low density, cost-effectiveness, high strain recovery, and biocompatibility. SLA, FFF, and DLP are the frequently explored methods for 4D printing SMPs. The preparation of specialized photocurable resins has become widespread due to the increasing use of SLA and DLP technologies. FFF and DIW processes are also under investigation for printing SMPs, SMCs, and hydrogels, where controlling the ink's rheological properties is crucial for effective DIW utilization. These materials can be actuated using thermal, electrical, and solvent stimuli.

As new engineering applications emerge, they will drive the development of innovative printing methods. Key abilities of 4D-printed smart materials include self-actuation, self-healing, self-diagnosis, and self-sensing, with ongoing research particularly focused on self-actuation and self-sensing. To ensure the sustainability of 4D printing technology, future needs must be considered. There are several critical issues that need attention:

Standardization: The review highlights that most SMMs are customized for specific applications. Standardization is necessary to make smart materials available as off-the-shelf engineering materials.

Interaction mechanism: The interaction between material and manufacturing process is vital for achieving reliable and durable structures. Therefore, to ensure long-term reliability and durability of 4D-printed structures, it is essential to develop more simulation and modeling methods.

Increasing research related to material standardization and optimizing mathematical models for smart materials can enhance and maximize the potential applications of 4D printing and smart materials.

Appendices and nomenclature

4D	four-dimensional
3D	three-dimensional
CNT	carbon nanotube
DART	Diels-Alder Reversible Thermoset
DLP	digital light projection
DPP	digital projection printing
CSL	direct ceramic stereolithography
DED	direct energy deposition
DIW	direct ink writing
DMD	direct metal deposition
EBAM	electron beam additive manufacturing
FFF	fused filament fabrication
GelMA	Gelatin-Methacrylate
LOM	laminated object manufacturing
LENS	laser engineered net shaping
LRM	laser rapid manufacturing
Ni-Mn-Ga	Nickel-Manganese-Gallium
Ni-Ti	Nickel-Titanium
PCL	polycaprolactone
PEEK	polyetheretherketone
PLA	poly-lactic acid
PBF	powder bed fusion
SLA	stereolithography
SLM	selective laser melting
SLS	selective laser sintering
SMA	shape memory alloy
SMC	shape memory composite
SME	shape memory effect
SMG	shape memory gel
SMH	shape memory hybrid
SMM	shape memory material
SMP	shape memory polymer
SRM	stimulus responsive material

Author details

Rajesh Pungampalayam Kannappan[1*] and Manikandan Nachimuthu[2]

1 PSG College of Technology, Coimbatore, India

2 Maxbyte Technologies Services Private Limited, Coimbatore, India

*Address all correspondence to: pkr.auto@psgtech.ac.in

References

[1] Hull CW. Apparatus for production of three-dimensional objects by stereolithography. U.S Patent 5556590. 1996. pp. 1-13

[2] Tibbits S. The Emergence of 4D Printing [Internet]. TED; 2014. Available from: https://www.ted.com/talks/skylar_tibbits_the_emergence_of_4d_printing

[3] Primer P. Print flexible solar cells. Nature. 2016;**539**:488-489

[4] Cui H, Miao S, Esworthy T, Lee S jun, Zhou X, Hann SY, et al. A novel near-infrared light responsive 4D printed nanoarchitecture with dynamically and remotely controllable transformation. Nano Research [Internet]. 2019;12(6):1381-1388. DOI: doi:10.1007/s12274-019-2340-9

[5] Sun L, Huang WM, Ding Z, Zhao Y, Wang CC, Purnawali H, et al. Stimulus-responsive shape memory materials: A review. Materials & Design [Internet]. 2012;**33**(1):577-640. DOI: 10.1016/j.matdes.2011.04.065

[6] Duerig TW, Melton KN, Stockel DWCM. Engineering Aspects of Shape Memory Alloys. London, United Kingdom: Butterworth-Heinemann; 1990. 499 p

[7] Huang WM, Ding Z, Wang CC, Wei J, Zhao Y, Purnawali H. Shape memory materials. Mater Today [Internet]. 2010;**13**(7-8):54-61. DOI: 10.1016/S1369-7021(10)70128-0

[8] Elahinia M, Shayesteh Moghaddam N, Taheri Andani M, Amerinatanzi A, Bimber BA, Hamilton RF. Fabrication of NiTi through additive manufacturing: A review. Progress in Materials Science [Internet]. 2016;**83**:630-663. DOI: 10.1016/j.pmatsci.2016.08.001

[9] Haberland C, Elahinia M, Walker JM, Meier H, Frenzel J. On the development of high quality NiTi shape memory and pseudoelastic parts by additive manufacturing. Smart Materials and Structures. 2014;**23**(10):1-13

[10] Meier H, Haberland C, Frenzel J. Structural and functional properties of NiTi shape memory alloys produced by selective laser melting. Innovative Developments in Virtual and Physical Prototyping [Internet]. 2011;**100**:291-296. Available from: http://www.crcnetbase.com/doi/10.1201/b11341-47

[11] Walker J, Elahinia M, Haberland C. An investigation of process parameters on selective laser melting of Nitinol. In: ASME 2013 Conference on Smart Materials, Adaptive Structures and Intelligent Systems, SMASIS. Utah, United States of America: ASME; 2013

[12] Bormann T, Müller B, Schinhammer M, Kessler A, Thalmann P, De Wild M. Microstructure of selective laser melted nickel-titanium. Materials Characterization [Internet]. 2014;**94**:189-202. DOI: 10.1016/j.matchar.2014.05.017

[13] Bormann T, Schumacher R, Müller B, Mertmann M, de Wild M. Tailoring selective laser melting process parameters for NiTi implants. Journal of Materials Engineering and Performance [Internet]. 2012;**21**(12):2519-2524. Available from: http://link.springer.com/10.1007/s11665-012-0318-9

[14] Meier H, Haberland C. Experimental studies on selective laser melting of metallic parts. Materialwissenschaft und Werkstofftechnik. 2008;**39**(9):665-670.

Available from: http://doi.wiley.com/10.1002/mawe.200800327

[15] Olier P, Barcelo F, Bechade LJ, Brachet CJ, Lefevre E, Guenin G. Effects of impurities content (oxygen, carbon, nitrogen) on microstructure and phase transformation temperatures of near Equiatomic TiNi shape memory alloys. Le Journal de Physique IV Fr. 1997;7: 143-148. DOI: 10.1051/jp4:1997522

[16] Frenzel J, Zhang Z, Somsen C, Neuking K, Eggeler G. Influence of carbon on martensitic phase transformations in NiTi shape memory alloys. Acta Materialia. 2007;**55**(4):1331-1341

[17] Krishna BV, Bose S, Bandyopadhyay A. Laser processing of net-shape NiTi shape memory alloy. Metallurgical and Materials Transactions A: Physical Metallurgy and Materials Science. 2007;**38**(5):1096-1103. Available from: http://link.springer.com/10.1007/s11661-007-9127-4

[18] Marattukalam JJ, Singh AK, Datta S, Das M, Balla VK, Bontha S, et al. Microstructure and corrosion behavior of laser processed NiTi alloy. Materials Science and Engineering: C. 2015;**57**:309-313. DOI: 10.1016/j.msec.2015.07.067

[19] Elahinia M. Shape Memory and Superelastic Alloys. Cambridge, United Kingdom: Woodhead Pub; 2014

[20] Hamilton RF, Palmer TA, Bimber BA. Spatial characterization of the thermal-induced phase transformation throughout as-deposited additive manufactured NiTi bulk builds. Scripta Materialia. 2015;**101**:56-59. DOI: 10.1016/j.scriptamat.2015.01.018

[21] Bandyopadhyay A, Krishna BV, Xue W, Bose S. Application of laser engineered net shaping (LENS) to manufacture porous and functionally graded structures for load bearing implants. Journal of Materials Science. Materials in Medicine. 2009;**20**:29-34

[22] Shiva S, Palani IA, Mishra SK, Paul CP, Kukreja LM. Investigations on the influence of composition in the development of Ni-Ti shape memory alloy using laser based additive manufacturing. Optics and Laser Technology. 2015;**69**(1):44-51. DOI: 10.1016/j.optlastec.2014.12.014

[23] Malukhin K, Ehmann K. Material characterization of NiTi based memory alloys fabricated by the laser direct metal deposition process. Journal of Manufacturing Science and Engineering [Internet]. 2006;**128**(3):691. Available from: http://manufacturingscience.asmedigitalcollection.asme.org/article.aspx?articleid=1450690

[24] Xu X, Lin X, Yang M, Chen J, Huang W. Microstructure evolution in laser solid forming of Ti-50 wt% Ni alloy. Journal of Alloys and Compounds. 2009;**480**(2):782-787

[25] Gu DD, Shen YF. Development and characterisation of direct laser sintering multicomponent Cu based metal powder. Powder Metallurgy [Internet]. 2006;**49**(3):258-264. Available from: http://www.tandfonline.com/doi/full/10.1179/174329006X95662

[26] Gustmann T, Neves A, Kühn U, Gargarella P, Kiminami CS, Bolfarini C, et al. Influence of processing parameters on the fabrication of a Cu-Al-Ni-Mn shape-memory alloy by selective laser melting. In: Additive Manufacturing. Vol. 11. Elsevier B.V; 2016. 23-31 p. DOI: 10.1016/j.addma.2016.04.003

[27] Caputo MP, Berkowitz AE, Armstrong A, Müllner P, Solomon CV.

4D printing of net shape parts made from Ni-Mn-Ga magnetic shape-memory alloys. Additive Manufacturing. 2010;**2018**(21):579-588. DOI: 10.1016/j.addma.2018.03.028

[28] Liu C, Qin H, Mather PT. Review of progress in shape-memory polymers. Journal of Materials Chemistry. 2007;**17**(16):1543-1558

[29] Yang B, Huang WM, Li C, Lee CM, Li L. On the effects of moisture in a polyurethane shape memory polymer. Smart Materials and Structures. 2004;**13**(1):191-195

[30] Huang W. On the selection of shape memory alloys for actuators. Materials and Design. 2002;**23**(1):11-19. Available from: http://linkinghub.elsevier.com/retrieve/pii/S0261306901000395

[31] Hao B, Li Y, Xiao X, Dai W, Chen H, Chen S. A facile photo-polymerization method for reconfigurable shape memory polymers. Materials Letters. 2019;**254**:214-217. DOI: 10.1016/j.matlet.2019.07.071

[32] Ge Q, Qi HJ, Dunn ML. Active materials by four-dimension printing. Applied Physics Letters. 2013;**103**(13):103-1-103-6

[33] Liu Y, Shaw B, Dickey MD, Genzer J. Sequential self-folding of polymer sheets. Science Advances. 2017;**3**(3):1-8

[34] Zarek M, Layani M, Cooperstein I, Sachyani E, Cohn D, Magdassi S. 3D printing of shape memory polymers for flexible electronic devices. Advanced Materials. 2016;**28**(22):4449-4454

[35] Zarek M, Layani M, Eliazar S, Mansour N, Cooperstein I, Shukrun E, et al. 4D printing shape memory polymers for dynamic jewellery and fashionwear. Virtual and Physical Prototyping [Internet]. 2016;**11**(4):263-270. DOI: 10.1080/17452759.2016.1244085

[36] Yu K, Dunn ML, Qi HJ. Digital manufacture of shape changing components. Extreme Mechanics Letters [Internet]. 2015;**4**:9-17. DOI: 10.1016/j.eml.2015.07.005

[37] Miao S, Zhu W, Castro NJ, Nowicki M, Zhou X, Cui H, et al. 4D printing smart biomedical scaffolds with novel soybean oil epoxidized acrylate. Scientific Reports. 2016;**6**(June):1-10. DOI: 10.1038/srep27226

[38] Wei H, Zhang Q, Yao Y, Liu L, Liu Y, Leng J. Direct-write fabrication of 4D active shape-changing structures based on a shape memory polymer and its nanocomposite. ACS Applied Materials & Interfaces. 2017;**9**(1):876-883

[39] Yang K, Grant JC, Lamey P, Joshi-Imre A, Lund BR, Smaldone RA, et al. Diels–Alder reversible thermoset 3D printing: Isotropic thermoset polymers via fused filament fabrication. Advanced Functional Materials. 2017;**27**(24):1-11

[40] Huang WM, Yang B, An L, Li C, Chan YS. Water-driven programmable polyurethane shape memory polymer: Demonstration and mechanism. Applied Physics Letters. 2005;**86**(11):1-3

[41] Yang B, Huang WM, Li C, Chor JH. Effects of moisture on the glass transition temperature of polyurethane shape memory polymer filled with nano-carbon powder. European Polymer Journal. 2005;**41**(5):1123-1128

[42] Havens E, Snyder EA, Tong TH. Light-activated shape memory polymers and associated applications. European Polymer Journal. 2005;**5762**:48. Available from: http://proceedings.spiedigitallibrary.org/proceeding.aspx?doi=10.1117/12.606109

[43] Razzaq MY, Anhalt M, Frormann L, Weidenfeller B. Thermal, electrical and magnetic studies of magnetite filled polyurethane shape memory polymers. Materials Science and Engineering A. 2007;**444**(1-2):227-235

[44] Du H, Zhang J. Solvent induced shape recovery of shape memory polymer based on chemically cross-linked poly(vinyl alcohol). Soft Matter. 2010;**6**(14):3370-3376

[45] Tokarev I, Minko S. Stimuli-responsive hydrogel thin films. Soft Matter. 2009;**5**(3):511-524

[46] Visser J, Peters B, Burger TJ, Boomstra J, Dhert WJA, Melchels FPW, et al. Biofabrication of multi-material anatomically shaped tissue constructs. Biofabrication. 2013;**5**(3):1-9

[47] Billiet T, Gevaert E, De ST, Cornelissen M, Dubruel P. Biomaterials the 3D printing of gelatin methacrylamide cell-laden tissue-engineered constructs with high cell viability. Biomaterials [Internet]. 2014;**35**(1):49-62. DOI: 10.1016/j.biomaterials.2013.09.078

[48] Xu T, Zhao W, Zhu JM, Albanna MZ, Yoo JJ, Atala A. Complex heterogeneous tissue constructs containing multiple cell types prepared by inkjet printing technology. Biomaterials. 2013;**34**(1):130-139. DOI: 10.1016/j.biomaterials.2012.09.035

[49] Soman P, Chung PH, Zhang AP, Chen S. Digital microfabrication of user-defined 3D microstructures in cell-laden hydrogels. Biotechnology and Bioengineering. 2013;**110**(11):3038-3047

[50] Raviv D, Zhao W, McKnelly C, Papadopoulou A, Kadambi A, Shi B, et al. Active printed materials for complex self-evolving deformations. Scientific Reports. 2014;**4**:1-8

[51] Naficy S, Gately R, Gorkin R, Xin H, Spinks GM. 4D printing of reversible shape morphing hydrogel structures. Macromolecular Materials and Engineering. 2017;**302**(1):1-9

[52] Bakarich SE, Gorkin R, In PM, Het SGM. 4D printing with mechanically robust, thermally actuating hydrogels. Macromolecular Rapid Communications. 2015;**36**(12):1211-1217

[53] Gao J, Wang H, Wang L, Wang J, Kong D, Yang Z. Enzyme promotes the hydrogelation from a hydrophobic small molecule. Journal of the American Chemical Society. 2009;**131**(32):11286-11287

[54] Sáez JA, Escuder B, Miravet JF. Supramolecular hydrogels for enzymatically triggered self-immolative drug delivery. Tetrahedron. 2010;**66**(14):2614-2618

[55] Zhao S, Lee J, Xu W. Supramolecular hydrogels formed from biodegradable ternary COS-g-PCL-b-MPEG copolymer with α-cyclodextrin and their drug release. Carbohydrate Research. 2009;**344**(16):2201-2208. DOI: 10.1016/j.carres.2009.08.017

[56] Verdejo B, Rodríguez-Llansola F, Escuder B, Miravet JF, Ballester P. Sodium and pH responsive hydrogel formation by the supramolecular system calix[4]pyrrole derivative/tetramethylammonium cation. Chemical Communications. 2011;**47**(7):2017-2019

[57] Fan K, Huang WM, Wang CC, Ding Z, Zhao Y, Purnawali H, et al. Water-responsive shape memory hybrid: Design concept and demonstration. Express Polymer Letters. 2011;**5**(5):409-416

[58] Wang CC, Huang WM, Ding Z, Zhao Y, Purnawali H. Cooling–/water-responsive shape memory hybrids. Composites Science and Technology [Internet]. 2012;**72**(10):1178-1182. DOI: 10.1016/j.compscitech.2012.03.027

[59] Huang WM, Yang B, Fu YQ. Polyurethane Shape Memory Polymers. Florida, United States of America: Taylor & Francis; 2012

[60] Mutlu R, Alici G, Panhuis MIH, Spinks G. Effect of flexure hinge type on a 3D printed fully compliant prosthetic finger. In: IEEE/ASME Int Conf Adv Intell Mechatronics. Busan, Korea (South): AIM; 2015

[61] Melchels FPW, Feijen J, Grijpma DW. A review on stereolithography and its applications in biomedical engineering. Biomaterials. 2010;**31**(24):6121-6130. DOI: 10.1016/j.biomaterials.2010.04.050

[62] Liu Y, Zhang W, Zhang F, Leng J, Pei S, Wang L, et al. Microstructural design for enhanced shape memory behavior of 4D printed composites based on carbon nanotube/polylactic acid filament. Composites Science and Technology. 2019;**181**:107692. DOI: 10.1016/j.compscitech.2019.107692

Chapter 7

Mechanical Properties of AISI 410L Stainless Steel Parts Manufactured by Laser Directed Energy Deposition in the as-Built and Heat-Treated States

Jurandir Marcos Sá de Sousa, Milton Pereira, Anselmo Thiesen Júnior, Henrique Santos Ferreira, Jhonattan Gutjahr and Juliane Ribeiro da Cruz

Abstract

Martensitic stainless steels (MSS) have extensive industrial applicability owing to their high hardness and resistance to wear and corrosion at milder temperatures. AISI 410L is a low-carbon MSS with better processability than the high carbon-content MSS grades. Additive manufacturing by laser directed energy deposition (AM L-DED) grants the benefits of geometry freedom, complex materials processability, and obtention of refined microstructures. Nonetheless, there are challenges because of the L-DED complex thermal history and brittle nature of martensite formed in MSS. Therefore, post heat-treatment are required to provide residual stress relief, martensite tempering, and overall mechanical properties enhancement. To synthesize the current knowledge on the factors affecting the mechanical properties of AISI 410L MSS processed by L-DED, while expanding the investigation on suitable heat-treatment routes, this work complies with original data and literature results to discuss how different carbon contents, building orientations, heat-treatments, and cladding procedures affect the microstructure, hardness, tensile, and Charpy impact mechanical properties of different MSS, in special AISI 410L, processed by L-DED.

Keywords: AISI 410L MSS, L-DED thermal history, post heat-treatment, carbon-content, microstructure refinement

1. Introduction

Martensitic stainless steels are cost-effective materials extensively used in industry because of their properties such as high hardness, mechanical strength, wear performance, and corrosion resistance at milder working temperatures. Because of this, industries including cutlery, mining, railroads, oil and gas, and transportation largely use martensitic stainless steels (MSS) to produce gears, bearings, exhaust systems, magnetic separators, turbine blades, punches and dies, plastic molds, rails, pumps, and valve parts, among other components [1, 2].

MSS components are usually manufactured by casting or wrought, and a mixture of martensite, ferrite, and carbides forms their microstructure. For the martensitic transformation to occur, it is necessary to carry out post heat-treatment (PHT) of solution annealing plus quenching from the austenitic field. Owing to the brittle nature of this phase, a tempering step must be performed after quenching to temper the martensite and improve the material ductility [3, 4]. Low-C MSS can offer better processability, such as weldability and machinability while exhibiting a property set comparable to higher-C MSS. In this context, for instance, the MSS 410L has found several applications [5, 6].

Additive manufacturing (AM) has brought to the industry some advantages to traditional processes, such as the manufacturing of customized and on-demand parts, the possibility of reducing stocks and lead-time, and the processability of complex alloys [7, 8]. Among the metal AM techniques, attention is drawn to the laser powder bed fusion (L-PBF) and laser directed energy deposition (L-DED) processes. In the first, the region of the pre-deposited powder layer that corresponds to the three-dimensional model is selectively melted by the laser's energy source [9]. This is more concerned with the manufacturing of smaller components with more complex geometry. L-DED, defined as a process in which focused thermal energy is used to melt materials as they are being deposited [10], is dedicated to manufacturing and repairing larger parts and less complex geometries compared to L-PBF. Besides, L-DED has the advantages of applying energy in a concentrated area, providing excellent heat-input control, low distortions, narrow heat-affected zone, and obtaining refined microstructures [7, 11].

In this context, the manufacturing of low C-content MSS parts by AM L-DED presents an interesting industrial alternative since it combines the attractive properties of this material and the technique advantages. However, there are challenges. On the material side, the brittle nature of martensite can compromise the component's performance. Conversely, the L-DED process exhibits a complex thermal history, leading to fast thermal cycles and high cooling speeds. During L-DED processing of MSS, residual stresses may result in cracks and cooling speed can generate carbide segregation at grain boundaries, resulting in localized chemical composition changes. Furthermore, the process's inherent heterogeneous microstructure may show a decrease in mechanical strength, particularly toughness [5, 12]. To mitigate these issues, the L-DED process must be properly parameterized to provide an as-built (AB), high-density component, free of geometric and metallurgical defects. Furthermore, it is required to map suitable post heat-treatment routes for the L-DED process thermal history, aiming to provide residual stress relief, microstructure homogenization, and martensite tempering [13–15].

Different techniques are employed in the literature to optimize L-DED parameters, including numerical simulation studies [16, 17] and experimental approaches. The search for operational maps varying the main L-DED parameters is commonly found for different materials. However, this is generally limited to the geometry of single beads, which restricts the observation of trends associated with subsequent layer and multilayer steps, which are necessary when thinking about AM [18]. Low C-content MSS processed by L-DED present, in the as-built (AB) state, microstructures formed by ferrite, martensite, and carbides, with characteristics of refined grains and multiple grain growth directions [5, 19, 20].

In this context, the present work aims to investigate the mechanical properties of a low C-content MSS AISI 410L additively manufactured by L-DED in the as-built state and to explore the effects of a post heat-treatment (PHT) route on these properties. For this, starting from a set of previously optimized L-DED parameters found in a full design of experiments method based on steps of single beads, single layers, and multilayers geometries [21–23], AISI 410L (0.02 wt.%C) multilayer samples were manufactured by L-DED for analysis of microstructure, hardness, tensile strength, and Charpy impact test. Four PHT routes were investigated, varying the tempering temperature. Results of the as-built and heat-treated conditions were benchmarked to those of the commercial quality solution annealed AISI 410 MSS (AISI 410 CQ-SA, with 0.15wt.%C), used as a reference material. An in-depth comparison to similar work in literature allowed a full understanding of how the carbon content, building orientations, heat-treatments, and cladding procedures affect the mechanical properties of this important alloy.

2. Methodology

2.1 Experimental steps

To perform this work, an AISI 410 MSS in the shape of square plates, received in a hot-rolled and annealed state, was used as the substrate. This commercial quality (CQ) and solution annealed (SA) substrate, hereafter referred to as AISI 410 CQ-SA, was used as a reference for all experimental analyses conducted in this work. The AISI 410L MSS gas atomized feedstock from Höganäs Inc., with particle size in the range of 53–150 μm, was used. The chemical composition of the substrate and powder feedstock, as supplied by the manufacturers, is shown in **Table 1**.

The L-DED 5-axis RPMI 535® system from RPM Innovations Inc., consisting of a build platform inside a chamber with oxygen-level control by adding argon gas, was used for the sample manufacturing. It made use of a 25° discontinuous coaxial nozzle. This system is connected to a 3000 W continuous Ytterbium-doped fiber laser source from IPG Photonics, with a 1064 nm beam wavelength, beam parameter product of BPP 10.0, and beam quality parameter of M^2 29.7.

Before the samples' production, the L-DED parameters were optimized using a design of experiments consisting of single beads, single layers, and multilayer geometry steps, following the methodology described by Thiesen Jr. [21]. Details of the L-DED process parameters optimization, which led to the parameters used in processing the AISI 410L MSS, can be found elsewhere [14, 22, 23]. Two sets of samples, each devoted to the assessment of density, microstructure, hardness, mechanical

Element (wt.%)	C	Cr	Mn	Si	Ni	Fe
410 MSS Substrate (CQ-SA)	0.15	13.5	1.0	1.0	0.7	Bal.
AISI 410L Powder	0.02	12.8	0.8	0.8	0.7	Bal.

Table 1.
Chemical composition of the AISI 410 commercial quality solution annealed substrate (AISI 410 CQ-SA reference material) and AISI 410L powder feedstock.

tensile strength [24], and Charpy impact toughness [25], were produced using the optimized L-DED parameters (**Figure 1**). These parameters and the number of tensile and Charpy impact specimens tested for each condition are shown in **Table 2**. A first set of samples was evaluated in their as-built (AB) state and a second set underwent the heat-treatment (HT) routes as shown in **Figure 2**.

After manufacturing, the 25 × 25 × 25 mm cubic samples (**Figure 3**) in the AB and HT states were metallographically prepared by wire electron discharge machining W-EDM, sanding #80–1200 mesh, and polishing in 0.3 μm diamond solid-solution suspension. In this preparation stage, the samples were analyzed by Archimedes density (METTLER ME204/A scale), porosity quantification by optical microscope OM (ZEISS AXIO M2M), and phase identification by X-ray diffraction XRD (RIGAKU MINIFLEX600 from LINDEN Laboratory UFSC, with Cu Kα radiation at a 2° per minute speed, 0.02° step, 40 kV, and 30 mA). These were then chemically etched using Ralph's reagent, and their microstructure was examined using an OM and scanning electron microscopy (SEM) (ZEISS SUPRA 55VP). Energy-dispersive spectroscopy (SEM-EDS) was used to evaluate the chemical composition. A WILSON 402 MVD tester was employed to assess Vickers hardness.

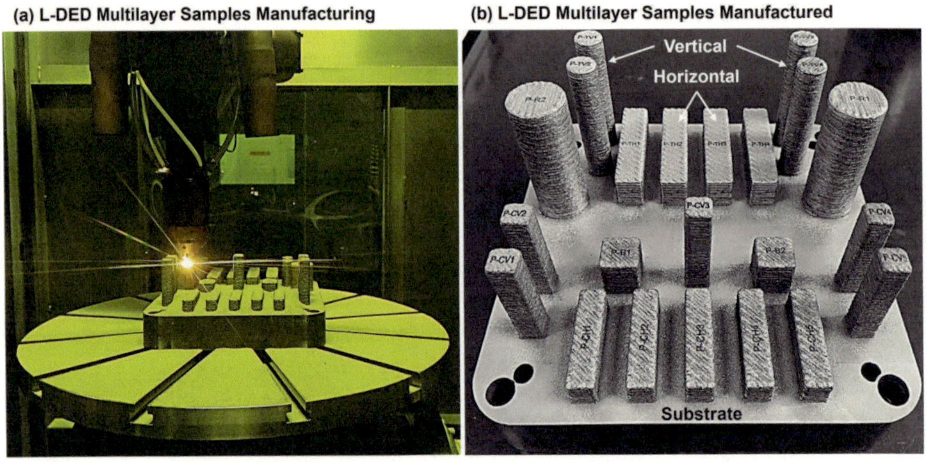

Figure 1.
L-DED samples obtention showing (a) L-DED samples manufacturing; (b) L-DED samples after manufacturing completion.

L-DED parameters		Samples	Quantity (per build)
Powder Feed rate	30 g/min	Cubes 25 × 25 × 25 mm^3	2
Laser Power	1300 W	Tensile Vertical Ø16 × 80 mm	4
Travel Speed	635 mm/min	Tensile Horizontal 80 × 20 × 20 mm	4
Hatch Spacing	1.93 mm	Charpy Vertical 10 × 10 × 60 mm	5
Layer Height	1.22 mm	Charpy Horizontal 60 × 10 × 10 mm	5
Slicing Strategy		45° rotation between all layers	

Table 2.
L-DED optimized parameters.

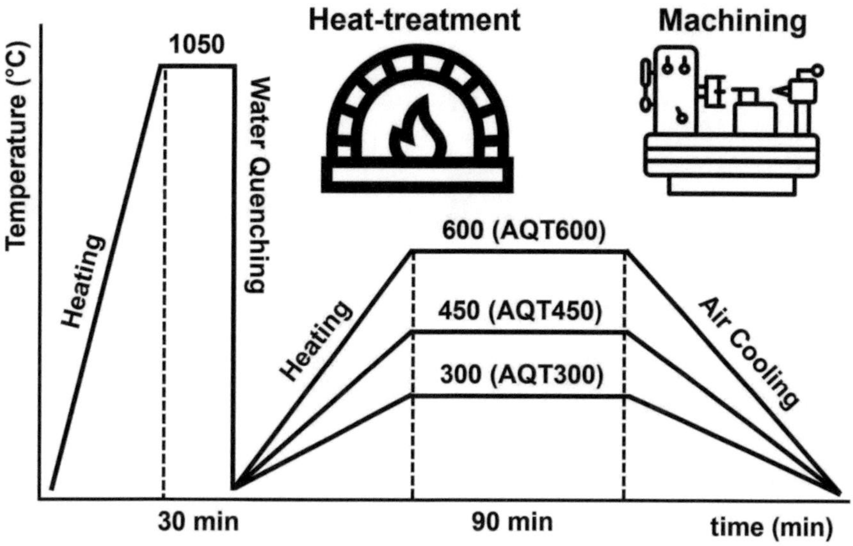

Figure 2.
Post heat-treatment route applied to the AISI 410L MSS samples.

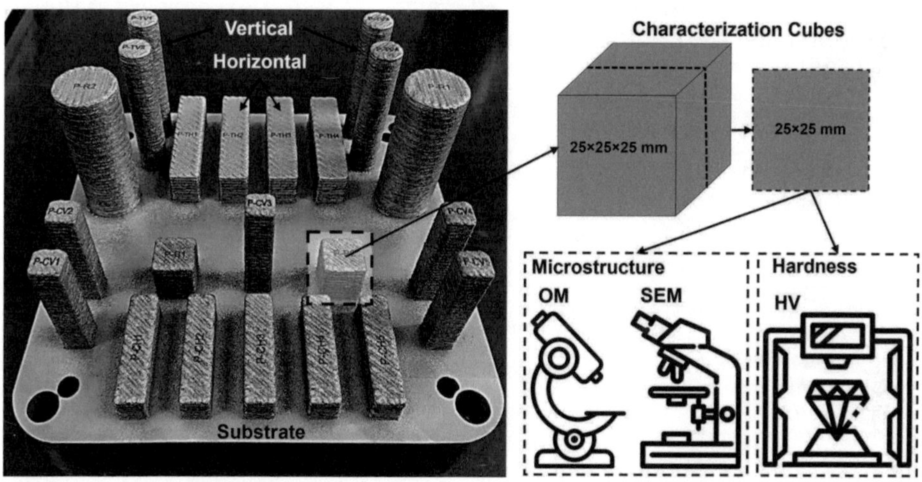

Figure 3.
Illustration of the samples for characterization steps of microstructure and hardness.

Samples intended for the mechanical tests (**Figure 4**) of both orientations and states were machined by a combination of W-EDM, turning, and milling. These were then tested following the recommendations for dimensions and parameters of the ASTM E8/E8M specimen 2 tensile [24] (INSTRON 5988 universal tester) and ASTM E23 Charpy impact [25] standards (INSTRON SI-1D3 tester). After being tested, the fractures of representative samples from each condition were verified by SEM.

2.2 Literature survey

The microstructure aspects and mechanical test results obtained in this work were benchmarked with the literature. As a comparative parameter, it was focused on

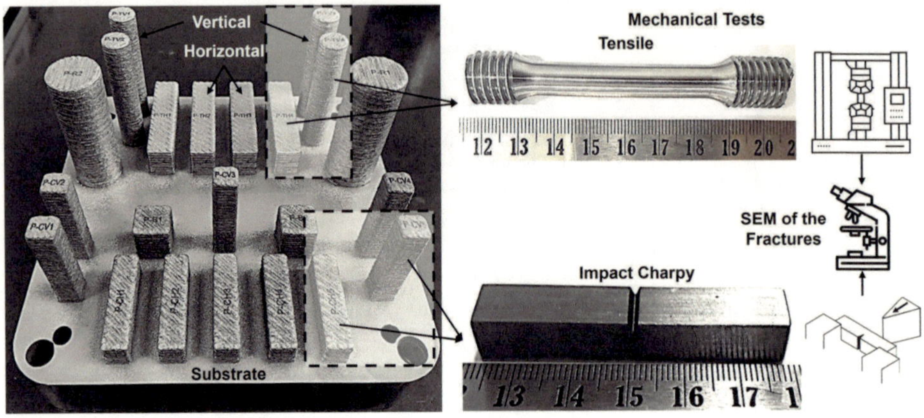

Figure 4.
Illustration of the samples for mechanical tests of tensile and Charpy impact.

works investigating the mechanical performance of MSS with different C-contents processed by L-DED. Aspects of microstructure refinement and post heat-treatment were also considered in this investigation. The results of the 410 MSS substrate used as reference material (CQ-SA), mentioned previously, were also employed in the comparison study. The primary information from the sources used for this survey is shown in **Table 3**.

Ref.	Material: Chemical composition wt.%	L-DED conditions	Purpose
[15]	AISI 410L: 0.004C, 12.20Cr, 1.00Mn, 1.00Si, 0.75Ni, Fe-Bal.	60 × 60 × 12 mm cubes	Evaluate the effects of the L-DED and heat-treatment on the mechanical performance.
[19]	AISI 410L: 0.01C, 12.70Cr, 0.51Mn, 0.47Si, 0.08Ni, 0.01Mo, Fe-Bal.	Single and double-layer coatings	Investigate the effects of pre-heating and pre- plus post-heating on the mechanical performance by shear-punch test.
[20]	410L: 0.01C, 12.70Cr, 0.51Mn, 0.47Si, 0.08Ni, 0.01Mo, Fe-Bal.	Laser cladding of 2 mm layer thickness of the rail sample and machining of tensile samples	Investigate the effects of pre- and post-heating on laser cladding rail steels.
[6]	AISI 410L: 0.02C, 12.20Cr, 0.72Mn, 0.35Si, 0.39Ni, Fe-Bal.	Five-layer coatings on a 70 × 40 × 2 mm area	Evaluate the effects of heat-treatment on microstructure, and hardness.
[5]	410L: 0.03C, 13.00Cr, 0.10Mn, 0.50Si, Fe-Bal.	Rectangular sample of 15 × 9 × 5 mm	Study the mechanical properties of 410L and Fe-Cr-B-Si-Mo alloys by laser cladding.
[13]	630 MSS: 0.05C, 15.80Cr, 0.72Mn, 0.60Si, 4.63Ni, 0.10Mo, Fe-Bal.	Rectangular multilayer sample	Study the effects of annealing and precipitation hardening on 630 MSS mechanical behavior.
[26]	410: 0.11C, 12.95, 0.82Mn, 0.78Si, 0.41Ni, Fe-Bal.	15 × 15 mm laser cladding area	Analyze the effects of TiC addition on 410 MSS performance.
[27]	AISI 410 SS: 0.13C, 13.20Cr, Fe-Bal.	Single layer cladding of 300 × 80 × 20 mm	Investigate the heat-treatment effects on AISI 410 plus Cr_3C_2 addition.

Ref.	Material: Chemical composition wt.%	L-DED conditions	Purpose
[12]	AISI 410L: 0.14C, 11.50Cr, 0.38Mn, 0.26Si, 0.14Ni, 0.03Mo, Fe-Bal.	Single walls of six layers and 150 mm length	Evaluate the mechanical properties of 410L manufactured by L-DED.
[28]	SC420: 0.14C,12.90Cr, 0.72Mn, 0.67Si, Fe-Bal.	70 × 13 × 13 mm vertical and 13 × 13 × 70 mm horizontal samples	Study the as-built tensile responses as a function of the build direction.
[29]	410 MSS: 0.15C, 12.50Cr, 1.00Mn, 1.0Si, 0.60Ni, Fe-Bal.	150 × 18 × 18 mm rectangular multilayer samples	Verify the effects of as-built and post heat-treatment on the microstructure and mechanical properties.
[30]	420: 0.23C, 13.50Cr, 0.31Mn, 0.59Si, Fe-Bal.	Single walls of 130 × 30 × 4 mm	Study the mechanical performance of 420 manufactured by L-DED.
[31]	AISI 420: 0.23C, 12.30Cr, 1.20Mn, 0.50Si, Fe-Bal.	45 × 45 × 25 mm cubic samples by laser cladding	Investigate the effects of different orientations and heat-treatments on mechanical properties.
[32]	SS420: 0.30C, 13.60Cr, 0.35Mn, 0.55Si, 0.01Ni, Fe-Bal.	25 × 25 × 25 mm cubic samples	Study the microstructure and mechanical properties of 420 MSS.

Table 3.
Information of the references employed in the comparative study survey.

3. Results and discussion

This section provides the main results of this work regarding macro aspects, density, microstructure, mechanical properties of hardness, tensile strength, and Charpy impact toughness. Furthermore, a discussion was made regarding the literature outcomes.

3.1 Effect of chemical composition on the microstructure and mechanical properties of AISI 410L processed by L-DED additive manufacturing

AISI 410L multilayer geometry samples did not exhibit deleterious defects like cracks, lack of fusion, or delamination. No significant geometric deviations were observed concerning the 3D CAD model. The densification level analyzed by the Archimedes method (**Figure 5**) was around 99.50%. The OM porosity quantification (**Figure 5**) corroborated this result and, besides, the detected porosities demonstrated a rounded shape, which suggests that they come from the gas blisters entrapment during processing [33, 34].

Figure 6 presents the microstructures of representative multilayer geometry samples of AISI 410L in the AB state captured by OM and SEM. The absence of harmful defects is a primary highlight, supporting the previously discussed macro aspects. Only a few rounded porosities were seen. It was possible to detect certain molten pool boundaries in the regions of hatch spacing between single beads and layer height between single layers, which is an inherent feature of metallic AM techniques like L-DED and L-PBF [8, 11]. This feature of heterogeneous microstructure accounts for process anisotropy since it impacts the solidification mode and grain growth direction, and it can provide various mechanical responses depending on the direction of effort applied [28, 31].

Figure 5.
Archimedes density analysis and porosity quantification by optical microscope.

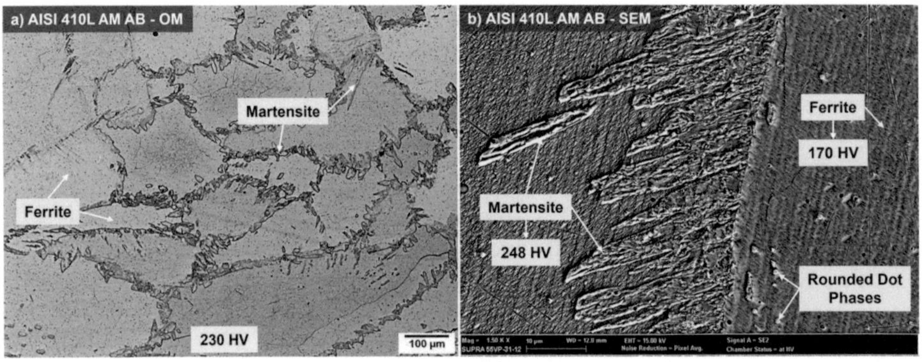

Figure 6.
Microstructure of the AISI 410L AB multilayer samples in the as-built state showing (a) OM and (b) SEM details.

Evaluating **Figure 6a**, the microstructure of the L-DED AB condition seen in the optical microscope was shown to be formed by columnar grains of ferrite with a lighter color, and refined martensite with a darker color, concentrated at the grain boundaries. **Figure 6b** confirms the ferritic grains' columnar appearance with the refined martensite phase concentrated only at the grain boundaries. Small, rounded dot phases were also observed at SEM higher magnification. The low C-content of AISI 410L powder employed as feedstock was one of those responsible for forming these microstructure aspects. Therefore, during solidification, there was insufficient C-amount to enrich the initial austenitic phase to provide a fully martensitic microstructure formation [5, 12, 15]. To investigate the chemical nature of the phases found, an EDS-map analysis was carried out. **Figure 7** showed no significant differences in the elements concerning the nominal levels. In other words, there were no substantial variations between ferrite grains and refined martensite concentration at the grain boundaries. However, the rounded dots found in both the OM and SEM analysis correspond to Si and O-rich particles, likely SiO_2, as reported by Nayebi et al. [35].

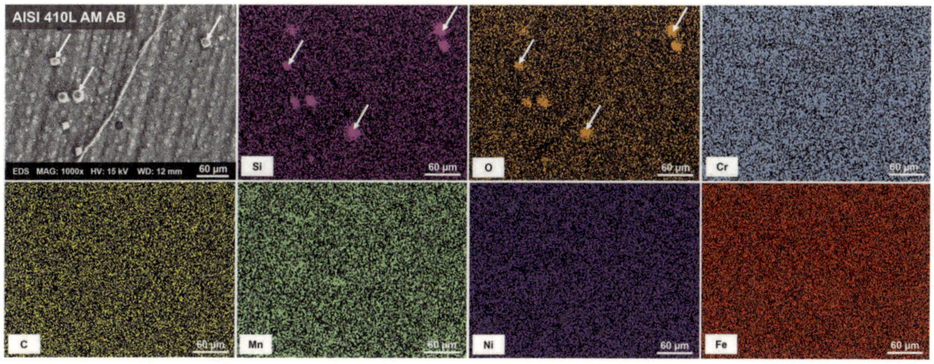

Figure 7.
SEM-EDS of the AISI 410L AB multilayer condition.

The substrate employed in this study also served as a reference material. An AISI 410 MSS CQ-SA (0.10 wt.%C) in the as-received state, rolled and annealed. **Figure 8** depicts the typical microstructure. OM produced **Figure 8a**, which shows some remaining marks from the rolling step. It was also able to emphasize a biphasic microstructure in two colors. **Figure 8b** shows that SEM examination enabled a more efficient distinction between the ferrite and tempered martensite phases. Smaller rounded phases were also found in the matrix and at the grain boundaries. The chemical composition of the ferrite and martensite phases did not differ in the **Figure 9** SEM-EDS analysis. The Cr and C-contents were greater in smaller rounded-shaped phases. Based on this discovery and the literature descriptions, these phases could be Cr-carbides.

An XRD analysis was performed to identify phases, and the resulting spectra are presented in **Figure 10**. For the AISI 410L feedstock, the AISI 410L additively manufactured samples, and commercial quality AISI 410 samples, peaks of the body-centered cubic BCC crystalline structure were found, corresponding to the existence of ferrite and tempered martensite phases. Near the 82° peak, a partial overlap slightly shifted to the right was observed for the AISI 410L additively manufactured samples and AISI 410 CQ-SA samples, which can be attributed to the body-centered tetragonal (BCT) structure, due to the formation of tetragonal refined martensite. During AISI 410 CQ-SA manufacturing, part of the carbon-supersaturated martensite was

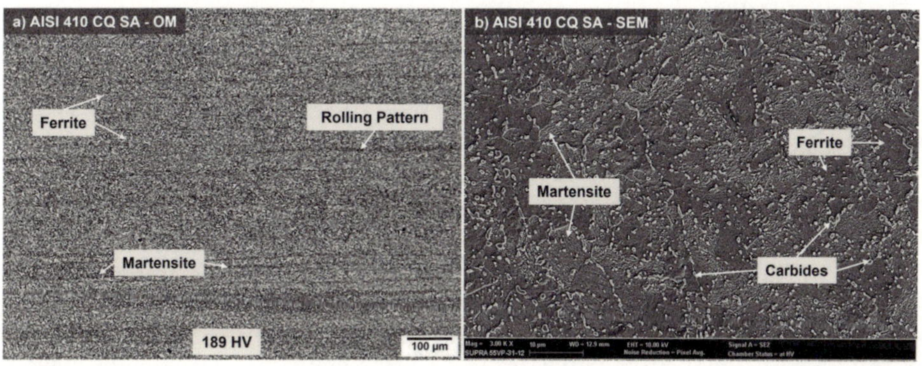

Figure 8.
Microstructure of the AISI 410 CQ-SA showing (a) OM and (b) SEM details.

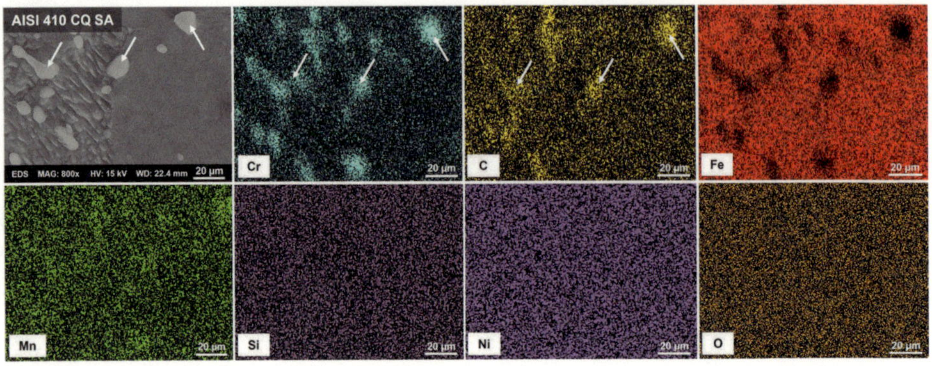

Figure 9.
SEM-EDS of the AISI 410 CQ-SA, showing high contents of Cr and C in the small, rounded phases.

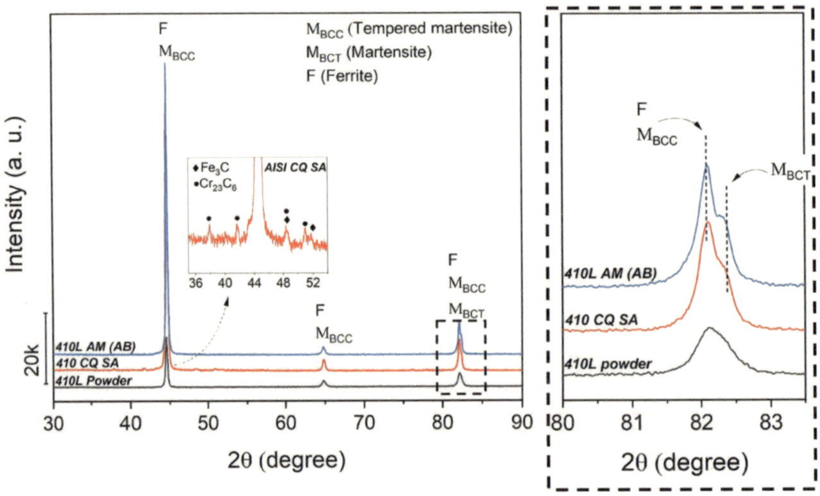

Figure 10.
XRD analysis of the 410L powder, AISI 410L samples processed by L-DED in the as-built condition (AISI 410L AM AB), and the AISI 410 commercial quality solution annealed reference samples (AISI 410 CQ SA).

transformed into BCC tempered martensite, Fe_3C cementite, and Cr-carbides, as seen in **Figure 9** SEM-EDS and **Figure 10** XRD. Based on the literature, the latter should be of the Cr_7C_3 type [4, 36]. These XRD responses are consistent with the microstructure features previously observed by OM and SEM, and the chemical composition analyzed by SEM-EDS. No indication of carbides was observed in the AISI 410L samples processed by L-DED, which was consistent with reports of other researchers, for the same material [12, 15, 19].

Analyzing hardness in the **Figure 11a** graph, the L-DED AB condition exhibited an average of 230 HV0.5. This value corresponds to the simultaneous response of the martensitic and ferritic regions as seen in **Figure 11b**. A hardness analysis using a reduced load (50 gf) revealed an average hardness of 170 HV0.05 for the ferritic phase and 250 HV0.05 for the grain boundary martensitic phase (**Figure 11b**), representing a 30% difference, thus corroborating the formation of harder martensite at the grain boundaries [5, 6, 14]. The literature describes how MSS C-content affects microstructure characteristics and mechanical properties. Observing authors who evaluated 410L

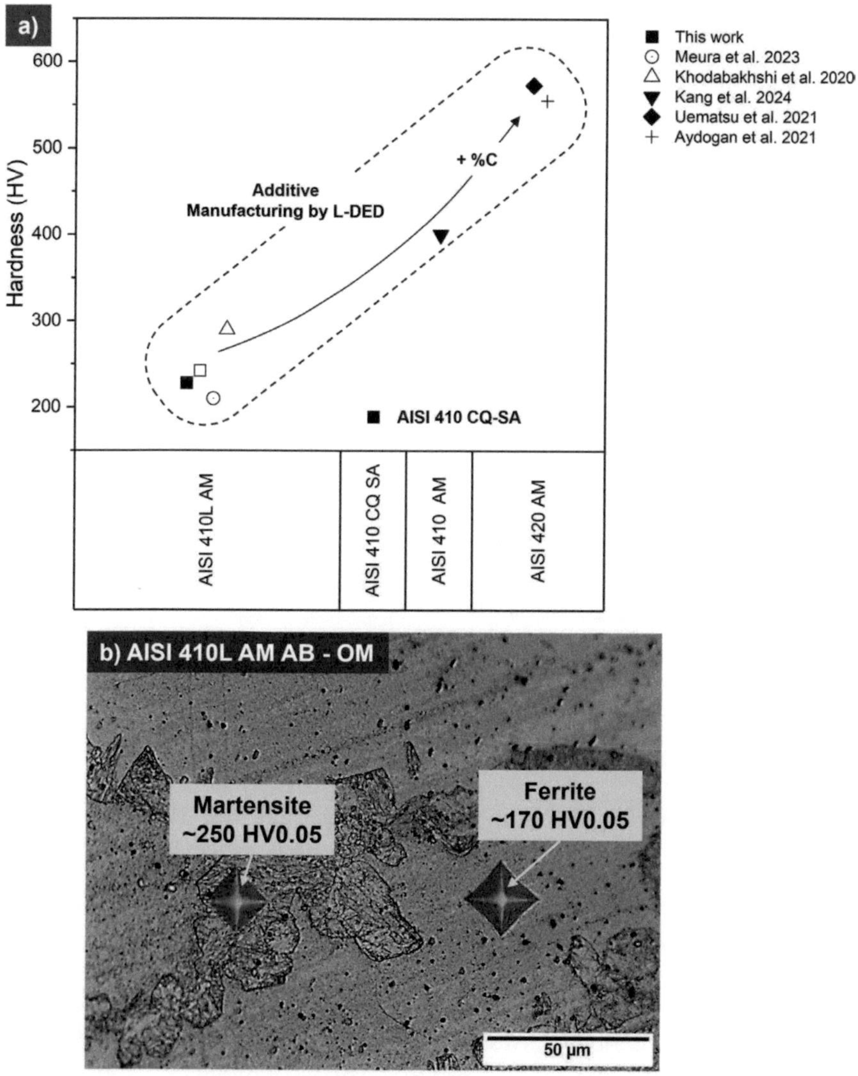

Figure 11.
Hardness (a) graph of the AISI 410L MSS additively manufactured by L-DED, the AISI 410 CQ-SA, and the literature survey with MSS of different C-contents; (b) hardness of the ferrite and martensite regions of the AISI 410L AM L-DED condition.

MSS with a C-content like that of the present work (0.01 wt.%C), they also found a microstructure composed of cellular or columnar ferrite grains and small amounts of martensite localized at the grain boundaries. Consequently, the average hardness was similar [6]. In **Figure 11a**, although the reference AISI 410 CQ SA had a higher C-content and the presence of carbides, the solution annealed condition in which the material was received led to the development of an average hardness lower than 200 HV. On the other hand, additively manufactured steels with even higher C-contents, such as the AISI 420 (0.14–0.30 wt.%), could be expected to reach values of about 550 HV [30, 32].

Figure 12 depicts the tensile results for YS and UTS. In the AB condition, the average YS and UTS were 326 and 528 MPa, respectively. As discussed in the previous paragraph, despite the higher C-content, YS and UTS of the reference AISI 410

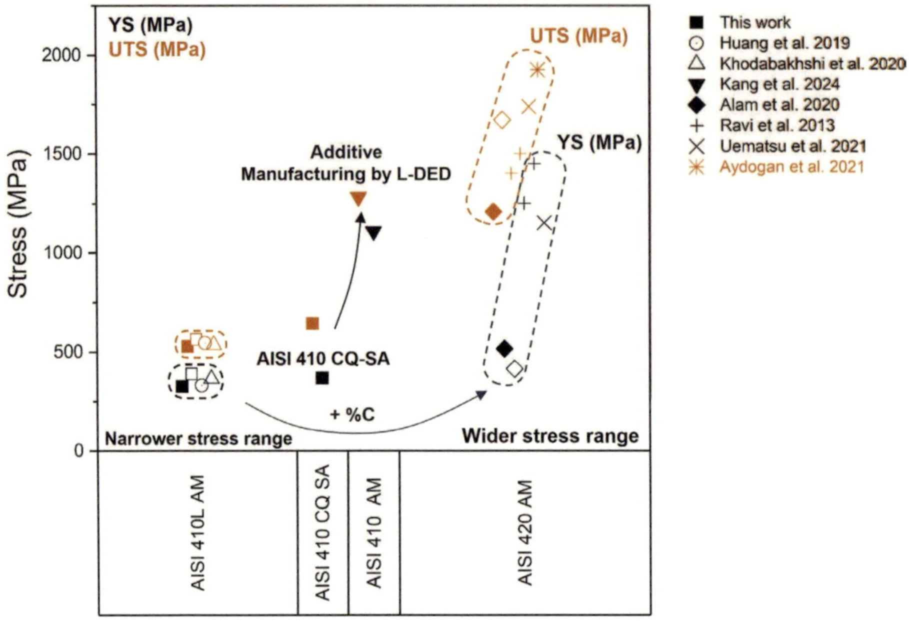

Figure 12.
Yield strength and ultimate tensile strength of AISI 410L MSS additively manufactured by L-DED, the AISI 410 CQ-SA, and the literature survey with MSS of different C-contents.

CQ SA samples were not significantly different, due to the material annealed state. Nevertheless, the YS and UTS of additively manufactured AISI 410 (0.15 wt.%) were found to reach values above 1000 MPa, associated with the formation of martensite during laser fast cooling [29]. L-DED processing of AISI 420, with even higher C-contents (0.14–0.30 wt.%), can lead to YS in the range of 500–1500 MPa, and UTS in the range of outstanding 1200–2000 MPa [28, 30–32].

On the other hand, such mechanical performance of YS and UTS was achieved at the expense of the EL reduction, which was much reduced for the high C-content MSS regarding the low C-content ones. AISI 420 processed by L-DED, as seen in **Figure 13**, showed values ranging from 4 to 16% [28, 30–32]. In the present work, the C-content of AISI 410L additively manufactured samples was around 27%. Maximum EL was seen in the reference AISI 410 CQ SA, whose annealed state granted an average of 32%.

Figure 14a illustrates the aspect of the tensile fracture of the AISI 410L MSS additively manufactured by L-DED, while **Figure 14b** presents the AISI 410 CQ-SA one. In the tensile test, the reference sample exhibited slightly higher YS and UTS (**Figure 12**) values regarding the AISI 410L AB condition. The AISI 410 CQ-SA showed significantly higher EL (**Figure 13**) than all L-DED conditions, which was corroborated by a fracture composed predominantly of dimples (**Figure 14b**). Although mostly ductile, the facture of the AISI 410L MSS additively manufactured samples showed cleavage signs, possibly associated with the martensitic grain boundaries embrittlement [12].

3.2 Effect of L-DED build orientation on the microstructure and mechanical properties of AISI 410L multilayer parts

The L-DED build orientation influences the component properties, as there is anisotropy due to the different thermal histories provided. In this sense, **Figure 15** presents the

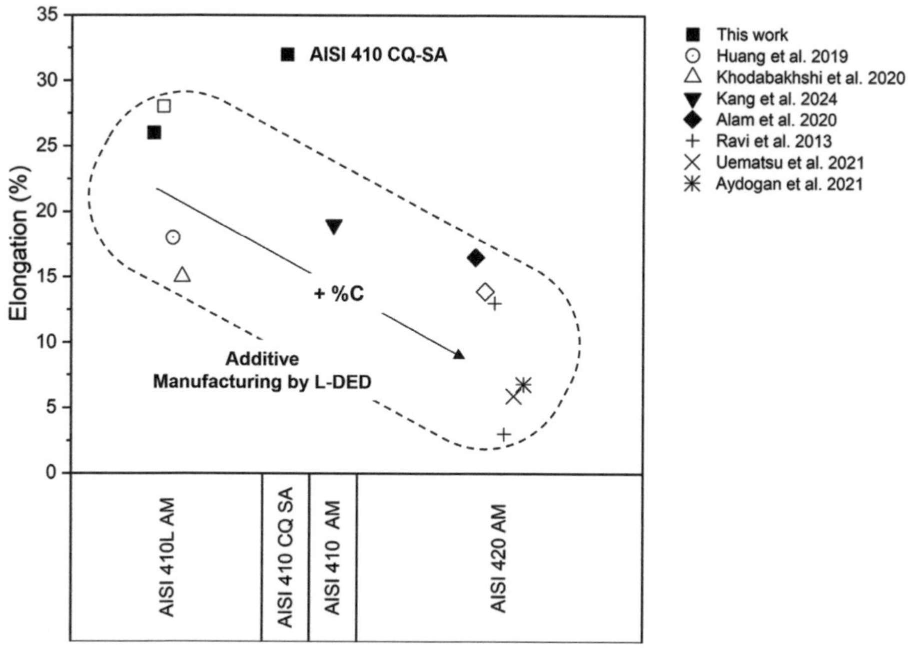

Figure 13.
Elongation of AISI 410L MSS additively manufactured by L-DED, the AISI 410 CQ-SA, and the literature survey with MSS of different C-contents.

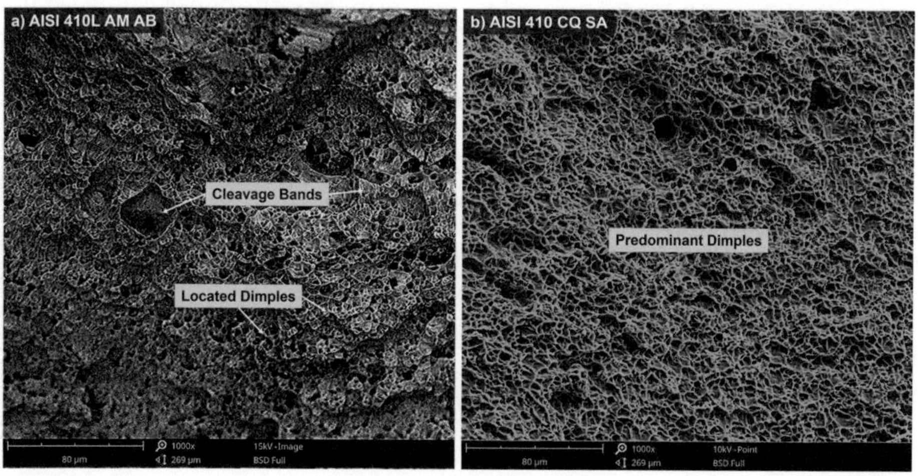

Figure 14.
SEM fractures of AISI 410L samples processed by L-DED in the as-built condition and AISI 410 reference CQ-SA.

microstructure of the AISI 410L MSS samples built by L-DED in vertical (**Figure 15a**) and horizontal (**Figure 15b**) orientations. The latter produced a microstructure that was more refined and had around 7% more area occupied by grain boundaries than the first. This feature is the outcome of a more efficient conduction heat-sink to the substrate.

Given that the samples were produced with identical L-DED parameters (travel speed, mm/min) and have the same volume (ASTM E8/E8M 2024 [24] tensile and

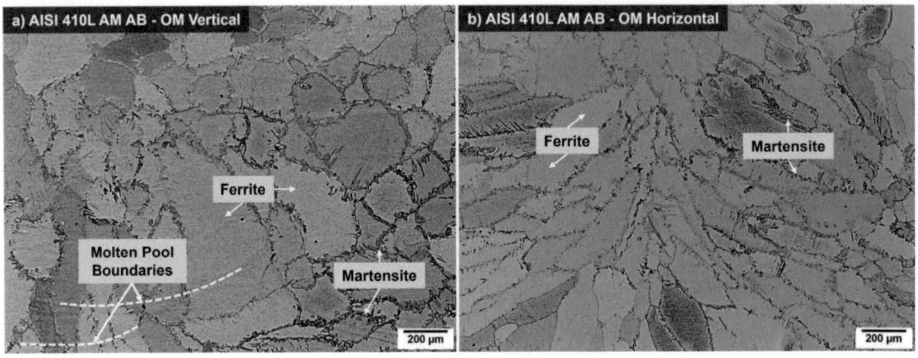

Figure 15.
Optical microstructure of the L-DED AB condition built in the (a) vertical and (b) horizontal orientations concerning the substrate.

ASTM E23 Charpy impact [25] standards), the only differences between them are their dimensions and the number of layers. This is because there are fewer layers, a larger contact area, a longer time needed to deposit a single layer, and a shorter distance between the upper layers and the substrate. The combination of these elements improves the conduction heat-sink mechanism action, generating more refined grains. In response, the hardness of the multilayer samples built in the horizontal orientation was higher (242 HV) than those processed in the vertical one (228 HV).

In the L-DED process, the relationship between the temperature gradient G and the solidification rate R is very high, which enhances the formation of refined grains [37, 38]. In this sense, **Figure 15** columnar ferrite grains grew along the build direction, and the preferential substrate conduction heat-sink. The L-DED fast cooling also contributes to the development of refined martensite laths. Consequently, the time needed for the dissolution of the grain boundary segregation into the matrix was unavailable [15].

Observing **Figure 16a**, when compared to the vertical orientation, the horizontal orientation provided a performance improvement in the YS and UTS properties for both the AISI 410L and AISI 420 MSS. This effect, however, was more noticeable for the 420 MSS owing to the higher number of carbides detected, and more sensitivity to

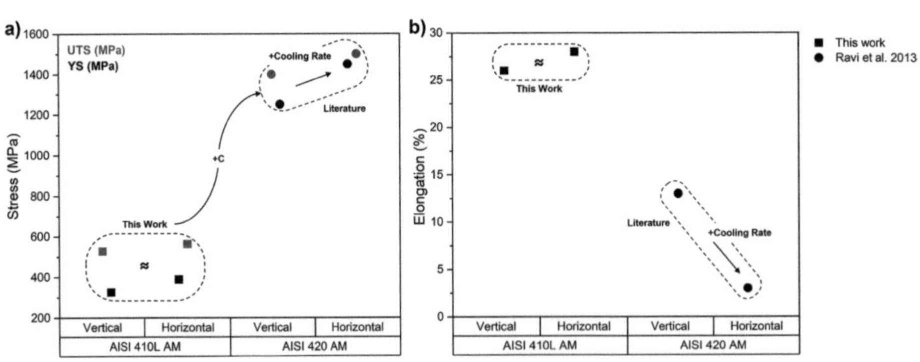

Figure 16.
Effect of vertical or horizontal build orientation on mechanical properties for AISI 410L and AISI 420 MSS additively manufactured by L-DED showing (a) YS and UTS; (b) elongation.

metallurgical transformations induced by fast cooling. Contrary to expectations, the Elongation of the AISI 410L AM sample seen in **Figure 16b** remained similar, indicating that there was no embrittlement when subjected to tensile stress. For the AISI 420 AM MSS, however, significant embrittlement was seen. The literature suggests a significant elongation decrease because of the brittleness and stress risers imposed by the martensite formed, particularly in C-richer MSS. Differences are associated with the grain refinement level because both orientations showed the same phases. In this sense, the time required for the deposition of a single horizontal layer is much longer than that of the vertical layer. Furthermore, in the latter, the greater number of layers provides multiple heating and cooling cycles, generating martensite tempering that leads to a significantly higher elongation (**Figure 13b**) in comparison to the horizontal orientation [28].

In **Figure 17**, SEM fractures show that both orientations exhibited a quasi-ductile fracture, with a similar amount of dimples and cleavage bands. However, the vertical orientation (**Figure 17a**) tended greater predominance of dimples, while in the horizontal one (**Figure 17b**), the cleavage bands occupied a larger area, which points to the higher ductility of the former [28, 39].

Besides the vertical and horizontal build orientations concerning the substrate, the MSS mechanical property can also be affected depending on whether the tensile samples were extracted from a longitudinal or transversal direction regarding the bead deposition. When comparing the direction of layers deposition (**Figure 18**), those deposited transversely to the mechanical effort direction generated greater anisotropy during cooling, which formed a higher number of grains and boundaries concerning the layers deposited longitudinally. In response, YS and UTS (**Figure 18a**) increased significantly at the expense of an EL (**Figure 18b**) reduction.

In summary, both the results of the present work and the literature denote the influence of the L-DED thermal history, such as refined grains, microstructure heterogeneity, precipitates, and residual stresses. To improve or correct these aspects, it is common to perform post-heat treatments.

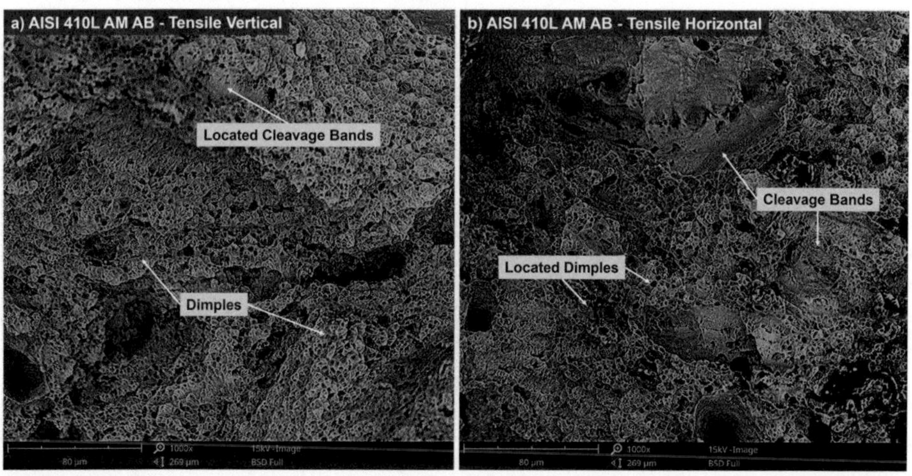

Figure 17.
SEM fractures of the AISI 410L additively manufactured samples in the AB conditions showing (a) vertical and (b) horizontal orientations.

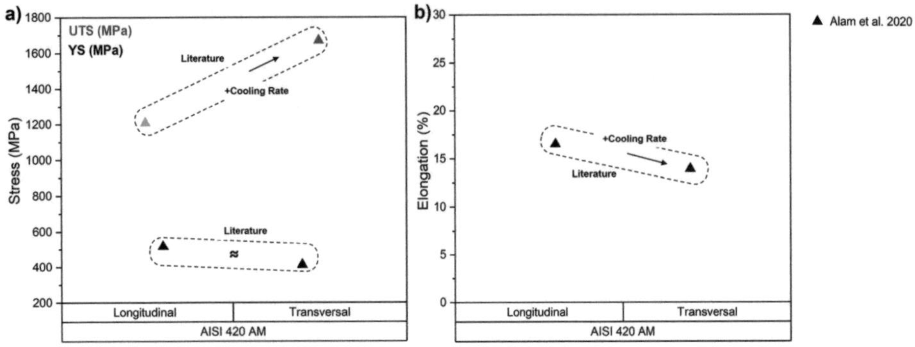

Figure 18.
Effect of the bead orientation on the layer deposition (transversal and longitudinal) on mechanical properties of AISI 420 MSS additively manufactured by L-DED showing (a) YS and UTS (b) elongation.

3.3 Effect of post heat-treatment on the microstructure and mechanical properties of AISI 410L multilayer parts

Figure 19 presents the microstructure captured by OM and SEM of the as-built and different heat-treated L-DED conditions. **Figure 19a–b** illustrates the effect of L-DED thermal history. The annealing plus quenching (AQ condition, **Figure 19c–d**) provided microstructure recrystallization so the AB start grain boundaries were dissolved and dispersed throughout the Fe-matrix, resulting in coarser grains. In response, the heterogeneous appearance resulting from the L-DED thermal history was removed, making it more difficult to distinguish between grains and boundaries. Thus, it is conceivable to conclude that the annealing temperature applied to the material was sufficient to achieve a fully austenitic state. There were no significant changes seen between the quenching plus tempering conditions at the lowest temperatures (AQT300 and AQT450) and the AQ condition. In this sense, only the AQ300 condition was represented in **Figure 19e–f**. In these conditions, the grains maintained their refinement level, and no evidence of other phases was seen. In contrast, tempering at the highest temperature (AQT600, **Figure 19g–h**) resulted in significant grain coarsening compared to the other conditions.

Evaluating **Figure 19a–b**, the biphasic microstructure resulting from the L-DED thermal history is highlighted. On the other hand, the AQ condition (**Figure 19c–d**) exhibited homogeneously distributed ferritic grains, while the martensite demonstrated distribution in packages with rounded morphology. Condition AQT300 (**Figure 19e–f**) presented these same aspects, without significant differences regarding AQ condition. The AQT600 (**Figure 19g–h**) condition, in turn, despite displaying these same phases, showed coarser grains. This indicates that this tempering temperature range was more suitable to provide residual stress relief from the L-DED process as well as from the first heat-treatment steps, which included annealing the material to put it in the austenitic state and quenching it to induce the martensitic transformation [6, 15]. Small, rounded phases were also seen in all heat-treated conditions, though in proportions that seemed to be lower, suggesting that some of them might have been dissolved during heat-treatment [35].

The L-DED AB condition and all heat-treated conditions showed the same peaks in the **Figure 20** XRD graph, all of which showed most of the BCC structure, which

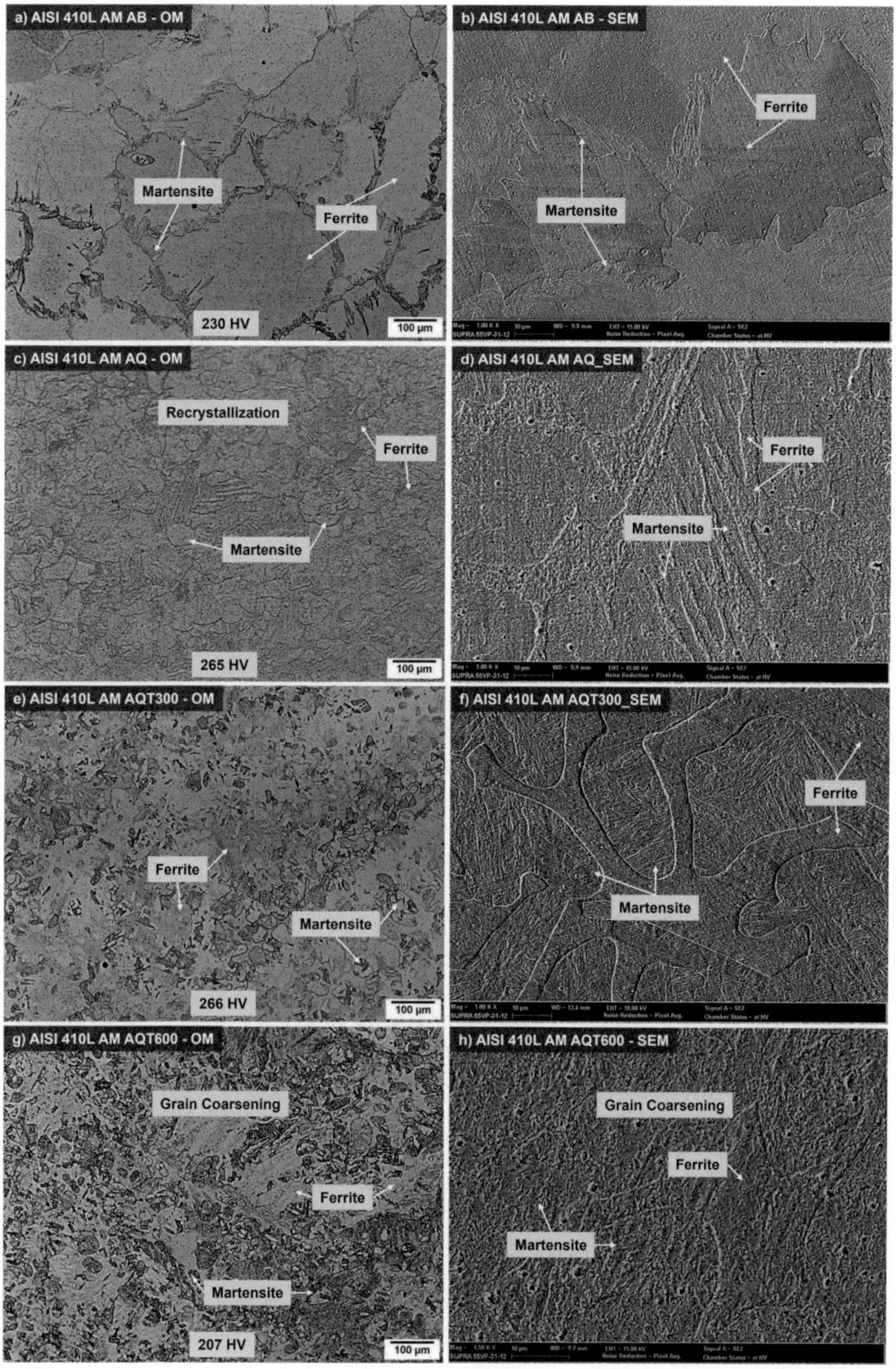

Figure 19.
Optical and scanning electron microscope showing (a) and (b) as-built condition; (c–d) annealed condition demonstrating the recrystallized microstructure; tempering at the lowest temperature; (e) and (f) 300°C did not significantly alter the microstructure; (g) and (h) tempering temperature increase provided grain coarsening.

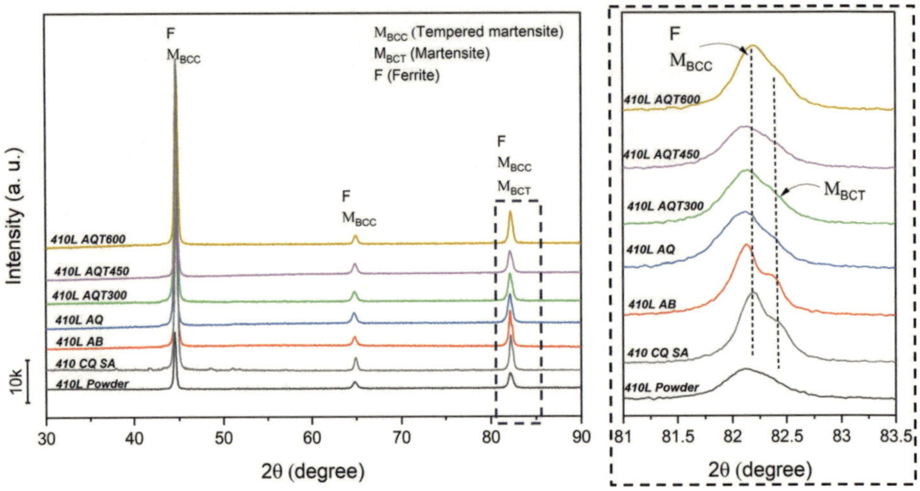

Figure 20.
XRD spectra highlighting the presence of ferrite and refined martensite peaks of BCC and BCT structure in the L-DED AB condition; ferrite and BCC tempered martensite in all L-DED HT conditions; ferrite, BCC tempered martensite, Cr-carbides, and cementite Fe_3C in the CQ-SA.

corresponds to the ferrite and tempered martensite phases. The microstructure characteristics seen by OM and SEM, as well as the chemical composition evaluated by SEM-EDS, are followed by the XRD responses. In all cases, the ferrite phase is present. Tempered martensite from heat-treated conditions is precisely due to the tempering step performed, which provided the controlled reheating and cooling cycles, thus forming a greater amount of tempered and homogeneously distributed martensite. No evidence of carbides was found in any of the L-DED conditions [6, 15].

In **Figure 21**, conditions AQ, AQT300, and AQT450 displayed comparable average hardness. These heat-treated conditions increased hardness by 15% regarding the AB

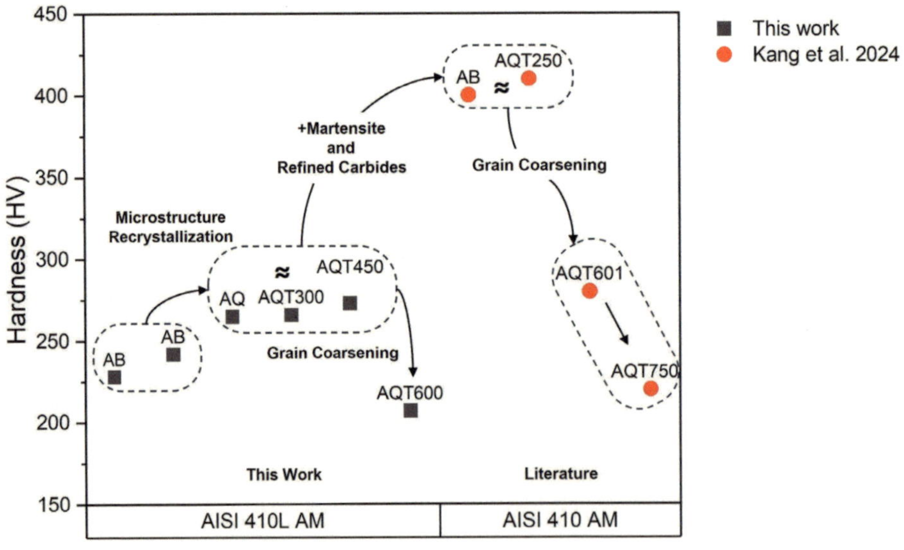

Figure 21.
Higher hardness values observed in as-built and tempering heat-treatment at lower temperatures conditions.

one, which can be attributed to the microstructure recrystallization. In comparison to the L-DED and all previous heat-treated conditions, the hardness decreased when the microstructure was tempered at a higher temperature (AQT600). The more thorough martensite tempering, grains coarsening, and microstructure homogenization are all responsible for this outcome [6]. In the higher C-content MSS, coarse Cr-carbides precipitation also contributes to hardness reduction [29].

In the **Figure 22** graph, it is possible to observe that YS and UTS increased with the performance of annealing and tempering at lower temperatures concerning AB condition. With the tempering carried out at the higher temperature (AQT600), YS and UTS showed a significant reduction regarding the other heat-treatment routes, demonstrating values closer to the AB condition. In the lower tempering temperatures, the microstructure refinement was kept, while the martensite was homogenized along with the ferritic matrix. Furthermore, the L-DED process's intense dislocation displacement and the annealing and quenching step that followed made the microstructure more stressed and, as a result, harder, and mechanically resistant. With the tempering temperature increase, the martensite was effectively tempered, the grains were coarsened and the residual stress from L-DED processing was relieved, which provided an elastic recovery and mechanical resistance decrease [15]. In the higher C-content MSS, the obtention of greater amounts of martensite after heat-treatment enhanced YS and UTS. A substantial mechanical performance improvement occurred with lower tempering temperatures owing to the refined Cr-carbides precipitation [29, 31].

Figure 23 graph presents a performance inversely proportional to that observed for hardness, YS, and UTS. The annealing heat-treatment (AQ) and maintenance at lower tempering temperatures (AQT300 and AQT450) deteriorate this property

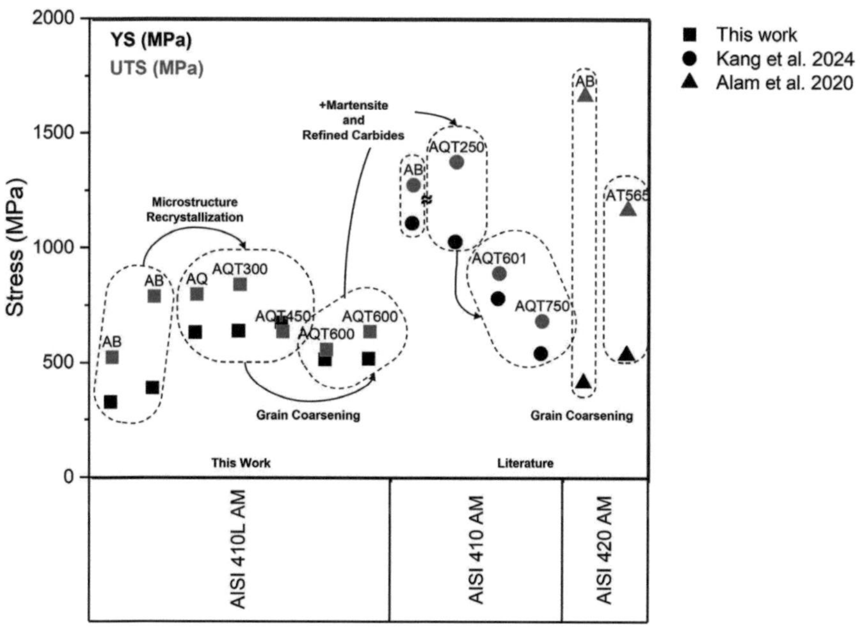

Figure 22.
YS and UTS were enhanced by the L-DED thermal history of the AB condition, and recrystallization of the annealing and tempering at lower temperatures. In higher C-content MSS, carbide precipitation induced at lower tempering temperatures increased hardness, which was reduced by the grain coarsening provided by the higher tempering temperatures.

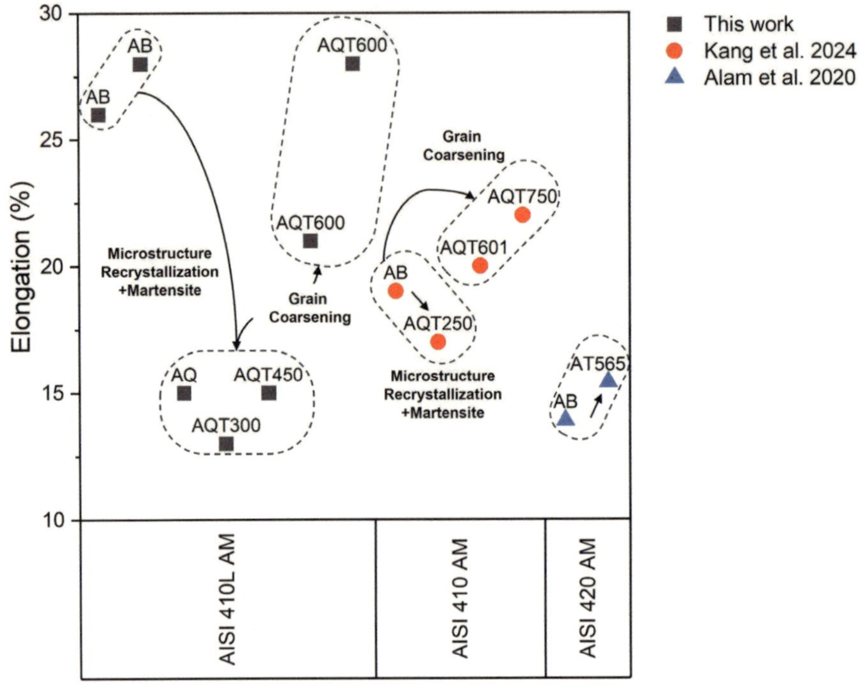

Figure 23.
Elongation performance vs. the trends of hardness, YS, and UTS.

regarding the AB condition. In the start microstructure, the L-DED samples showed large ferritic bands of high ductility, which were reduced with the imposed recrystallization. Only the AQT600 condition with a higher tempering temperature was able to recover this property, given the more intense martensite tempering and, consequently, more effective residual stress relief. In the higher C-content MSS, ductility was damaged with lower tempering temperatures, both by the formation of a greater amount of refined martensite and by the hard and brittle carbides precipitated. Tempering temperature increase helped to recover the elongation through the same aspects highlighted for the low C-content MSS [13], as well as through the carbides coarsening and the formation of retained austenite bands [29, 31].

Figure 24, which depicts the distinctive fractures of heat-treated L-DED conditions, supports the mechanical results. In the AB condition (**Figure 24a**), a sizable number of dimples localized on the high-ductility ferritic regions was feasible. Along the grain boundaries, some cleavage bands were seen. AQ (**Figure 24b**), AQT1, and AQT2 (**Figure 24c**) all showed very similar characteristics, with most cleavage bands present and few localized dimples. Conversely, even more dimples were present in the AQT600 (**Figure 24d**). In the literature, however, tempering temperature increase impaired elongation due to the precipitation of refined Cr-carbides [13]. This effect was more pronounced in higher C-content MSS, owing to sensitization embrittlement [31].

3.4 Charpy impact outcomes

The biphasic AISI 410L AB AM microstructure generated satisfactory results in terms of mechanical tensile strength. However, the toughness investigated by the

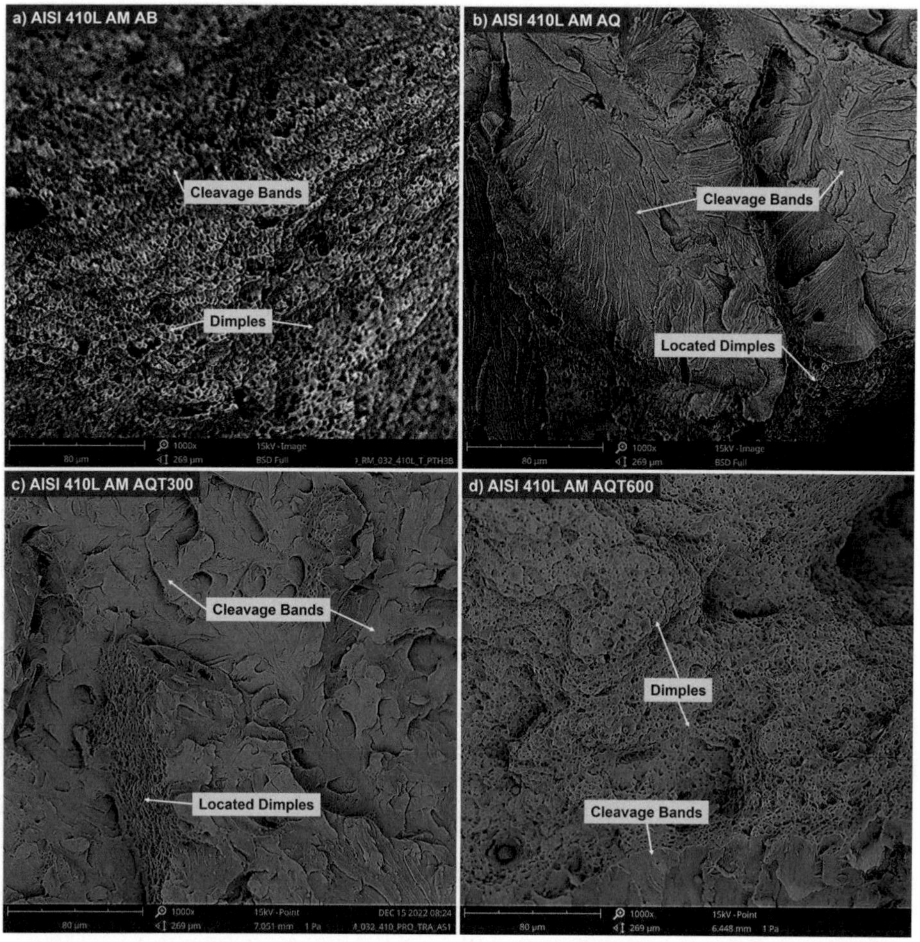

Figure 24.
Effect of heat-treatment on tensile fractography, showing (a) cleavage and dimples; (b) and (c) predominance of cleavage bands and located dimples; (d) greater number of dimples detected in the higher tempering temperature heat-treatment.

Charpy impact test exhibited low performance. **Figure 25** shows Charpy impact energy absorption values of AISI 410L MSS additively manufactured by L-DED, in the as-built and heat-treated conditions. When the Charpy impact was analyzed, the results showed very low values, between 5 and 7 J for all conditions except for samples tempered at the highest temperature of 600°C. In the AB condition case, this poor performance can be attributed to the embrittlement imposed by the refined martensite segregated at the grain boundaries so that these regions served as stress risers and crack propagators during the test shock. For the conditions only annealed AQ and tempered at the lowest temperatures AQT300 and AQT450, the recrystallization and maintenance of the refined, hard, and brittle martensite throughout the ferritic matrix were the factors responsible for the low performance observed. On the other hand, grain coarsening, martensite tempering, and effective residual stress relief provided by tempering at the higher temperature AQT600 significantly improved the Charpy impact toughness. In the literature, similar highlights were presented. However, in the heat-treatment performed at higher temperatures, the precipitation of carbides at the grain boundaries reduced this property owing to the sensitization phenomena [13].

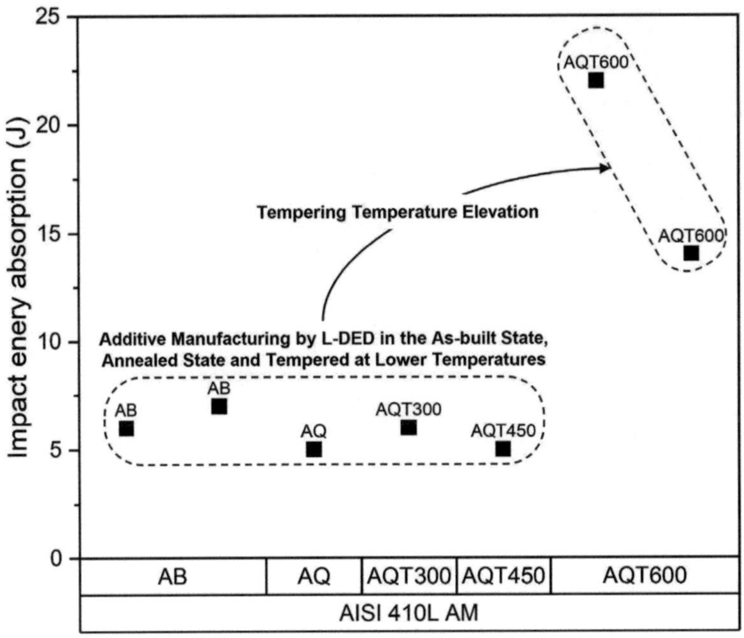

Figure 25.
Charpy impact energy absorption of AISI 410L MSS additively manufactured by L-DED. The as-built microstructure showed low performance. This property could not be altered by the annealing and tempering at lower temperatures. The only conditions that enhanced Charpy impact energy absorption were the tempering at higher temperatures.

Figure 26 shows that Charpy impact SEM fractures were mostly composed of large cleavage bands in the AB condition (**Figure 26a**), annealing AQ (**Figure 26b**), and lower tempering temperatures AQT300 (**Figure 26c**). A toughness recovery was observed only in the tempering condition at the highest temperature AQT600 (**Figure 26d**), which can be attributed to the elimination of grain boundaries with refined martensite compared to the AB condition, and the grain coarsening and effective residual stress relief concerning the heat-treated conditions of annealing and tempering at lower temperatures. In response, located ductile dimples were perceived [40, 41].

Owing to the brittle martensite nature and the MSS application field, there are few studies in the literature exploring this property in this class of material. The only reference found on this research topic explored a 630 MSS, with 0.05 wt.%C, and higher contents of the alloy elements Cr, Mn, Si, Mo, and mainly Ni. The Charpy impact showed a significantly greater AE (138 J) in the as-built condition. Nevertheless, a significant AE drop to 27 J was noted when the material underwent a heat-treatment of annealing plus tempering, which is like the 20 J of the AQT600 condition of the current work. This change was attributed to the transition from an as-built microstructure rich in ferrite and retained austenite bands of high ductility to a fully martensitic microstructure of high hardness and brittleness after heat-treatment [13]. On the other hand, the AISI 410 CQ-SA reference material exhibited at least two times higher AE (40 J) concerning all L-DED conditions and the low C-content MSS. Its annealed and tempered martensite microstructure, with homogeneously distributed

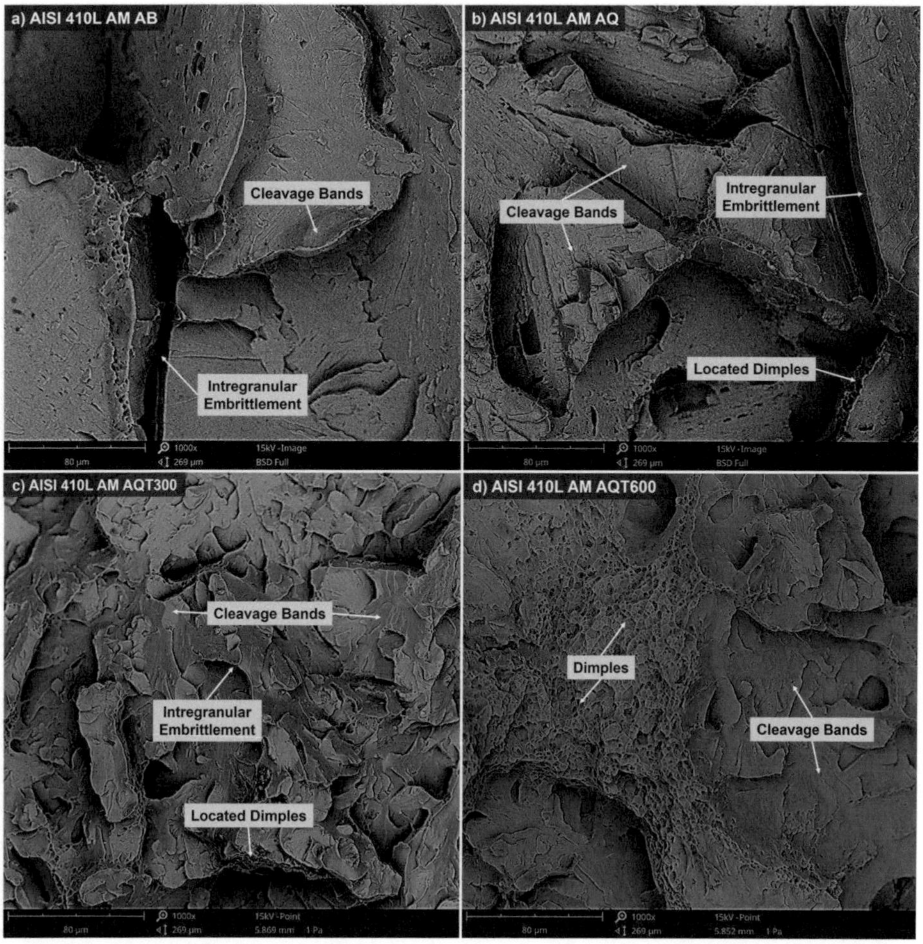

Figure 26.
Charpy impact fractography of the L-DED as-built and heat-treated conditions. All conditions (a) AB, (b) AQ, and (c) AQT300 showed large cleavage bands. Higher tempering temperature (d) AQT600 demonstrated located dimples.

grains and tempered martensite. **Figure 27** shows that Charpy impact fractures justify this behavior since multiple cleavage bands were identified in AISI 410L AM AB (**Figure 27a**), while AISI 410 CQ-SA (**Figure 27b**) demonstrated a predominance of dimples. This variation is due to the different thermal history of the traditional manufacturing process [3, 41].

3.5 Effect of substrate material on the microstructure and mechanical properties of AISI 410L MSS in the laser cladding configuration

In the L-DED setup of just a few deposited layers, such as for coating applications, it is interesting to investigate the substrate's influence on the microstructure and hardness. **Figure 28** depicts the typical cross-section of an AISI 410L single layer deposited using the optimized L-DED parameters, captured by OM and SEM. Single layer microstructure was formed by ferrite (lighter color) and refined martensite

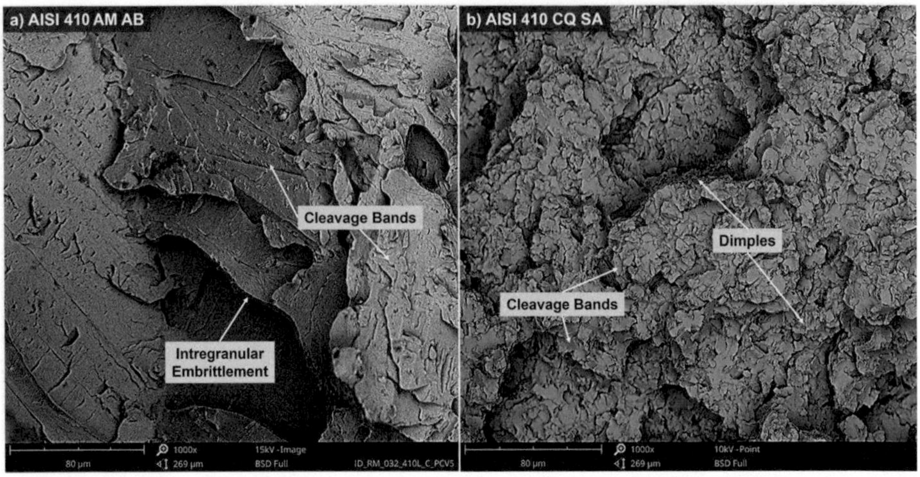

Figure 27.
Charpy impact fractography showing (a) predominance of cleavage bands and intergranular fracture in the L-DED AB AM condition; (b) a higher number of dimples in the AISI 410 CQ SA reference material.

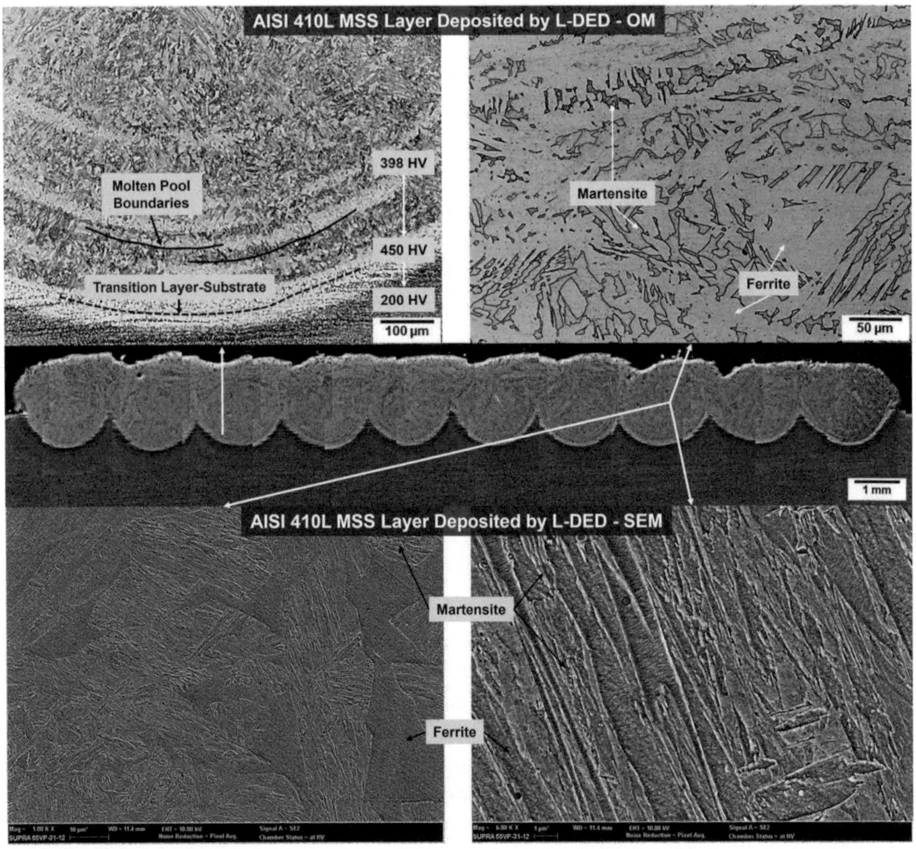

Figure 28.
Characterization of the AISI 410L single layer: OM highlighting the transition between layer and substrate, molten pool boundaries, ferrite, and refined martensite phases; SEM showing the ferrite and refined martensite details in higher magnifications; microhardness sweeping the single layer, transition region, and substrate.

laths (darker color). When compared to multilayer geometries, in the single layers, martensite laths formed more massively due to the C flotation from the AISI 410 C-richer substrate (0.10 wt.%C) regarding the AISI 410L (0.01 wt.%C) powder feedstock. The average hardness (398 HV0.5) was also much higher than that of multilayer geometries, which was linked to the C-content increase and the L-DED thermal history. Continuing the hardness profile from the top to bottom of the single layer, a peak of 450 HV0.5 was detected in the transition region near the substrate, indicating the production of a narrow martensitic band that is refined and harder. The substrate's greater C-content also improved this. The indentations showed decreasing values until stabilizing around 200 HV0.5, suggesting that substrate was reached [42, 43].

As observed in the present work, authors who explored MSS 410L coatings deposited on C-rich MSS substrates (0.13–0.93 wt.%) reported ferritic microstructures with refined martensite laths. The primary factor displayed is the C that enters the layers through convection currents from the substrate. Therefore, the more turbulent the process, especially regarding the energy density level, the more intense the effect. It is typical to find a maximum hardness peak at the layer interface and heat-affected zone, specifically provided by the more refined and C-enriched martensite, which exhibits a more tetragonal structure. When deposited on hypoeutectoid substrates, lower hardness (**Figure 29** graph) values are reported, around 300 HV, which rise to close to 400 HV in the case of heat-treated hypoeutectoid or hypereutectoid steels. The intense formation of refined martensite was the factor attributed to this response in both cases [19, 20].

Figure 30 shows that the yield strength (YS) and ultimate tensile strength (UTS) of low carbon-content AISI 410L martensitic stainless steel (MSS) increased approximately twofold compared to the initial material. The difference is considerably more noticeable when deposited on hypereutectoid substrates. The primary cause is suggested to be the production of a significant amount of refined martensite. The YS and UTS values of hypoeutectoid steel doubled with heat-treatment performance, a phenomenon ascribed to the precipitation of refined carbides [15, 20].

Figure 31 graph shows that elongation exhibited a trend opposite to that of hardness, YS, and UTS. Deposition on hypoeutectoid substrates resulted in greater elongation compared to deposition on heat-treated hypoeutectic and hypereutectoid

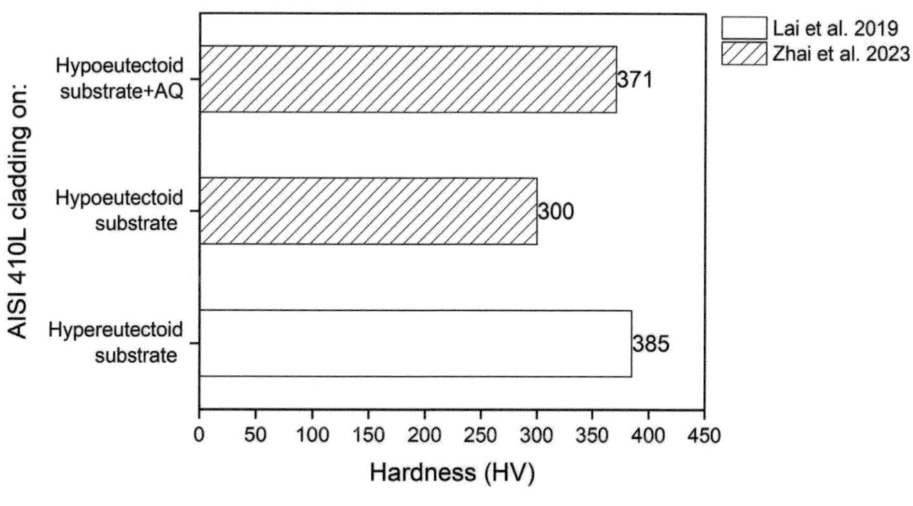

Figure 29.
Hardness performance of AISI 410L MSS deposited by L-DED on different C-content MSS.

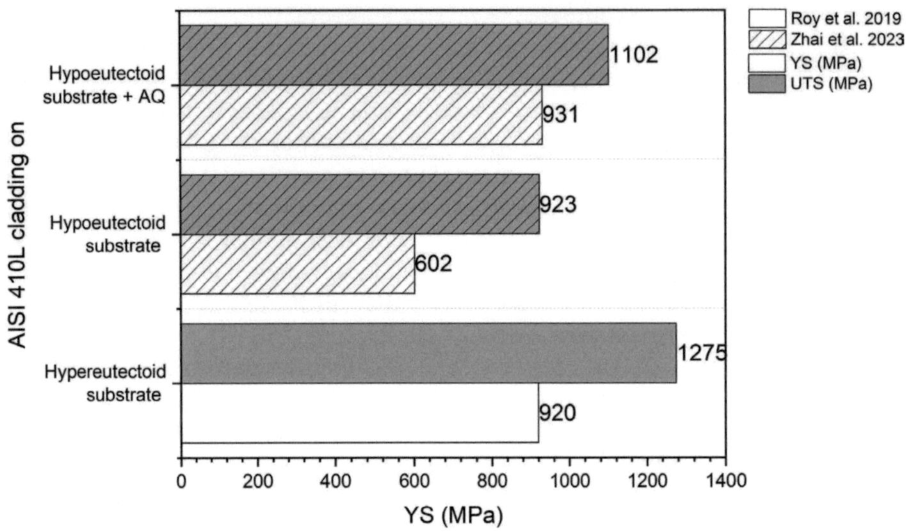

Figure 30.
YS and UTS of AISI 410L MSS deposited by L-DED on different C-content MSS.

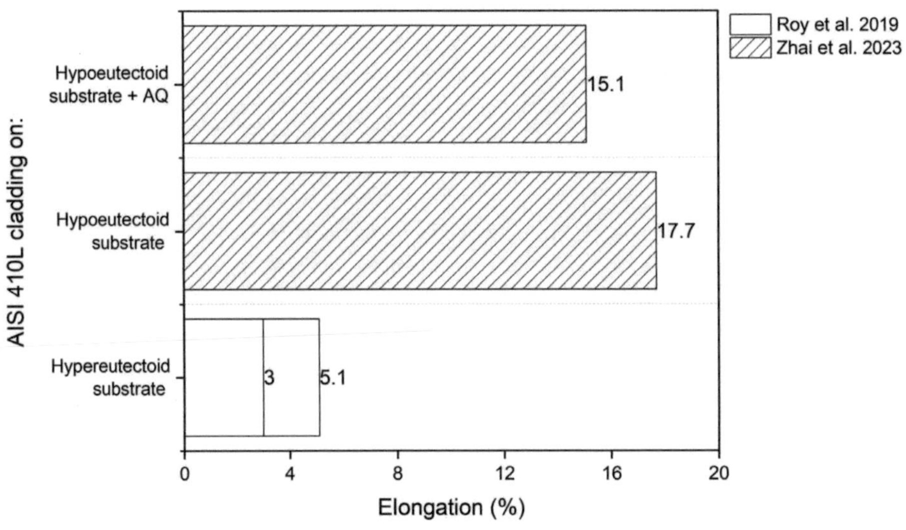

Figure 31.
Elongation of AISI 410L MSS deposited by L-DED on different C-content MSS.

substrates. In the first case, the large amount of ductile ferrite can be attributed to this performance. In the latter, the establishment of a microstructure predominantly composed of martensite and refined carbides embrittles the material [15, 20].

4. Conclusions

This work presented a study of AISI 410L MSS processed by additive manufacturing using the L-DED technique, in the as-built state and after heat-treatment performance. Furthermore, a literature survey was performed. In general, it was possible to infer:

- It is essential to optimize L-DED parameters for additive manufacturing, to obtain as-built multilayer geometries free from defects and with high density.

- The L-DED thermal history, added to the low C-content of AISI 410L MSS, generated a ferritic matrix with refined martensite laths at the grain boundaries, providing heterogeneity in hardness, yield strength, and ultimate tensile strength. However, the martensite brittleness led to reduced elongation and impact toughness.

- Grain refining is linked to L-DED parameters and build strategies. The layers' orientation concerning the substrate provides variations in the heat-sink efficiency, affecting the precipitation behavior. More refined microstructures generated higher hardness, yield strength, and ultimate tensile strength at the expense of reductions in elongation and impact toughness.

- Post heat-treatment applied to MSS processed by L-DED provides recrystallization of the start heterogeneous microstructure. Grain coarsening in lower C-content MSS increased elongation and toughness at the trade-off of decreased mechanical strength. In higher C-content MSS, increased dislocation displacement, martensite formation, and carbide precipitation enhanced mechanical resistance while reducing elongation and impact toughness.

- A ferrite plus refined martensite was produced by the C-flotation from the substrate to the layer, making the deposition of low-C MSS coatings on C-rich substrates an interesting alternative. Higher hardness and mechanical resistance with acceptable ductility were found.

- Elongation and toughness are weak points of MSS processed by L-DED. Post-heat treatments are required. Higher tempering temperatures enable recrystallization, grain coarsening, and residual stress relief, eliminating the L-DED AB microstructure heterogeneity, resulting in elongation and impact energy absorbed compatible with MSS from traditional processes.

Acknowledgements

The authors acknowledge the Precision Engineering, and LINDEN Laboratories from the Federal University of Santa Catarina, and acknowledge the Senai Innovation Institute for Manufacturing Systems and Laser Processing for their technical scientific support.

Conflict of interest

The authors declare no conflict of interest.

Author details

Jurandir Marcos Sá de Sousa[1,2*], Milton Pereira[1], Anselmo Thiesen Júnior[2], Henrique Santos Ferreira[2], Jhonattan Gutjahr[3] and Juliane Ribeiro da Cruz[4]

1 Precision Engineering Laboratory, Federal University of Santa Catarina, Campus Reitor João David Ferreira Lima, Florianópolis, Brazil

2 Senai Innovation Institute for Manufacturing Systems and Laser Processing, Joinville, Brazil

3 The Welding Institute Ltd. - Laser Additive Manufacturing, Rotherham, South Yorkshire, United Kingdom

4 Department of Mechanical Engineering, Polytechnic School of Engineering, University of São Paulo, São Paulo, Brazil

*Address all correspondence to: jurandirmarcos37@gmail.com

References

[1] Batista MN, Marinelli MC, Alvarez-Armas I. Effect of initial microstructure on surface relief and fatigue crack initiation in AISI 410 ferritic-martensitic steel. Fatigue Fracture of Engineering Materials and Structures. 2018;**42**:61-68. DOI: 10.1111/ffe.12868

[2] Gonçalves KAMB, de Faria GL, de Siqueira RHM, de Oliveira TR, de Lima MSF. Heat treatment effects on the hardness and wear behavior of laser-welded AISI40 martensitic steel plates. The International Journal of Advanced Manufacturing Technology. 2021;**114**:1155-1163. DOI: 10.1007/s00170-021-06942-6

[3] Chakraborty G, Das CR, Albert SK, Bhaduri AK, Thomas Paul V, Panneerselvam G, et al. Study on tempering behaviour of AISI 410 stainless steel. Materials Characterization. 2015;**100**:81-87. DOI: 10.1016/j.matchar.2014.12.015

[4] Mabruri E, Syahlan ZA, Sahlan PS, Anwar MS, Chandra SA, et al. Influence of austenitizing heat treatment on the properties of the tempered type 410-1Mo stainless steel. Materials Science and Engineering. 2017;**202**:012085. DOI: 10.1088/1757-899X/202/1/012085

[5] Huang S, Li D, Zhang L, et al. Tailoring the mechanical properties of laser cladding-deposited ferrous alloys with a mixture of 410L alloy and Fe–Cr–B–Si–Mo alloy powders. Materials. 2019;**12**:1-12. DOI: 10.3390/ma12030410

[6] Meura VH, Pacheco JT, Veiga MT, Rabelo A, da Silva LJ, Milan JC, et al. Effect of post-heat treatment on microstructure and tribological performance of AISI 410L coatings processed by laser-directed energy deposition (L-DED). Progress in Additive Manufacturing. 2023;**8**:1367-1377. DOI: 10.1007/s40964-023-00404-z

[7] Gradl P, Tinker DC, Park A, Mireles OR, Garcia M, Wilkerson R, et al. Robust metal additive manufacturing process selection and development for aerospace components. Journal of Materials Engineering and Performance. 2022;**31**:6013-6044. DOI: 10.1007/s11665-022-06850-0

[8] Cunha Â, Marques A, Silva MR, Bartolomeu F, Silva FS, Gasik M, et al. Laser powder bed fusion of the steels used in the plastic injection mould industry: A review of the influence of processing parameters on the final properties. The International Journal of Advanced Manufacturing Technology. 2022;**121**:4255-4287. DOI: 10.1007/s00170-022-09588-0

[9] ISO/ASTM 52900. Additive Manufacturing—General Principles—Fundamentals and Vocabulary. Standard: International Organization for Standardization/American Society for Testing and Materials; 2021

[10] ASTM F3187. Standard Guide for Directed Energy Deposition of Metals. Standard: American Society for Testing and Materials; 2021

[11] Cooke S, Ahmadi K, Willerth S, Herring R. Metal additive manufacturing: Technology, metallurgy and modelling. Journal of Manufacturing Processes. 2020;**57**:978-1003. DOI: 10.1016/j.jmapro.2020.07.025

[12] Khodabakhshi F, Farshidianfar MH, Gerlich AP, Nosko M, Trembošová V, Khajepour A. Effects of laser additive manufacturing on microstructure and

crystallographic texture of austenitic and martensitic stainless steels. Additive Manufacturing. 2020;**31**:100915. DOI: 10.1016/j.addma.2019.100915

[13] Oh WJ, Son Y, Cho SY, Yang SW, Shin GY, Shim DS. Solution annealing and precipitation hardening effect on the mechanical properties of 630 stainless steel fabricated via laser melting deposition. Materials Science and Engineering: A. 2020;**794**:139999. DOI: 10.1016/j.msea.2020.139999

[14] Sousa JMS, Pereira M, Cruz JR, Thiesen Junior A, Ferreira HS, Gutjahr J. Influence of post-processing heat-treatment on the mechanical performance of AISI 410L stainlees steel manufactured by L-DED process. Journal of Laser Applications. 2023;**35**:042056. DOI: 10.2351/7.0001135

[15] Zhai W, Wu N, Zhou W. Laser metal deposition of low carbon 410L stainless steel and heat treatment. Materials Science and Engineering: A. 2023;**872**:144987. DOI: 10.1016/j.msea.2023.144987

[16] Li Y, Chen Y, Hou L, Ye C, Chen S, Huang X, et al. Process parameter optimization of K477 and GH4169 for defect control in laser-directed energy deposition using Taguchi design of experiments. The International Journal of Advanced Manufacturing Technology. 2024;**131**:4447-4467. DOI: 10.1007/s00170-024-13277-5

[17] dos Santos Paes LE, Pereira M, Xavier FA, Weingaertner WL, D'Oliveira ASCM, Costa EC, et al. Understanding the behavior of laser surface remelting after directed energy deposition additive manufacturing through comparing the use of iron and Inconel powders. Journal of Manufacturing Processes. 2021;**70**:494-507. DOI: 10.1016/j.jmapro.2021.08.061

[18] Aghili SE, Shamanian M. Investigation of powder fed laser cladding of NiCr-chromium carbides single-tracks on titanium aluminide substrate. Optics and Laser Technology. 2019;**119**:105652. DOI: 10.1016/j.optlastec.2019.105652

[19] Lai Q, Abrahams R, Yan W, Qiu C, Mutton P, Paradowska A, et al. Influences of depositing materials, processing parameters and heating conditions on material characteristics of laser-cladded hypereutectoid rails. Journal of Materials Processing Technology. 2019;**263**:1-20. DOI: 10.1016/j.jmatprotec.2018.07.035

[20] Roy T, Abrahams R, Paradowska A, Lai Q, Mutton P, Soodi M, et al. Evaluation of the mechanical properties of laser cladded hypereutectoid steel rails. Wear. 2019;**432-433**:202930. DOI: 10.1016/j.wear.2019.202930

[21] Thiesen A Jr. Selection of processing parameters for the laser directed energy deposition process applied to additive manufacturing: A methodological proposal [master thesis]. BR: University of Santa Catarina; 2021

[22] Sousa JMS, Gutjahr J, Thiesen Junior A, Ferreira HS, Pereira M, da Silva LJ. Laser directed energy deposition (L-DED) of AISI 410L: Development of an operational map and the effect processing parameter on the bead geometry. In: Proceedings of the 26th International Congress of Mechanical Engineering—COBEM 2021. Florianópolis: Brazil; 2021

[23] Sousa JMS, Pereira M, Ferreira HS, Thiesen Junior A, Cruz JR. Laser directed energy deposition process optimization parameters for AISI 410L single layers. In: Proceedings of the 27th International Congress of Mechanical Engineering—COBEM 2023. Florianópolis: Brazil; 2023

[24] ASTM E8/E8M. Standard Test Methods for Tension Testing of Metallic Materials. Standard: American Society for Testing and Materials; 2022

[25] ASTM E23. Standard Test Methods for Notched Bar Impact Testing of Metallic Materials. Standard: American Society for Testing and Materials; 2018

[26] Zhu H, Ouyang M, Hu J, Zhang J, Qiu C. Design and development of TiC-reinforced 410 martensitic stainless steel coatings fabricated by laser cladding. Ceramics International. 2021;**47**:12505-12513. DOI: 10.1016/j.ceramint.2021.01.108

[27] Fenech M, Mallia B, Grech M, Betts JC. Post-deposition heat treatment of co-deposited Cr3C2 and AISI 410 stainless steel using the coaxial laser deposition technique. Journal of Materials Science. 2012;**48**:2224-2235. DOI: 10.1007/s10853-012-6998-6

[28] Ravi GA, Hao XJ, Wain N, Wu X, Attallah MM. Direct laser fabrication of three dimensional components using SC420 stainless steel. Materials and Design. 2013;**47**:731-736. DOI: 10.1016/j.matdes.2012.12.062

[29] Kang HK, Lee H, Oh CS, Yoon J. Microstructure and mechanical properties of laser direct energy deposited martensitic stainless steel 410. Micromachines. 2024;**15**:837. DOI: 10.3390/mi15070837

[30] Uematsu Y, Sasaki R, Kakiuchi T, Yamamoto S, Zensho A, Sano R, et al. Fatigue design curves for laser-metal-deposited type 420 stainless steel and effect of an interval during deposition process. The International Journal of Advanced Manufacturing Technology. 2021;**116**:2917-2927. DOI: 10.1007/s00170-021-07605-2

[31] Alam MK, Mehdi M, Urbanic RJ, Edrisy A. Mechanical behavior of additive manufactured AISI 420 martensitic stainless steel. Materials Science and Engineering: A. 2020;**773**:138815. DOI: 10.1016/j.msea.2019.138815

[32] Aydogan B, O'Neil A, Sahasrabudhe H. Microstructural and mechanical characterization of stainless steel 420 and Inconel 718 multi-material structures fabricated using laser directed energy deposition. Journal of Manufacturing Processes. 2021;**68**:1224-1235. DOI: 10.1016/j.jmapro.2021.06.031

[33] Reddy L, Preston SP, Shipway PH, Davis C, Hussain T. Process parameter optimisation of laser clad iron based alloy: Predictive models of deposition efficiency, porosity and dilution. Surface and Coatings Technology. 2018;**349**:198-207. DOI: 10.1016/j.surfcoat.2018.05.054

[34] Aversa A, Saboori A, Librera E, de Chirico M, Biamino S, Lombardi M, et al. The role of directed energy deposition atmosphere mode on the microstructure and mechanical properties of 316L samples. Additive Manufacturing. 2020;**34**:101274. DOI: 10.1016/j.addma.2020.101274

[35] Nayebi B, Najafi H, Farnia A. Microstructural, hardening, and wear characteristics of surface re-melted AISI 410S stainless steel via fiber laser process. Journal of Mechanical Science and Technology. 2021;**35**:4419-4426. DOI: 10.1007/s12206-021-0912-6

[36] Lu SY, Yao KF, Chen YB, Wang MH, Liu X, Ge X. The effect of tempering temperature on the microstructure and electrochemical properties of a 13wt.% Cr-type martensitic stainless steel. Electrochimica Acta. 2015;**165**:45-55. DOI: 10.1016/j.electacta.2015.02.038

[37] Kou S. Welding Metallurgy. Book. 2nd ed. Hoboken, New Jersey: John

Wiley and Sons, Inc; 2003. 466 p. ISBN 0-471-43491-4

[38] Hemmati I, Ocelík V, De Hosson JTM. The effect of cladding speed on phase constitution and properties of AISI 431 stainless steel laser deposited coatings. Surface and Coatings Technology. 2011;**205**:5235-5239. DOI: 10.1016/j.surfcoat.2011.05.035

[39] Shen LC, Yang XH, Ho JR, Tung PC, Lin CK. Effects of build direction on the mechanical properties of a martensitic stainless steel fabricated by selective laser melting. Materials. 2020;**13**:5142. DOI: 10.3390/ma13225142

[40] Deev A, Kuznetsov P, Zhukov A, Bobyr V. The structure and properties of the samples produced by selective laser melting of 410L steel-based metal powder. Physics Procedia. 2017;**89**:31-38. DOI: 10.1016/j.phpro.2017.08.013

[41] Moreira DC, Furtado HC, Buarque JS, Cardoso BR, Merlin B, Moreira DDC. Failure analysis of AISI 410 stainless-steel piston rod in spillway floodgate. Engineering Failure Analysis. 2019;**97**:506-517. DOI: 10.1016/j.engfailanal.2019.01.035

[42] Luo G, Li H, Li Y, Mo J. Microstructures and properties of a low-carbon-chromium ferritic stainless steel treated by a quenching and partitioning process. Materials. 2019;**12**:1704. DOI: 10.3390/ma12101704

[43] Bonagani SK, Vishwanadh B, Tenneti S, Naveen Kumar N, Kain V. Influence of tempering treatments on mechanical properties and hydrogen embrittlement of 13 wt% Cr martensitic stainless steel. International Journal of Pressure Vessels and Piping. 2019;**176**:103969. DOI: 10.1016/j.ijpvp.2019.103969

Chapter 8

Additive Manufacturing of Ferroelectric and Piezoelectric Composites Based on Eco-Circular Salts

Etienne Lemaire and Atilla Atli

Abstract

Modern societies are increasingly challenged to adopt resilient, circular manufacturing practices. The integration of 3D printing with liquid-phase epitaxy offers a promising path for creating eco-circular piezoelectric transducers and groundbreaking ferroelectric components. This chapter delves into innovative additive manufacturing techniques for these advanced pseudo-composites and investigates the revival of historically significant piezoelectric salts. We evaluate these options based on their affordability, durability, and strong piezoelectric or, in some cases, ferroelectric properties. We also explore their potential for eco-circularity, assessing the life cycle impacts, the properties achieved, and the opportunities they present for environmentally friendly multi-physics components and applications. Depending on the chosen polymer and salt, these components could either be designed for long-term use or be made biodegradable. In this work, the low-cost fabrication of pseudo-composite transducers using a combination of polymer and Rochelle is achieved using 3D printing and liquid-phase epitaxy. The ferroelectric and piezoelectric performance of these new samples is presented, including improvements over previous studies.

Keywords: 3D printing, eco-design, ferroelectricity, piezoelectricity, manufacturing

1. Introduction

3D printing has opened up new avenues for the manufacture of piezoelectric materials, offering unprecedented design freedom and customization. This technology makes it possible to create complex, porous, multi-material structures, paving the way for innovative applications in a variety of fields, including biomedicine. Piezoelectric materials, capable of converting mechanical energy into electrical energy and vice versa, are of particular interest for implantable medical devices, self-powered sensors, and energy harvesting applications. However, 3D printing of these materials presents specific challenges in terms of material selection, piezoelectric sample preparation, polarization techniques, and mechanical properties. This chapter explores recent advances, challenges, and future prospects for the 3D printing of piezoelectric materials.

Several 3D printing techniques have been adapted to manufacture the piezoelectric materials, each having distinct advantages and disadvantages. The choice of a technique depends on the properties of the piezoelectric material, the complexity of the desired geometry, and the intended application. The most widely used 3D printing technique is fused deposition printing (FDM). FDM is a widely used technique that extrudes a molten thermoplastic material to build, layer by layer, three-dimensional objects. The method is appreciated for its simplicity, speed, and ability to create complex structures. The most commonly printed plastics are ABS (acrylonitrile butadiene styrene), PET (polyethylene terephthalate), and PLA (polylactic acid). The latter has the advantage of being agro-sourced and biodegradable, which is an important factor in reducing the environmental impact of printed objects. FDM enables the printing of composites incorporating these plastics; for example, PLA filament filled with wood particles can be purchased commercially. However, the resolution of FDM can be limited, as it is constrained by the molten plastic filament. This can affect the precision of the produced devices. In addition, the choice of materials is limited to thermoplastics compatible with the melting process. Numerous works have used FDM to print various piezoelectric materials such as PVDF-TrFE [1], barium titanate ($BaTiO_3$) [2], or other materials including those based on lead [3]. In their study, Sikder et al. [4], have used FDM to develop piezoelectric scaffolds built on polycaprolactone (PCL) and $BaTiO_3$ particles. These structures, designed for orthopedic applications, feature controlled porosity and a specific architecture. The researchers highlight the effectiveness of FDM in creating these kinds of scaffolds with interconnected pores of uniform size, thus promoting osteogenesis and bone integration.

Stereolithography (SLA) is another 3D printing technique. It uses light, usually a laser or projector, to polymerize a photosensitive thermoset resin printed layer by layer. This technique offers high resolution and precision, making it ideal for manufacturing complex medical devices and microfluidic applications. However, the choice of materials for SLA is more limited compared to FDM. The light-sensitive resins used in SLA must have specific properties to ensure efficient polymerization and adequate mechanical properties. Wang et al. [5] used microstereolithography to produce lead-free potassium sodium niobate (KNN) piezoelectric ceramics. They succeeded in fabricating a segmented annular lattice with a pitch of 1 mm on a hollow hemispherical structure, demonstrating the ability of SLA to create piezoelectric components with complex architectures. They also highlighted the importance of using low-viscosity, high-solids ceramic/polymer suspensions to achieve high-quality prints. This complex, multi-physics technique requires the right combination of polymer and piezoelectric materials to be effective.

Selective laser sintering (SLS) is a technique that uses a laser to selectively sinter powders, including piezoelectric ceramics. This method enables the creation of dense, complex parts without the need for material support during the printing process. However, SLS can have lower accuracy than SLA, and post-processing of printed parts can be more complex. Yang et al. [6], for example, used SLS to fabricate scaffolds based on poly(L-lactic acid) (PLLA) with graphene and barium titanate. They found that graphene increased the electrical conductivity of material, while barium titanate conferred piezoelectric properties. The 3D-printed scaffold showed an enhanced piezoelectric response, suggesting its potential for bone stimulation applications.

The inkjet printing technique deposits photopolymerizable material droplets onto a substrate and then cures them using UV light. It is appreciated for its capacity to print in a variety of materials and colors, making it useful for the creation of realistic prototypes and customized medical devices. Numerous inks based on piezoelectric polymers such as polyvinylidene fluoride-trifluoroethylene copolymer (PVDF-TrFE) have made it possible to print piezoelectric devices using inkjet [7] and especially PVDF-TrFE-based inks [8]. However, resolution can be limited compared with other techniques such as SLA. Similar to inkjet, syringe-based (3D) printing allows great flexibility in structural design and is compatible with a wide range of materials, including piezoelectric composites. It is particularly well suited to the manufacture of small-series devices and complex structures. As an example, Chang et al. [9] used a commercial Digital Light Processing (DLP) 3D printer to fabricate a functional, flexible, tactile sensor by integrating piezoelectric, PZT-based, conductive, and structural polymer layers into a single auxetic and stretchable structure [9]. Others preferred to use a $BaTiO_3$-based ink with a similar approach [10].

Although 3D printing of piezoelectric materials offers many possibilities, it also presents technical challenges that need to be overcome. A polarization step, i.e., the application of an electric field to align the dipoles, is essential to achieve optimum piezoelectric properties in a piezoelectric ceramic or polymer. The integration of polarization into the 3D printing process is an active area of research, with techniques such as corona polarization, contact polarization, and additive manufacturing-assisted electrical polarization under development. For example, Lee and Tarbutton [11] have developed an electrical polarization-assisted additive manufacturing (EPAM) process that combines additive manufacturing and electrical polarization to build continuous free-form piezoelectric devices. However, the need to individually polarize each sample is an additional key step that should not be overlooked. This is non-necessary in piezoelectric single crystals (which are already oriented by nature).

Resolution and precision are also two key parameters in 3D printing. Improving these parameters is necessary to manufacture small-sized, complex piezoelectric devices. Advances in printing techniques, such as high-resolution stereolithography, will enable the creation of miniaturized, more precise devices. For example, in their study, Tiller et al. [12] used a commercial Digital Light Processing (DLP) 3D printer to fabricate a functional piezoelectric microphone with a horizontal resolution of the order of 30 μm.

3D-printed piezoelectric materials have enormous potential, particularly in the biomedical field. Applications involving ultrasound devices are just one example. 3D-printed piezoelectric ceramics, such as $BaTiO_3$, are used to manufacture ultrasound transducers for imaging and energy recovery. The ability of 3D printing to create complex geometries enables to improve the resolution and sensitivity of these devices. For example, Zeng et al. (2020) [13] used 3D printing to fabricate $BaTiO_3$-based honeycomb structures for ultrasonic sensing applications. Ultrasound therapy and imaging devices now account for a large share of the potential market for printable piezoelectric transducers [14, 15]. Other portable devices are also being developed. 3D-printed piezoelectric polymers, such as PVDF-TrFE, are being integrated into wearable devices for health monitoring (heart rate, respiration, etc.) and energy harvesting from body movements. Their flexibility and biocompatibility make them ideal for prolonged use. Pei et al. [16] have demonstrated the use of 3D printing to fabricate a polarization-free PVDF-based piezoelectric energy harvesting device, simplifying the manufacturing process and paving the way for self-powered wearable devices.

Medical implants dedicated to bone are also being considered. 3D-printed piezoelectric and bioactive composites, such as $BaTiO_3$/hydroxyapatite, are being investigated for their potential to stimulate bone growth and promote bone regeneration. These intelligent implants could revolutionize the treatment of fractures and bone defects. Polley et al. [17] have 3D-printed barium titanate and bioactive glass composites for bone tissue engineering applications, demonstrating the capacity of 3D printing to create piezoelectric and bioactive scaffolds that promote cell growth and proliferation. Printing self-repairing and recyclable devices is also possible in some cases. He et al. [18] developed a strategy for growing piezoelectric, recyclable, and self-repairing Rochelle salt crystals in 3D-printed structures inspired by cuttlefish bone. These 3D-printed composites have shown remarkable mechanical and piezoelectric performance, as well as excellent properties. These partially 3D-printed materials can be used to create smart protective devices. These partially 3D-printed materials can be employed to create smart protective devices, such as armor for athletes and knee pads with fall detection for the elderly [18]. These devices offer mechanical protection and electrical sensitivity for impact monitoring and fall detection. This work uses Rochelle salt (RS) as a piezoelectric material. The latter is edible and biodegradable. Indeed, the development of new 3D-printable, biocompatible, and biodegradable piezoelectric materials may be crucial for biomedical applications.

Researchers are therefore exploring new materials such as silk-based composites [19], cellulose nanocrystal-based materials [20], and piezoelectric salts [21] to meet these requirements.

Indeed, 3D printing of piezoelectric materials represents a significant advance in the field of additive manufacturing, opening up new possibilities for the creation of innovative and, in some cases, sustainable devices [22]. Advances in printing techniques and the integration of eco-compatible materials will make it possible to create piezoelectric devices that are more efficient, more versatile, and, in this case, more environmentally friendly. The future of this technology is promising, with potential applications that could dramatically improve quality of life and open up new avenues for regenerative medicine, responsible wearable electronics, and other fields.

Here, the latest works cited are taken up, illustrated, and extended in this chapter with new elements and using more advanced characterizations than in the past.

In particular, the measurement frequency band is extended (>1 MHz). These features allow for the manufacture of affordable piezoelectric transducers using 3D printing. Several improvements and prospects are also discussed. The transducer manufacturing protocol is first described in detail, including problems encountered. Different sample design versions are also illustrated. Characterization elements are then presented, followed by a discussion of the eco-circular manufacturing technique developed recently. The latter is of great interest in a pedagogical context, making scratch-built piezoelectric transducers accessible to all, which are also partly characterizable with simple means.

2. Experimental protocol for manufacturing the transducers

Rochelle salt is soluble in water, and its solubility increases with temperature [23]. At 25°C, RS dissolves in water with almost equal mass, i.e., 1 g of RS in 1 g of water. However, the RS dissolution in water is endothermic, leading to a temperature drop in the mixture.

We prepared solutions with 500 g of RS and 500 mL of demineralized water in a glass beaker. We first added about a third of RS powder into demineralized water. At this stage, the temperature drop can reach around 10°C, provoking a strong decrease in dissolved RS amount. The baker was slightly heated, without exceeding 30°C to face the temperature drop. We then gradually poured RS powder into the water while keeping the temperature stable (**Figure 1a**). Once all the salt is dissolved, we add some water (15% of the mass of the solution) in order not to be too close to the saturation concentration that would induce fast, numerous crystallization starts. There is always at least some spontaneous crystallization at the bottom of the beaker. When a large mass of crystals is removed from the solution, this mass of salt must be added again to compensate and maintain the correct concentration.

The beaker was left at room temperature for a few hours or days. Small RS crystals were formed at the bottom of the beaker. These small crystals (a few mm) were correctly placed at the bottom of the beaker in order to orient the RS crystal in the appropriate growth direction. We then removed RS crystals from the beaker when they were around 1–3 cm long.

We picked up these RS crystals and tied them with a wire in suspension in a beaker filled with a 90% RS-saturated solution. The beaker is kept for 12–48 hours at room temperature. The RS crystals should grow without any further branched crystallization; otherwise, it is a sign that there are particles and impurities in the solution. Then we redissolved the solution, refiltered, and rinsed the crystal. Alternatively, we placed the beaker in a fridge (at around 8–10°C) and waited for the growth of RS crystal to reach the desired size. At lower temperatures, RS crystal growth is faster. If necessary, we renewed the growing solution. Then we attached a 3D-printed matrix to one side of the crystal and followed the same protocol as for growing the crystals (**Figure 1b**). We used a wire to attach the crystal and matrix together and to suspend the assembly. If the hemp wire is too fragile, it may break when attaching to the matrix or because of the weight of the crystal as it grows. The good rigidity of the wire is also useful for holding the crystal more effectively. For beakers to be stored in the fridge, it is preferable to use a plastic beaker. Rochelle salt has a coefficient of thermal expansion of between 30 and 60 10^{-6}°C^{-1}, while glass has a coefficient of thermal expansion of 4 10^{-6}°C^{-1}. If the bottom of the beaker is totally covered by RS crystals, as it warms

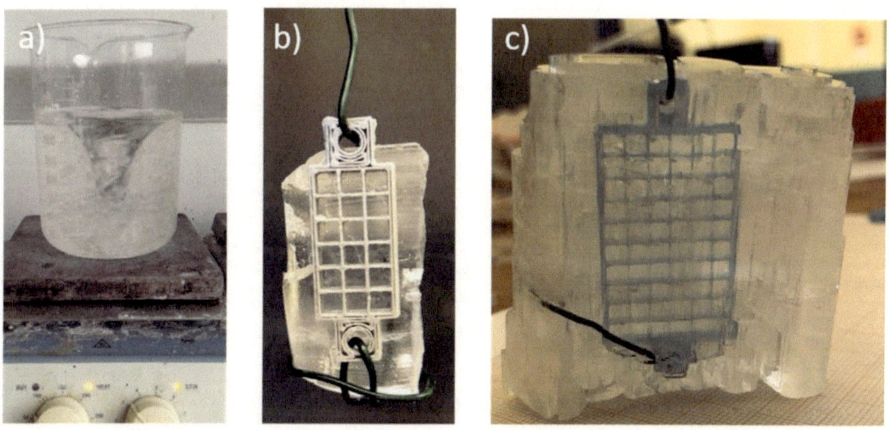

Figure 1.
a) Preparation of RS solution on a heating plate with a magnetic stirrer, b) 3D-printed PLA matrix attached to RS seed, c) 3D-printed polymer matrix embedded into an RS single crystal.

up when the beaker is taken out of the fridge, the glass beaker can break since the thermal expansion of RS is around one order of magnitude bigger than that of glass. If a glass beaker is placed in the fridge, it is important to break the crystal layer formed at the bottom of the beaker quickly. This can be done with a knife or by pouring a little tap water over the crystal, allowing it to dissolve. When a big RS single crystal is tightly attached to a polymer matrix, we repeat the crystal growth process until the polymer matrix is totally embedded into the RS crystal (**Figure 1c**). The excess of RS around the polymer matrix is first roughly removed by careful cutting using a handsaw followed by a polishing step with different grades of sandpaper. Finally, we obtained a polymer matrix where all the cells were correctly filled by RS crystals oriented in the same direction.

2.1 Problems encountered

We've experienced a few issues, and here are the main ones. Sometimes the Rochelle salt (RS) seed crystals detached from the matrix and fell down. One potential reason would be the wire has not been sufficiently tightened, and the vibrations have caused the crystal to fall down when the beaker was moved (**Figure 2a**). The RS seed crystal dissolved because the brine was too hot for the RS concentration. It is possible that air bubbles may be present in the cells in the matrices when immersed in brine. They produce holes in the crystal after crystallization (**Figure 2b**). It was sometimes difficult to obtain single crystals used as seeds. If the concentration is too high for the temperature, and if there are impurities in brine, these impurities will be initiators for crystallization. If the solution has not been properly filtered in order to remove the impurities, the polycrystalline structures with very poor piezoelectric properties were obtained (**Figure 2c**).

It took us about 1 hour and 30 minutes a day, and we alternated the person in charge of redissolving and filtering the brine, but not systematically.

2.2 3D printing of the matrix: Role, design, and tools

The matrices in which the RS crystals are incorporated have been prepared as samples. The use of matrices is not for their piezoelectric capabilities, but, as with all composites, they bring other useful characteristics to our samples. Rochelle salt

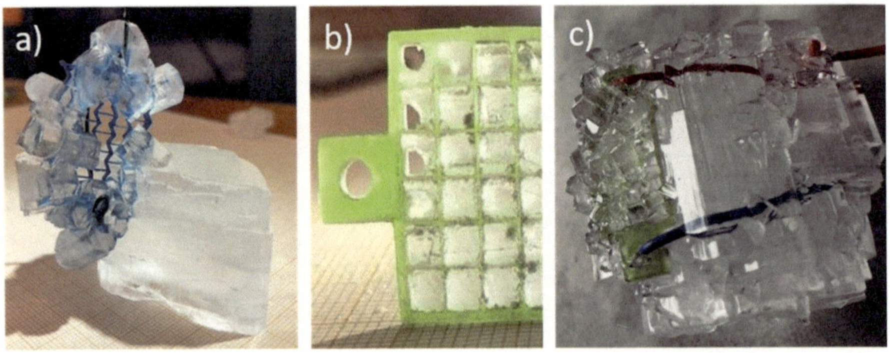

Figure 2.
a) Matrix with fallen RS crystal, b) holes in the crystals caused by air bubbles, c) Matrix in RS polycrystalline structures.

is not particularly resilient. A polymer matrix makes it less sensitive to external mechanical stresses and shocks. RS is water-soluble and degrades when exposed to UV light. A polymer matrix in which RS is enclosed is therefore a way to improve sample lifetime [24]. Last but not least, the matrix allows us to design samples with the desired shape and size, on which we can apply a load thanks to the fixing lugs on the sides of the matrix.

We created four rectangular-shaped cellular model matrices: two with 3 mm square meshes and two with 2.5*5 mm auxetic meshes. All four dies are 2.5 mm thick. Auxetic materials or structures are those with a negative Poisson's ratio. This means that when a force is applied in the right direction, instead of a striction phenomenon reducing the cross-section, the cross-section increases. The other effect we are interested in here is that the applied loads are much greater than those with a non-auxetic form. We have simulated the effect of an identic tensile stress on both matrices with a finite element method (ANSYS) (**Figure 3**). Since the voltage at the electrodes of a piezoelectric crystal is related to stress, we wanted to find out whether these shapes were more efficient than a more conventional square mesh. We were also concerned that sharp corners, where surfaces are close together, might lead to spontaneous crystallization. This was not a problem in the end.

For the additive manufacturing of the matrix, we used two 3D printing methods and therefore two materials. First, we produced the matrices on a Creality CR-10 3D printer. The PLA dies were 3D printed using a 0.4 mm diameter brass nozzle and a 1.75 mm diameter PLA filament. The second option was to use the resin-based stereolithography printing. We used an Anycubic Photon Mono 3D printer. This printer uses Masked Stereolithography (MSLA), also called LCD 3D printing, as MSLA printers, like this one, typically use Liquid Crystal Display (LCD) screens to mask the light and only cure the desired shape. The wavelength of light used is 405 nm to cure the UV resin, and this printer has a maximum resolution of about 40 μm. The resin used is the one recommended by Anycubic (Standard resin+). It should be noted that a durable resin can also be used (eg: bio-resin, Anycubic). Die dimensions are identical for both printing methods. We compared two printing methods, i.e., stereolithography and filament printing, to estimate their reliability for our application. The scrap rate of filament-printed dies in printer output is around 30%. Filament printing is not suitable for making single-wire thick walls, causing holes in walls and brittle

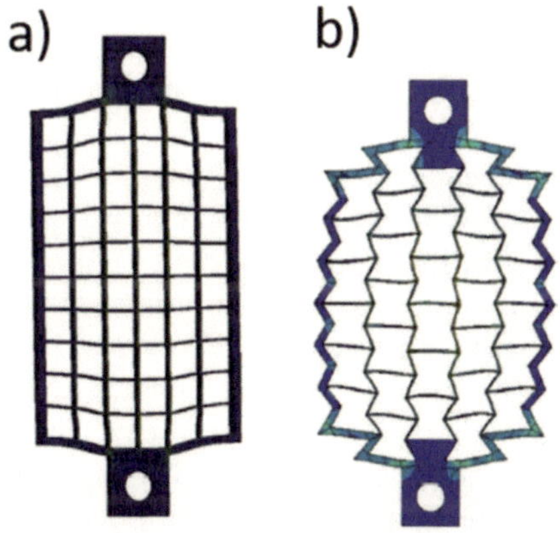

Figure 3.
3D-printed polymer matrices subjected to the same tensile stress simulated by ANSYS: a) a square-mesh matrix and b) an auxetic-mesh matrix.

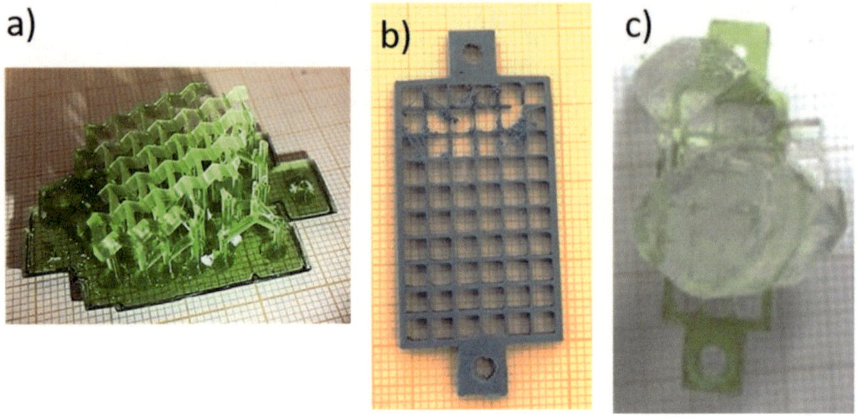

Figure 4.
a) Broken resin matrix with attached support below, b) PLA matrix with printing defects, c) resin matrix on which crystals have grown randomly.

structures. Such problems have not been noticed with photolithography printing. It is also highly recommended to insert a support "under" the matrix to avoid deforming printed parts. These supports must then be removed. A simple solution found was to cut out carefully the supports with a multi-purpose rotary tool (Dremel® brand) fitted with a cutting disk; then we removed the supports and polished the die to flatten it, eliminating thus all the problems. Parts printed in PLA require no modification after the printing process.

The surface finish of resin-printed parts is better than that of filament-printed PLA parts, resulting in observing a difference in RS crystallization quality in matrices as expected. In the figure we could see two matrices on which crystals have grown spontaneously. The solutions in which they were placed were identical, coming from the same beaker. Less crystallization starts on the resin (**Figure 4c**). On average, there is a higher rate of anarchic crystallization starts on PLA. Nevertheless, PLA filament printing is more attractive than resin printing: (1) PLA costs less. (2) PLA is biodegradable under the right conditions. (3) PLA can be bio-sourced from corn starch. (4) ePLA printing is also slightly faster. Sometimes, it is possible to obtain very good PLA composites.

2.3 Matrix extraction, polishing, and electrode deposition

Once the RS crystal completely covered the polymer matrix, the whole block (matrix entirely embedded into RS crystal) was removed from the RS brine. The block was kept in ambient conditions for a few days to dry out. Then the RS crystal was cut down by saw from the midst in order to remove the exceeded RS (**Figure 5a**). The aim was to keep one to two centimeters of RS crystal from each side of the matrix, preserving the RS-incorporated matrix intact. The RS crystal cut removed from the block was reused for further growth as a seed. The thickness of the crystal around the matrix was then progressively reduced with a rasp and then different grades of sandpapers. The polishing process was stopped as soon as we reached the wet RS crystal. For the last few millimeters of thickness, we used gradually finer grade sandpapers to remove the excess RS around the matrix in order to avoid tearing off any crystal splinters in the matrix.

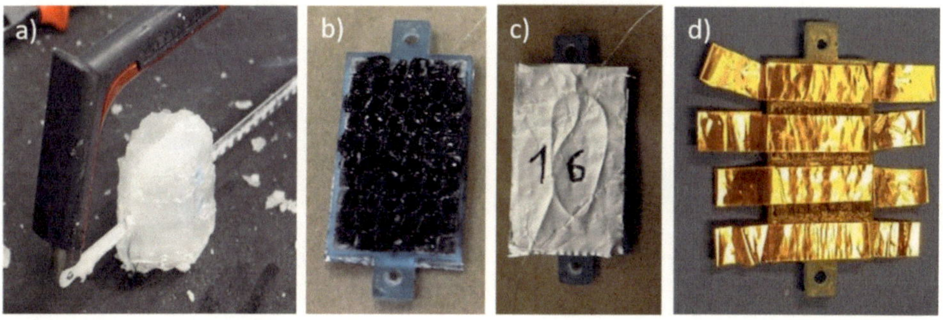

Figure 5.
a) Crystal cut with a saw, b) sample after conductive graphite ink deposition, c) finished samples with aluminum foil fixed on top, d) multiple electrodes for i.e., RCA configuration.

The samples were wiped with slightly humidified textile to clean their surfaces by removing RS powder and scratches left by sanding and such, thereby improving the surface finishing. Samples require electrodes, made of aluminum or copper foil, to which a wire is attached for connection to the measuring equipment. Graphite-based ink is brushed onto the surface of the sample (**Figure 5b**). A metallic wire is put directly on the graphite ink by looping it to increase the contact surface area of the wire in the sample.

The other end of the metallic wire is used to attach the clamps for measurements. An aluminum foil, cut to the same size as the sample, is then fixed on top of the sample surface, with the most reflective side against the glue. We waited ten seconds and ensured that the entire electrode was glued in place. Samples can have a single electrode covering the entire surface (**Figure 5c**), or a single electrode per square (or per group of squares) using a Row column Addressed (RCA) configuration (**Figure 5d**). It should be noted that the characterization of a single crystal will give the best results compared to the performance of several crystals connected together.

3. Electrical characterization

PLA/RS pseudo-composite samples were mainly characterized by using four measurement techniques: (1) impedancemetry, (2) laser vibrometry, (3) direct piezoelectric current measurement produced by the impact of a falling object of known mass on the sample, and (4) ferroelectric characterizations. These are the characterizations classically made in works dealing with the same type of sample [22]. Here, more details are given on the experimental approach, which can also be used as a confirmation of previous works with different samples having variable geometries.

The impedance analyzer (Agilent 4294a) is used to measure the complex impedance or admittance of a component over a wide frequency range (typically 40 Hz–110 MHz). These parameters on our samples were measured between 10 kHz and 4 MHz. Most of them showed the piezoelectric coupling modes below 1 or 2 MHz, with the most pronounced modes lying in the kilohertz band. As an example, **Figure 6** shows three identifiable modes on a rectangular PLA/RS pseudo-composite sample. A first mode is identifiable around 35 kHz, two others are around 500 kHz, and then 1 MHz.

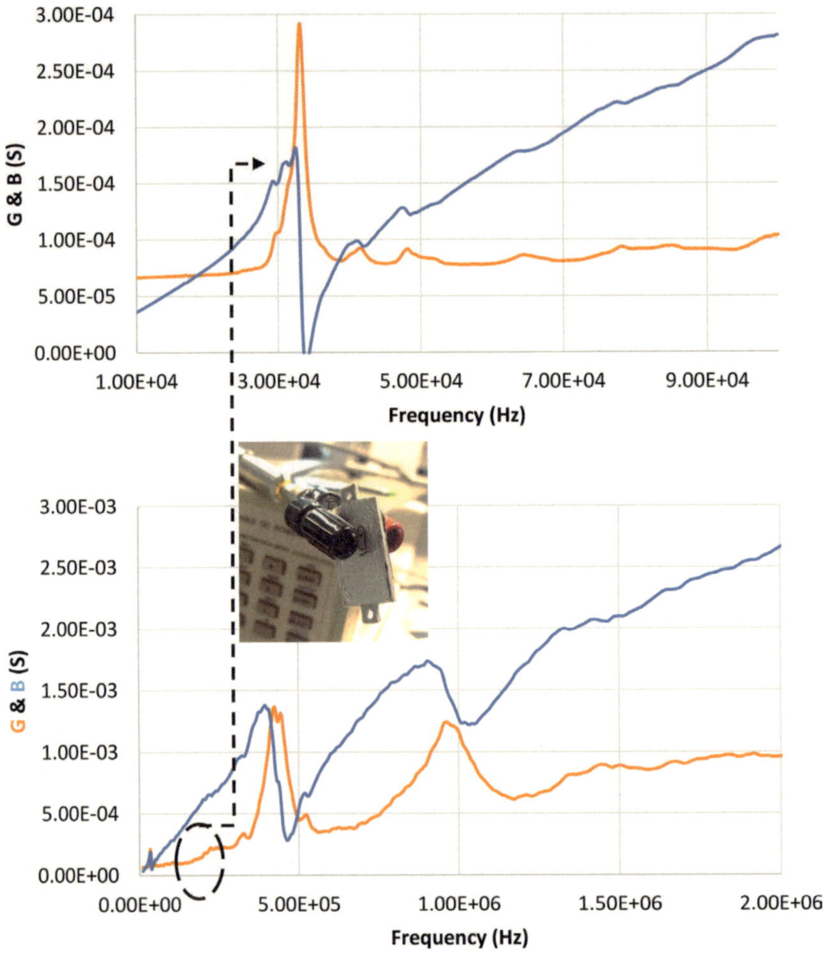

Figure 6.
Conductance (G) and Susceptance (B) of a rectangular-shaped sample manufactured by PLA/RS.

These coupling modes, or resonances, observed in these structures are a function of the key dimensions of the piezoelectric crystals and their arrangement. Parameters such as the frequency of the first resonance mode and the associated quality factor have been compiled for several samples of different formats (**Figure** 7). The observed frequencies may be roughly grouped by the design dimensions of samples. However, the frequency variations observed suggest that the reproducibility of thickness and crystallization quality from one sample to another can largely be improved. Indeed, the variation of quality factors shows a strong disparity in the observed coupling factors, which will lead to a strong variation in measured properties. This simple manufacturing method, designed to be accessible to all, is still homemade and enables us to produce samples that are functional but relatively different from one another. This can be clearly seen in **Figure** 7. In addition, it was impossible to pinpoint any particular advantage or disadvantage of auxetic-shaped meshes over regular square-shaped meshes. The latter perform neither better nor worse than the others. On the other hand, there is an advantage for smaller samples achieving sometimes better performance. Statistically, they are likely less affected by crystallization than a sample with a larger surface area.

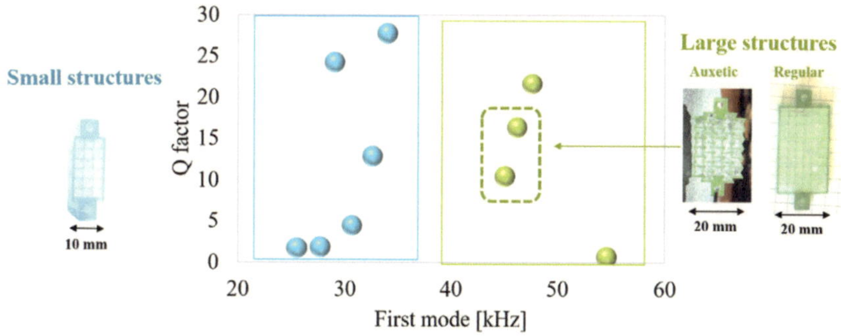

Figure 7.
Tested first resonant mode (f_0 [kHz]) with respect to Quality factors (Q) extracted from conductance and susceptance spectra of RS/Resin structures (left plot: small size sample with square meshes, right plot: large size samples with auxetic or square meshes).

The samples were also characterized several years after their manufacturing (up to 4 years) using a laser Doppler vibrometer (MSA 500 from Polytech). This type of measurement provides information on the vibration spectrum (in terms of speed or displacement) of the structure. In this case, the sample is electrically actuated, since it is piezoelectric. This measurement technique highlights the inverse piezoelectric effect. The out-of-plane vibration modes can be measured predominantly with respect to the surface and should correspond to the modes measured by impedancemetry, which vibrate in the right spatial direction.

Two low-frequency vibration modes (<100 kHz) and two higher-frequency modes around 300 kHz obtained on an auxetic PLA/RS sample are shown in **Figure 8** as an example. Note that the modes around 1 MHz observed on an impedance spectrum do not correspond to any significant out-of-plane vibration (although they do exist, as flatness is never perfect).

Vibrometer measurements correspond to the out-of-plane displacement of the largest sample surface. In other words, measured displacement is in the direction of the applied electric field (perpendicular to the two electrodes). From this normalized displacement spectrum (in direction 3), knowing the applied voltage (in direction 3), we can roughly estimate the effective d33 coefficient for these samples. This calculation requires a few assumptions: The measured first vibration mode corresponds to

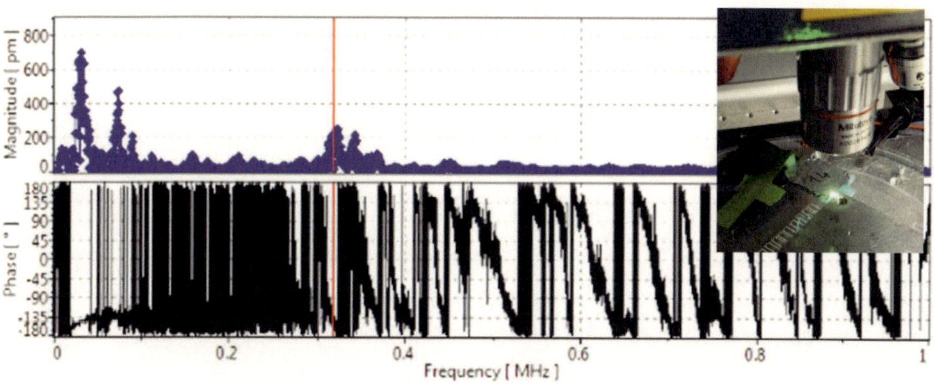

Figure 8.
Displacement magnitude and phase measured by laser vibrometer on an auxetic PLA/RS sample.

a second-order resonant system with damping factor ζ and natural frequency f_0 (or resonant frequency f_r and quality factor Q). D_0 is the static displacement and the 2nd order approximation of dynamic one $D(f)$ in Eq. 1:

$$D(f)=\frac{D_0}{1+2\zeta\frac{f}{f_0}+\frac{f^2}{f_0^2}}D \tag{1}$$

For $f=f_0$

$$D_0=2\zeta.D(f_0)\approx\frac{D(f_r)}{Q}\approx d_{33}\ if\ D[m/V] \tag{2}$$

Although this indirect method is a rough approximation for calculating the effective piezoelectric coefficient d33 (Eq. 2), it provides a very good order of magnitude for this coefficient. In principle, it relies on the proper calibration of the laser vibrometer with the standardized data given in [m/V]. The latter is normally well calibrated. However, it can be checked by repeating the experiment over different frequency ranges, optical signal intensity, and value of applied voltage. This is indeed the case, and the displacement magnitude is always consistent. In addition, to subtract the measurement noise from the equipment, we can plot the static displacement D_0 as a function of applied voltage (**Figure 9**). A reliable quality factor value can be extracted (recalculated in damping $\zeta = (2Q)^{-1}$), either using a narrow-band vibrometer or impedance meter measurement on the concerned mode, depending on the noise in measurements. The impedance meter gives an overall value with very little noise over the whole sample, whereas the vibrometer measurement is more local (although it can be averaged over measurement points using a spatial mesh) and sometimes noisier on this type of sample. Here, the measured effective coefficient d33 appears to be around 30 pm.V^{-1} (slope of the straight line in **Figure 9**). This result is in agreement with the values on similar samples manufactured by the same method [22].

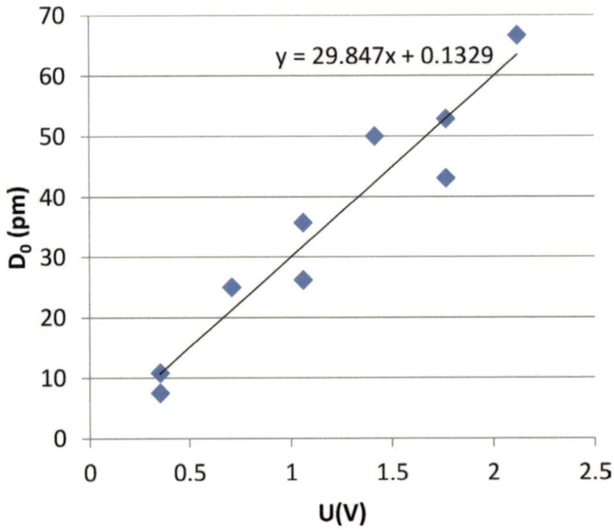

Figure 9.
Calculated displacements ($D_0 = 2\zeta.\ D(f_0)$) from vibrometer measurements as a function of applied voltage U.

If advanced characterization equipment is not available, a simple oscilloscope can be used to visualize the direct piezoelectric effect. By tapping on a piezoelectric transducer connected to an oscilloscope, the generated voltage can be measured (**Figure 10**). We can also attempt to control the force of the impact applied to the sample by dropping an object of known mass and initial height. The speed of approach to the point of impact and the mass of the object will determine the force applied to the piezoelectric transducer. At the same time, we can record the electrical signal produced by the impact and calculate both the energy generated and the charge produced.

The signal recorded during the impact of a 1.7 g mass on a sample can be converted into a current, which can then be digitally integrated to calculate charges (**Figure 11**). The difference between the slope corresponding to the integration of the measurement noise and the local charge maximum slope represents the amount of charge generated by the impact of the object. This quantity in [pC] can be divided by

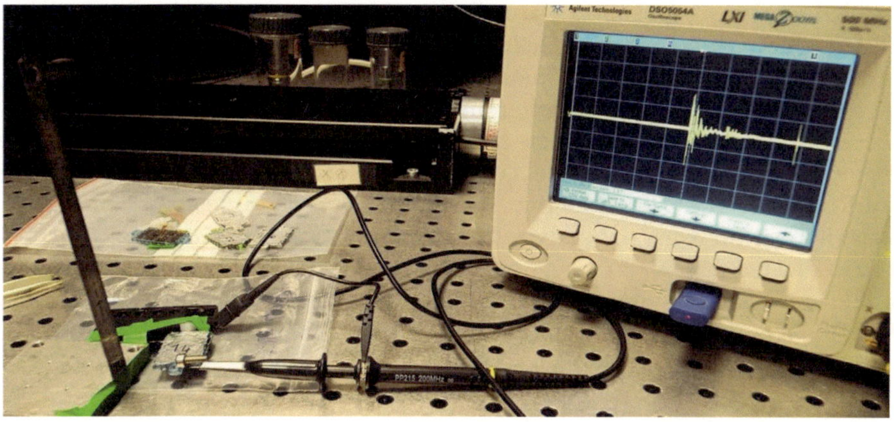

Figure 10.
Image of a piezoelectric transducer connected to an oscilloscope using an x1/x10 probe. The signal generated by the impact of an object dropped from a height on a transducer is visible on the oscilloscope.

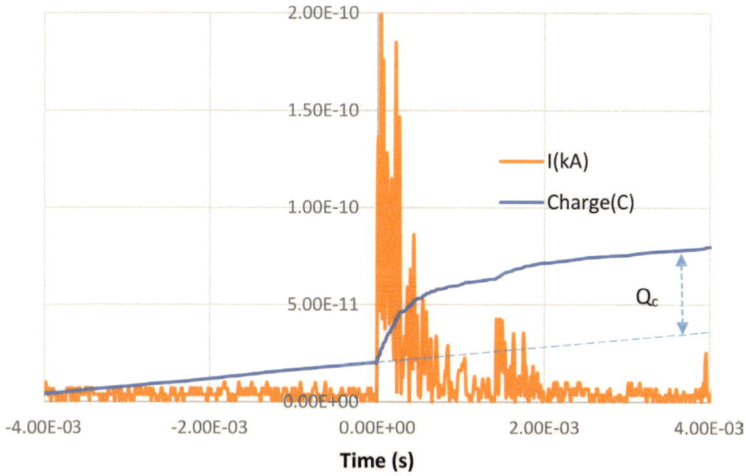

Figure 11.
Current generated and charge produced at the terminals of the piezoelectric element by the impact of a 1.7 g mass.

the force applied during the impact in [N]. Here, the force applied is around 2 N. The experiment can be repeated several times to obtain several values. We obtained values ranging from 21 to 57 $pC.N^{-1}$ with an average of around 31 $pC.N^{-1}$ for effective d33, further confirming the previously obtained values. Here, the same d33 piezoelectric coefficient is obtained using the direct piezoelectric effect. This very simple method is subject to several measurement biases: the location where the object falls on the sample, rebounds, uncertainties in release heights, the way in which the current is measured (impedance matching), etc. This technique, however, has the merit of being highly accurate. It is functional and provides results in a good order of magnitude. It can be performed anywhere with a simple oscilloscope, without requiring high bandwidth.

Since polymer/RS matrices are structured as single-crystal-filled cells, ferroelectricity measurements were carried out on a few of them at 22°C. As a result, ferroelectricity on Rochelle salt was confirmed. Although the material is quite conductive, a hysteresis cycle is easily identified when an AC voltage of +/– 380 V is applied at 20, 200, and 1000 Hz, respectively (**Figure 12**). It should be noted that if an electrode covers several single crystals (as done for previous experiments), the ferroelectricity becomes difficult to measure since all the single crystals do not necessarily respond coherently from a ferroelectricity point of view, and contributions from one crystal can be canceled by another one. However, as it is possible to individually address each crystal square mesh of the matrix with a set of RCA electrodes (Row Column Addressed), it is possible to exploit the ferroelectricity of these crystal matrices. Here, the characterization was carried out with a dedicated ferroelectricity measurement tool (Aixacct PES tool), but it is possible to measure it with a good impedance meter by measuring the capacitance at low frequency, which must evolve as a function of the DC voltage applied to the sample, as was done in our previous work [22].

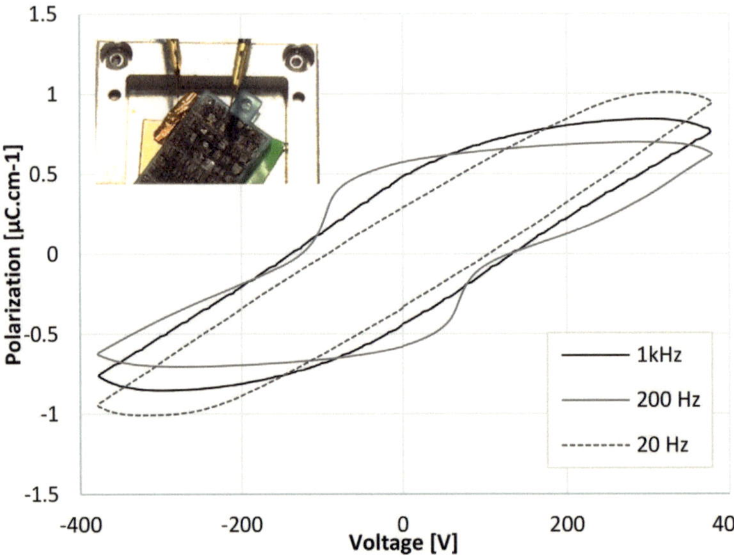

Figure 12.
P(V) loop of one square mesh of RS crystal in the RS/PLA matrix at 20 Hz, 200 Hz, and 1 kHz frequencies.

4. Evaluation of eco-circular manufacturing technology

The proposed technique is perfectly eco-circular (**Figure 13**). In other words, the manufactured samples can be fully dissolved: the salt can return to the solution, and the PLA can be processed into chips and then into filament if needed. A new structure can then be produced from the same resources. Additionally, the structure can be repaired: liquid-phase epitaxy can be repeated to make up for defects or repair a broken crystal tip. This liquid-phase epitaxy process can be repeated indefinitely. Moreover, all the materials used in the samples, including aluminum electrodes, are recyclable or agro-sourced. This includes PLA and potassium bitartrate, which is derived from winemaking and serves as a chemical precursor to Rochelle salt. If magnesium, graphite, or copper electrodes are used, the device becomes fully biodegradable, as all these materials (PLA, salt, copper, and/or carbon) have no environmental impact. We have documented the production of a transducer that can be made by anyone, with minimal resources, and is fully eco-friendly.

This type of transducer, with its highly interesting piezoelectric and ferroelectric properties, offers exceptional eco-circularity compared to other 3D-printed piezoelectric composite structuring techniques. Indeed, based on simple bio- or agro-sourced chemistry, this solution significantly surpasses, from an eco-design perspective, alternatives that use fluorinated compounds (such as PVDF-TrFE), lead and zirconium (such as PZT), barium titanate ($BaTiO_3$), aluminum nitride (AlN), or potassium niobate (KNN). While offering better piezoelectric and mechanical properties than some of these alternatives. What's more, the first two options are more problematic as they require much more energy-intensive synthesis and

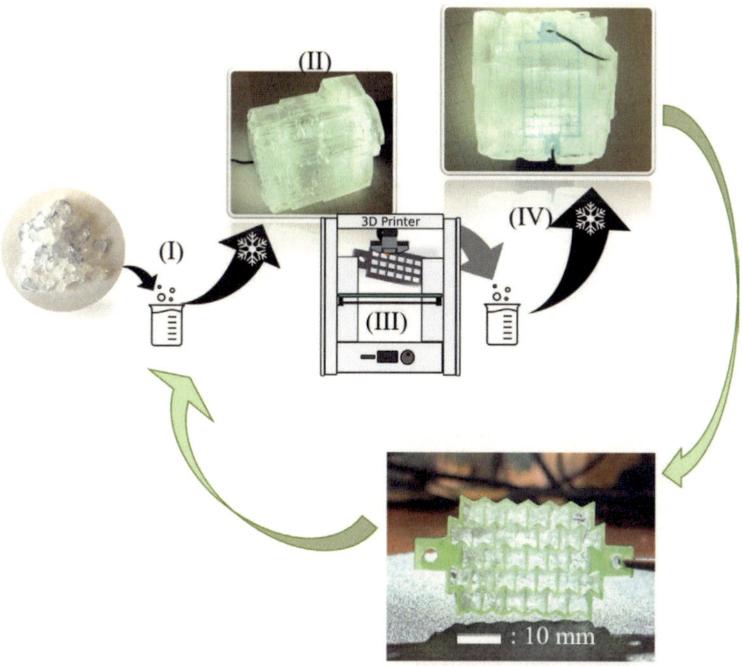

Figure 13.
Diagram illustrating some of the key steps of the additive manufacturing process, with the possibility of infinite cycling.

processing compared to the one presented here. These synthetic polymers or ceramics generally require a critical polarization step for each sample, which is not necessary with the technique presented here. Moreover, some piezoelectric ceramics contain rare elements. Niobium, found in KNN, and Scandium, found in the best-performing doped AlN, are problematic because of their relative rarity, respectively 20 ppm and 22 ppm in the earth's crust. More complex products meet much stricter specifications for resistance to humidity and temperature. It's true that samples made from Rochelle salt exhibit better aging resistance compared to unpackaged salt crystals, thanks to the polymer matrix and electrodes. While the former remains functional for several years, a single crystal can degrade in less than a year. On the other hand, since temperature resistance is intrinsic to Rochelle salt, the samples will be piezoelectric only between −18°C and + 45°C. This temperature range might be too narrow for many industrial applications. Nevertheless, this type of eco-designed transducer could still be suitable for a range of indoor applications, refrigerated environments, or in contact with living organisms. Additionally, it can be rapidly regenerated if a temperature threshold is exceeded. Moreover, this type of additive manufacturing, combined with the characterizations documented here, offers an excellent teaching scenario that integrates low-cost manufacturing with scientific experimentation.

5. Conclusion

We have successfully manufactured composite ferroelectric and piezoelectric transducers using a combination of 3D-printed PLA and Rochelle salt, which is grown by liquid-phase epitaxy. This manufacturing process is significantly more environmentally friendly compared to industrial standards of modern technologies that use lead-based ceramics, potassium niobate, or fluorinated polymers. Additionally, the piezoelectric performance of these new samples has improved compared to previous results.

This approach allows us to produce biodegradable, cost-effective piezoelectric transducers in any location, achieving a quality factor of up to 110 in certain specific cases, such as addressing a single crystal square. These elements hold great potential for integrating into smart devices for energy harvesting or detection applications, all while being environmentally friendly. Indeed, the energy consumption and environmental impact of manufacturing these composite piezoelectric transducers are extremely low while the performance achieved is satisfactory. The approach presented is also highly formative in a pedagogical context, using modest resources. An additional advantage of this manufacturing protocol is the flexibility permitting it to provide various shapes of the final samples. Although we have not yet concluded the specific benefits of using auxetic shapes, they proved to be as easy to manufacture and as effective as other shapes. Moreover, any 2D shape can be envisioned: a square, a hollow rectangle, or even a triangle is entirely possible, allowing a wide range of integrations. The miniaturization of structures is another possibility worth exploring. Beyond a few hundred nanometers, the Rochelle salt crystal remains piezoelectric, and microstructures inspired by this process could be developed.

Finally, this manufacturing technique can be applied to any crystalline piezoelectric salt, including monoammonium phosphate (ADP) or monopotassium phosphate (KDP), which are also piezoelectric and significantly more temperature-stable (withstanding temperatures up to around 190°C). However, growing these crystals by liquid-phase epitaxy would take a much longer time. It has therefore not been carried out for this type of transducer with a volume of the order of one cubic centimeter, which would require several weeks of growth at room temperature (without dedicated growth devices). With moderate resources or slightly more advanced industrial resources, with its good eco-circularity, this type of composite based on 3D printing and epitaxy could make sense for ADP or KDP. The KDP's electro-optical properties could also lead to low-cost, low-environmental-budget electro-optical modulation applications.

Acknowledgements

The author would like to thank the two master's students, Benjamin Dubois and Théo Luirard, who contributed to the manufacturing of the samples and wrote their observations in a French report that partially feeds the manufacturing part of this chapter. This manufacturing part has been done at ECAM Lyon during a master project supervised by the authors. This work also benefited from the CERTeM technological platform in Tours, especially for the measurements. The authors received no specific funding for this study.

Conflict of interest

The authors declare no conflict of interest.

Author details

Etienne Lemaire[1*] and Atilla Atli[2]

1 GREMAN, UMR-CNRS7347, Polytech Tours, Université de Tours, Tours, France

2 Universite de Lyon, ECAM LaSalle, LabECAM, Lyon, France

*Address all correspondence to: etienne.lemaire@univ-tours.fr

References

[1] Sun D, Lu Y, Karaki T. Review of the applications of 3D printing technology in the field of piezoelectric ceramics. Resources Chemicals and Materials. 2023;**2**(2):128-142

[2] Jiang Z, Cheng L, Zeng Y, Zhang Z, Zhao Y, Dong P, et al. 3D printing of porous scaffolds BaTiO3 piezoelectric ceramics and regulation of their mechanical and electrical properties. Ceramics International. 2022;**48**(5):6477-6487

[3] A review on 3D printed piezoelectric energy harvesters: Materials, 3D printing techniques, and applications. ScienceDirect [Internet]. [cited 2024 Sep 3]. Available from: https://www.sciencedirect.com/science/article/pii/S2352492823002313

[4] Sikder P, Nagaraju P, Naganaboyina HP. 3D-printed piezoelectric porous bioactive scaffolds and clinical ultrasonic stimulation can help in enhanced bone regeneration. Bioengineering. 2022;**9**(11):679

[5] Wang W, Sun J, Guo B, Chen X, Ananth KP, Bai J. Fabrication of piezoelectric nano-ceramics via stereolithography of low viscous and non-aqueous suspensions. Journal of the European Ceramic Society. 2020;**40**(3):682-688

[6] Yang Y, Peng S, Qi F, Zan J, Liu G, Zhao Z, et al. Graphene-assisted barium titanate improves piezoelectric performance of biopolymer scaffold. Materials Science and Engineering: C. 2020;**116**:111195

[7] Fully-printed piezoelectric devices for flexible electronics applications - Aliqué - 2021 - Advanced Materials Technologies, Wiley Online Library [Internet]. [cited 2024 Sep 4]. Available from: https://onlinelibrary.wiley.com/doi/full/10.1002/admt.202001020

[8] Piezoelectric P(VDF-TrFE) film inkjet printed on silicon for high-frequency ultrasound applications. Journal of Applied Physics. [cited 2024 Sep 4]. Available from: https://pubs.aip.org/aip/jap/article/129/19/195107/849112

[9] Chang SM, Hur S, Park J, Lee DG, Shin J, Kim HS, et al. Optimization of piezoelectric polymer composites and 3D printing parameters for flexible tactile sensors. Additive Manufacturing. 2023;**67**:103470

[10] Liu K, Zhou C, Hu J, Zhang S, Zhang Q, Sun C, et al. Fabrication of barium titanate ceramics via digital light processing 3D printing by using high refractive index monomer. Journal of the European Ceramic Society. 2021;**41**(12):5909-5917

[11] Lee C, Tarbutton JA. Electric poling-assisted additive manufacturing process for PVDF polymer-based piezoelectric device applications. Smart Materials and Structures. 2014;**23**(9):095044

[12] Tiller B, Reid A, Zhu B, Guerreiro J, Domingo-Roca R, Curt Jackson J, et al. Piezoelectric microphone via a digital light processing 3D printing process. Materials and Design. 2019;**165**:107593

[13] Micromachines | Free Full-Text | 3D-Printing Piezoelectric Composite with Honeycomb Structure for Ultrasonic Devices [Internet]. [cited 2024 Sep 4]. Available from: https://www.mdpi.com/2072-666X/11/8/713

[14] Flexible large-area ultrasound arrays for medical applications made using

embossed polymer structures. Nature Communications [Internet]. [cited 2024 Sep 4]. Available from: https://www.nature.com/articles/s41467-024-47074-1

[15] Emerging ultrasonic bioelectronics for personalized healthcare. ScienceDirect [Internet]. [cited 2024 Sep 4]. Available from: https://www.sciencedirect.com/science/article/pii/S0079642523000427

[16] Pei H, Xie Y, Xiong Y, Lv Q, Chen Y. A novel polarization-free 3D printing strategy for fabrication of poly (Vinylidene fluoride) based nanocomposite piezoelectric energy harvester. Composites. Part B, Engineering. 2021;**225**:109312

[17] Polley C, Distler T, Scheufler C, Detsch R, Lund H, Springer A, et al. 3D printing of piezoelectric and bioactive barium titanate-bioactive glass scaffolds for bone tissue engineering. Materials Today Bio. 2023;**21**:100719

[18] He Q, Zeng Y, Jiang L, Wang Z, Lu G, Kang H, et al. Growing recyclable and healable piezoelectric composites in 3D printed bioinspired structure for protective wearable sensor. Nature Communications. 2023;**14**(1):6477

[19] Chiesa I, De Maria C, Ceccarini MR, Mussolin L, Coletta R, Morabito A, et al. 3D printing silk-based bioresorbable piezoelectric self-adhesive holey structures for *In vivo* monitoring on soft tissues. ACS Applied Materials & Interfaces. 2022;**14**(17):19253-19264

[20] Maity K, Mondal A, Saha MC. Cellulose nanocrystal-based all-3D-printed pyro-piezoelectric nanogenerator for hybrid energy harvesting and self-powered cardiorespiratory monitoring toward the human–machine Interface. ACS Applied Materials & Interfaces. 2023;**15**(11):13956-13970

[21] Lemaire E, Moser R, Borsa CJ, Briand D. Green paper-based piezoelectronics for sensors and actuators. Sensors and Actuators A: Physical. 2016;**244**:285-291

[22] Lemaire E, Thuau D, De Vaulx JB, Vaissiere N, Atilla A. Rochelle salt-based ferroelectric and piezoelectric composite produced with simple additive manufacturing techniques. Materials. 2021;**14**(20):6132

[23] Shyju TS, Anandhi S, Gopalakrishnan R. Comparative studies on conventional solution and Sankaranarayanan–Ramasamy (SR) methods grown potassium sodium tartrate tetrahydrate single crystals. CrystEngComm. 2012;**14**(4):1387-1396

[24] Lemaire E, Atli A, Certon D. Rochelle salt revisited for eco-designed ultrasonic transducers. IEEE Open Journal on Ultrasonics, Ferroelectrics, and Frequency Control. 2024;**4**:171-176

Chapter 9

Optimizing Metal AM Potential through DfAM: Design, Performance, and Industrial Impact

Jose M. Costa, Elsa W. Sequeiros, Daniel Figueiredo, Ana R. Reis and Manuel F. Vieira

Abstract

Design for Additive Manufacturing (DfAM) is a transformative approach designed explicitly to optimize designs using metal additive manufacturing (AM). Exploring core DfAM principles, the chapter highlights the advantages of geometric freedom, material selection, and aligning designs with the capabilities of specific metal AM processes. It examines advanced optimization techniques like topology optimization and lattice structures to achieve high performance on lightweight metallic components. The transformative impact of metal DfAM is shown through real-world applications encompassing aerospace, healthcare, and automotive domains. The chapter acknowledges challenges inherent in metal DfAM, such as geometric limitations, surface finish considerations, and cost implications. Finally, it emphasizes the critical role of sophisticated software tools in driving design efficiency and explores future trends in AM metallic materials, technologies, and research.

Keywords: Design for Additive Manufacturing, metal additive manufacturing, topology optimization, lattice structures, lightweighting, performance enhancement

1. Introduction

Metal AM encompasses a range of technologies that build complex three-dimensional (3D) components directly from digital models by adding material layer by layer, unlike traditional subtractive manufacturing methods. AM offers unparalleled design freedom, reduced material waste, and the potential for rapid prototyping and small-batch production. Each metal AM technology offers unique characteristics, making them suitable for different applications [1, 2]. **Table 1** presents a comparison between processes regarding materials, advantages, limitations, and typical applications; it can be a valuable reference for decision-making.

1.1 Laser Powder Bed Fusion (L-PBF)

L-PBF employs a high-powered laser to melt thin layers of metal powder selectively. It is compatible with various materials, including steel, titanium, aluminum,

Technology	Advantages	Limitations	Typical applications
Laser Powder Bed Fusion (L-PBF)	High-detail resolution can produce complex geometries without specific orientation.	Support structures for overhangs are required, and post-processing is often needed.	Aerospace, medical implants, tooling industry.
Electron Beam Melting (EBM)	Can cut very thin holes of large aspect ratio, precise and distortion-free.	High capital and maintenance cost, and requires vacuum.	Aerospace, automotive, and medical industries.
Fused Filament Fabrication (FFF)	Low-cost, easy to use, wide range of materials.	Lower detail resolution compared to PBF methods and support structures often needed.	Prototyping, educational purposes, hobbyist projects.
Direct Energy Deposition (DED)	Can repair existing components, compatible with a range of materials.	Requires post-processing, less precise than PBF.	Repair and maintenance of structural components, aerospace.
Binder Jetting (BJT)	Fast print speeds, low operating costs per component.	Post-processing is often required, and less durable components.	Full-color prototypes, large sand-casting cores and molds, low-cost metal component

Table 1.
Comparative analysis of advantages, limitations, and typical applications across different AM processes for metallic materials [1–20].

and nickel-based superalloys [21]. L-PBF is recognized for its high precision, excellent detail reproduction, and the ability to produce components with superior mechanical properties [22]. However, the process can be relatively slow and requires support structures for complex geometries in customized components due to its unparalleled design freedom and material efficiency [23]. L-PBF is widely used in aerospace, medical devices, tooling, and complex production [24].

1.2 Electron Beam Melting (EBM)

EBM operates on a principle similar to L-PBF but uses an electron beam instead of a laser as an energy source. The process occurs within a vacuum chamber, particularly suitable for reactive metals like titanium and alloys [25]. EBM offers faster build speeds than L-PBF for certain materials, excellent properties for high-temperature and reactive materials, and the ability to produce components with slightly lower residual stress. However, EBM typically has a rougher surface finish compared to L-PBF, involves high equipment costs, and requires the added complexity of a vacuum environment [26]. Despite these limitations, EBM finds applications in aerospace, medical implants, and the manufacture of large or highly reactive metal components due to its precision and distortion-free nature [27, 28].

1.3 Direct Energy Deposition (DED)

DED is a versatile AM process that melts feedstock (powder or wire) as it is deposited using a laser or electron beam [29]. DED can process various metals, composites, and graded materials [30]. Its advantages include the ability to create very large components, the option to repair or add features to existing components, and faster build speeds than PBF processes for large components [31]. However, DED often has lower accuracy and surface finish than powder bed processes and usually necessitates post-processing [32]. Despite these challenges, DED is commonly used for large-scale aerospace components, repair work, and coating applications due to its compatibility with various materials and ability to repair existing components [33].

1.4 Binder Jetting (BJT)

Binder Jetting Technology (BJT) operates by precisely depositing a liquid binder onto a metal powder bed to join the particles. After manufacturing, the created components are submitted to a debinding to remove the binder, and a sintering process is performed to achieve total density. This method is mainly used with stainless steel, tool steel, and bronze [34]. Binder jetting is a cost-effective option for larger production runs, as it removes the need for support structures and permits the production of porous structures. Nonetheless, compared to fully melted techniques, binder jetting typically results in components with lower density and weaker mechanical properties. Additionally, the finishing steps involved (polishing, surface finishing, heat treatments to improve mechanical properties) can be time-consuming and require significant effort [34, 35]. Notwithstanding these limitations, BJT is utilized in prototyping, sand-casting molds, and the production of porous medical implants due to its rapid print speeds and low operating costs per component [36].

2. Design for AM (DfAM)

DfAM has revolutionized the production of metallic components by allowing intricate and complex shapes that traditional manufacturing methods cannot achieve [4]. DfAM refers to a specialized approach in engineering and design that focuses on optimizing and tailoring designs specifically for AM processes [37–40]. DfAM is centered around harnessing the unique capabilities of AM by considering the technology's constraints, opportunities, and intricacies during the design phase. There are several reasons to optimize components: lightweight design, performance increase, efficiency improvement, and decreased cost [37, 41–43]. Designing and developing components for AM can be more accessible or tougher if the designer/engineer has the experience, the requirements of the component, and the specificity of the type of component industry [44, 45]. As an example of this, **Figure 1** shows three intertwined hollow pyramids and one ring produced with FFF.

The significance of DfAM lies in its ability to leverage AM technology's capabilities thoroughly [39, 43, 46]. It enables designers and engineers to capitalize on the inherent advantages of AM, such as the freedom to create complex shapes, reduced material waste, and the production of intricate internal structures that enhance functionality [26, 47]. DfAM allows for optimizing component designs to achieve improved performance, lighter weight, increased strength, and enhanced functionality while minimizing the need for assembly [48]. Moreover, it streamlines the production process, reducing lead times and costs and making it a pivotal approach in advancing the potential and applicability of AM across various industries. By considering design intricacies specific to AM, DfAM unlocks a new realm of innovative possibilities that traditional manufacturing methods cannot match [11, 49].

It is required to evaluate which products will benefit the most from applying optimization plus AM. If companies develop optimization and AM strategies to correctly assess traditional products and processes (fabricated with traditional

Figure 1.
Three intertwined hollow pyramids with a ring were produced with FFF.

manufacturing), if they can become more efficient and achieve—at least—the same level of performance after optimization, at the same time, costs are reduced (material, manufacturing, production process, among others), companies will gain competitive advantage regarding competitors [20, 29, 50, 51].

One of the most known advantages of optimization, and one of the biggest challenges, is designing products with lower mass and similar (or better) performance under work conditions [41, 52, 53]. The manufacturing of optimized components is also a challenge, and they can often be manufactured by something other than traditional processes (like trimming, milling, drilling, and grinding). Two methods used and where DfAM can create a shape advantage for AM are topology optimization (TO) and lattice design. Both methods reduce the weight of components [41, 42], maintaining mechanical characteristics and properties. Both these strategies are widely used and are executed through Computer-Aided Design (CAD), Computer-Aided Engineering (CAE)—including Finite Element Analysis (FEA)—and Computer-Aided Manufacturing (CAM).

DfAM starts with a simple geometric shape in CAD software, ensuring it meets basic functional requirements. Distinctive design tools then optimize the components with TO and lattice structures for weight and material reduction [41, 42]. Finally, simulation software evaluates the design's performance under stress, loads, and specific operating conditions. Different methods are used to manufacture complex or expensive shapes, and traditional manufacturing is used to redesign them. The optimization software aims to create a mathematical idealization of an actual physical system; depending on the shape, material, series volume, and other criteria, series production is economically possible using metal AM [42]. It is required to determine analysis model properties, select appropriate response quantities, and choose a suitable finite element mesh to calculate the wanted responses within acceptable

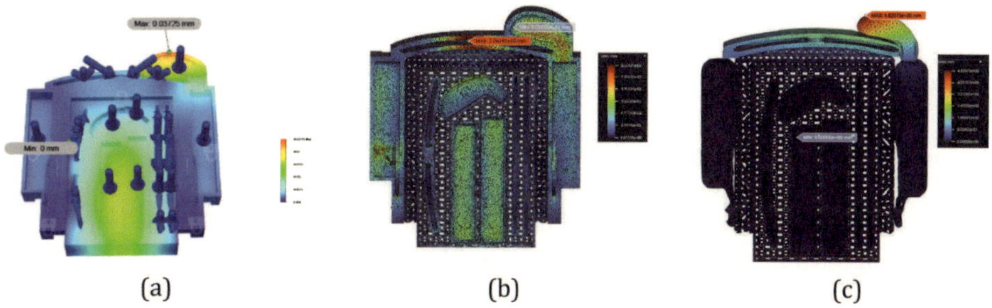

Figure 2.
Analysis of a box to store augmented reality glasses for different processing stages: (a) initial model, (b) small optimization model and (c) fully optimized model [54].

accuracy. The results achieved depend on the analysis type performed but include elements such as element stresses and modal frequencies and define materials, element thicknesses, element meshes, and other quantities needed to complete the analysis, as shown in **Figure 2** [55].

Two distinctive design methods exist. The first, process-driven shape, focuses on reducing manual interaction with a human designer to reduce design time and improve the design performance [42]. The second is the designer-driven shape, a process where a human designer controls the shape, contributing with manufacturing knowledge to avoid cost increase in fabrication. Both methods consider the AM aspect, balancing performance increase and per-component cost in series production [42, 56].

The optimization methodologies and simulation can be repeated unlimited times, based on the design methods, until a component that suits requirements is found. The only constraint is the software's time to process information and run algorithms. The previous figure shows the design strategies for redesigning AM components to improve performance and reduce costs.

3. Optimization techniques

The pursuit of optimization techniques in AM involves a multifaceted approach to maximizing the benefits of this transformative technology. This chapter investigates advanced methods such as TO, which enables the creation of structures with optimized material distribution for enhanced performance and reduced weight. Additionally, it explores the realm of designing lightweight structures to capitalize on the inherent freedom provided by AM processes. Furthermore, the study delves into generative design techniques, which leverage algorithms to iteratively generate and refine designs based on specified constraints and objectives. This research aims to unlock new avenues for innovation and efficiency in AM applications by delving into these sophisticated methodologies [57, 58].

According to Gibson et al., Design for Manufacturing and Assembly (DfMA) can be defined as the "practice of designing products to reduce, and hopefully minimize, manufacturing and assembly difficulties and costs" [59]. Theoretically, it looks easy to apply. However, it can become complex and time-consuming to apply [10], especially when the design team does not control variants well. DfMA also uses traditional manufacturing processes, which can represent significant costs for manufacturing optimized components. When AM becomes an alternative, we move from DfMA to DfAM, where the purpose is to utilize the AM fabrication capabilities best to achieve desired performance and lifecycle objectives through the combination of shapes, sizes, geometric mesostructures (such as lattice unit cells), material compositions, and microstructures [60–62].

3.1 Topology optimization (TO)

TO is a numerical approach that identifies where material should be placed in a given domain to achieve a desired functionality (e.g., stiffness) for a given set of loads and constraints while optimizing for qualities such as minimal material usage/weight or uniform stress distribution [43]. TO algorithmically determines the most structurally efficient design within a defined space and under given constraints. This process results in organic, optimized shapes that use material only where necessary for structural integrity. With AM, TO is compelling since it can create complex, load-bearing structures that are difficult or impossible to achieve through traditional manufacturing methods. These optimized designs enhance strength-to-weight ratios, reducing material usage while maintaining structural integrity, as shown in **Figure 3**, where an office stapler is shown with reduced weight while maintaining the mechanical properties and functionality [38]. TO is a process based on FEA where an algorithm is used to determine the space that provides the optimal behavior, removing material that is not required to manage the applied load to meet defined requirements and fulfill a specified optimization goal. The process requires initial inputs to determine optimization, usually boundary conditions (external loads and optimization criteria) plus component geometry [41, 42]. TO methods are commonly applied in components developed for traditional manufacturing processes, like casting and machining, which are limited—in terms of design—by the manufacturing process since it is required to compromise between an optimal form and ease of manufacture. These approaches have significant manufacturing constraints that must be considered during the design stage to ensure a feasible design [63].

TO enables mass reduction and stiffness optimization of a maximum allowed volume and a load set for a specific volume reduction goal [41]. Considering AM's ability, combining AM with TO will enable a compromise between weight and shape.

There are several software for TO, and a big part of them already take into consideration additive manufacturability. Hexagon, Siemens NX, Dassault Systèmes, and

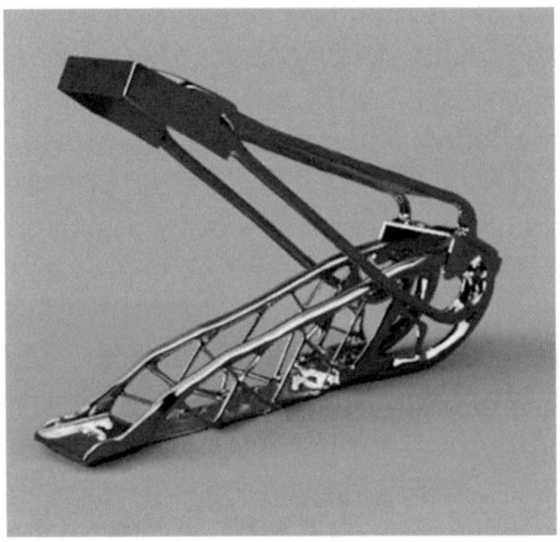

Figure 3.
Topologically optimized desk stapler [38].

Autodesk, among others, are starting to present well-developed software solutions, not only for TO but also including DfAM's perspective.

3.2 Lattice and lightweight structures

Lattice structures mimic nature by replicating the organic shapes and structures found in natural materials like wood, bone, sponge, and coral. It is made of a pre-defined external geometry and internal architecture, similar to cellular materials structures, referred to as lattice structures due to being inherently non-stochastic [64]. Lattice structures, likewise the TO, are an excellent way of producing light-weight, robust performance, reducing fabrication time and overall costs, which are important in industries such as aerospace and transportation. Lattice structures are intricate, mesh-like designs that feature a network of beams or struts [65]. Some examples are presented in **Figure 4**.

These structures provide strength and support while significantly reducing weight. A lattice is a cellular structure of repeated unit cells to form a larger vol-ume, and it is like trying to mimic nature by creating organic shapes and forms like

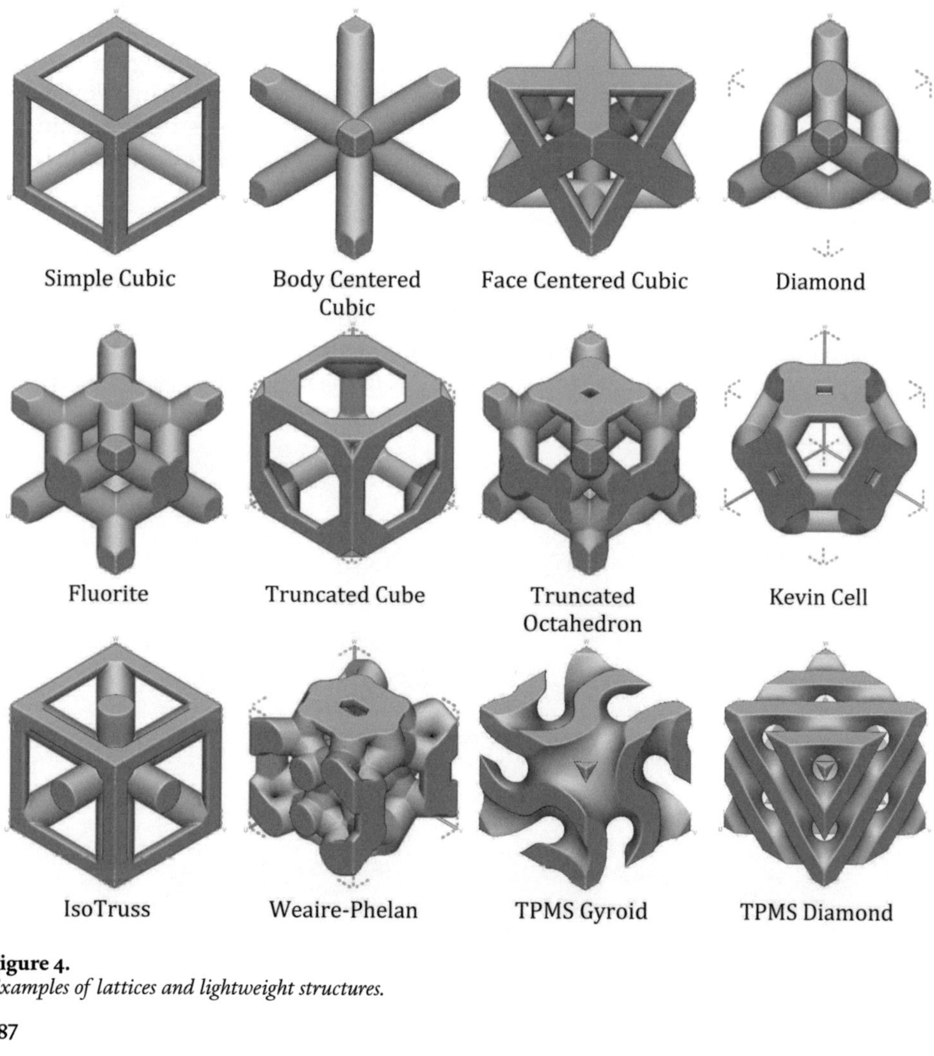

Figure 4.
Examples of lattices and lightweight structures.

wood, bone, sponge, and coral. It can be used as lightweight support structures, to increase robustness against overloading, decrease mass (less material used), and reduce production time (smaller costs) [26, 66]. There are many options for the shape and size of such lattice cells and for the pattern in which they are repeated in a conforming pattern (due to being inherently non-stochastic). Countless examples of lattice structures are used to reduce the material to improve its strength-weight ratio or replace support material in components. AM allows for the creation of these complex lattice structures easily. These lightweight structures are advantageous in aerospace, automotive, and biomedical applications where weight reduction without compromising strength is critical [67, 68]. **Figure 5** shows a gear bearing with lattice structures, which is used for the weight reduction of components, fully optimized for production through AM.

Lattice design is a designer-guided approach involving intersecting or cutting components of the design space with a consistent pattern of smaller strut-like structures and used as lightweight support structures [41], and they decrease mass and fabrication time [42]. Lattice designs can be used within components, increase robustness against overloading of the final components, and enable the utilization of less material than a solid representation of the same design space and may thus cost less to fabricate, which makes it a possible way of improving manufacturability and to lower series components cost [68]. The design-driven shape method uses lattice structures to reduce mass and decrease AM fabrication time, maintaining the performance of the components by creating self-supporting structures in the interior of the components [42]. The placement and quantities of these—lattice—structures and where they are built and disposed of are done by specific software, with appropriate algorithms running, confirmed in simulation software.

Implementing the lattice structure of AM design strategies can be beneficial when considering both the mechanical performance and manufacturing aspects [64, 69]. According to Panesar et al. [64], lattice structures are utilized in AM for various advantages, such as facilitating the production of components with intermediate densities, reducing component distortions by minimizing residual stresses through inherent porosity, and needing fewer supports due to the inclusion of self-supporting

Figure 5.
Gear bearing optimized with lattice structures.

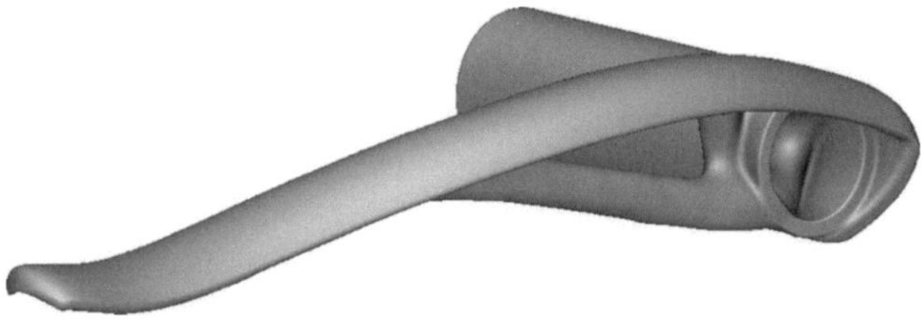

Figure 6.
Generative design of a door handle [40].

unit cells. This approach also enhances design robustness. As mentioned in the previous paragraph, the lattice structure creation is built using software where CAD files are structurally analyzed to introduce selected lattice patterns and analyzed to support stresses when an external load is applied. Again, FEA is crucial in evaluating whether the designed components support loads and stresses under work conditions [70, 71].

3.3 Generative design

Generative design utilizes advanced algorithms to explore numerous design permutations based on specified parameters and constraints, creating innovative designs by allowing optimized, efficient, creative designs that capitalize on AM's capabilities to develop superior, economically viable products [72]. This iterative approach tests and refines possibilities to optimize the final design for its intended application. Producing complex and organic shapes that traditional methods might overlook is precious in AM. This capability leverages AM's geometric freedom to encourage the development of novel, efficient designs uniquely suited to its processes. The door handle in **Figure 6** is a good example of this [40].

Generative design can refine designs in a multi-material context to meet performance criteria while minimizing material use [57]. By highlighting its role in optimizing multi-scale structures, which is crucial for applications needing varying material properties within a single component. This iterative testing ensures that designs are innovative, practical, and manufacturable [58]. Generative design in AM supports creating structures that enhance performance while reducing material waste. It enables intricate lattice structures, conformal cooling channels, and complex geometries, contributing to significant cost savings and sustainability through reduced material use and post-processing [45, 73].

4. Fundamentals of DfAM

The fundament of DfAM lies in understanding and implementing its basic principles and guidelines. This thoroughly explores various facets, including material choices and selecting appropriate AM processes [45]. As mentioned above, designing products optimized for AM entails a deeper discussion of fundamental principles, encompassing design considerations, exploitation of geometric freedom, and the reasonable selection of materials. By delving into these foundational aspects, practitioners gain insights into enhancing the efficiency and effectiveness of AM processes, leading to advancements in product design and manufacturing methodologies [54, 74]. DfAM is challenging since he must change the paradigm in which he learned and worked with familiar, safe, and proven design theories [75]. To get a new and innovative perspective and improve product performance, four areas should be considered: integrated design, individualization, lightweight design, and efficient design [69, 75, 76]. The first step identifies components and assemblies where AM offers a benefit, which—after deciding on a suitable design strategy—develops a design concept and proceeds to the design of a detailed component [75, 77]. Mass reduction assumes relevance when deciding to design for AM. Items like reduction of trimming operations, feedstock saving, complexity, and economic savings, among others, favor AM and DfAM, compared to traditional manufacturing processes [41, 42, 69]. It is not a matter of AM being able to produce components since any components—independent of design shape—are suitable to be designed and fabricated through AM; the limitations are physical (build volume enough to produce components), processability, and efficiency (time, cost, finishing), as well as post-processing operations required, functionality.

One of the most significant advantages of metal AM is that, unlike traditional methods, it allows for complex designs without driving up production costs. Traditional processes often cannot handle complex designs, and the costs are much higher when they can produce them. AM technology brought a wide range of new possibilities by allowing the production of complex components at a minor manufacturing cost [40, 63]. The production cost does not vary with the complexity since it will only increase costs marginally; however, the current fixed costs are higher when compared to the traditional process. Thus, as soon as the break-even point is reached, LPBBF will enable total cost and cycle time reduction. AM enables quick single-component manufacturing, and compared to traditional manufacturing, it is more cost-effective and flexible. Small batch sizes, highly complex forms, and designs with integrated cooling or tempering channels are not problematic for metal AM and L-PBF but are critical in traditional processes [78]. With the end of some Original Equipment Manufacturing (OEM) patent protection, plus the attention that the market is starting to show, shortly, technology will become more effective and independent of additional processes to get the same finishing as traditional manufacturing processes [77, 79].

4.1 Design considerations for complex geometry utilization

As mentioned, DfAM enables the creation of intricate and complex shapes that traditional manufacturing methods cannot achieve. Designers should leverage this capability by fully utilizing AM's layer-by-layer construction, which may involve consolidating components, creating internal structures, or optimizing designs for lightweight [38, 80]. **Table 2** provides a comparative analysis of feature types across different AM processes to enable the creation of designs fully utilizing AM's layer-by-layer construction, which involves consolidating components, creating internal structures, and optimizing lightweight, among others. DfAM allows designers and engineers to incorporate features that enhance performance and efficiency.

4.2 Design considerations for performance

The design for improved performance depends on the product. For that, the designer needs to know and understand the customer's perceived quality, performance, utilization rate, and the value that the customer gives to that component/product. The strategy depends on what is valued: performance, cost, and

Feature type	L-PBF	EBM	FFF	DED	BJT
Orientation	Allows construction of complex geometries without specific orientation.	Similar to L-PBF, it can build complex geometries without particular orientation.	Orientation affects surface finish.	Allows for multi-axis deposition, meaning the orientation can be varied.	Orientation affects surface finish.
Support angle	Supports needed for overhangs >45°.	Supports needed for overhangs >45°.	Supports needed for overhangs >45°.	Supports are not typically required due to high deposition rates.	Supports unnecessary as the unbound powder acts as a support.
Wall thickness	It can produce thin walls down to 0.2 mm.	It can produce thin walls down to 0.2 mm.	The minimum wall thickness depends on the nozzle diameter.	Wall thickness is dependent on nozzle diameter.	It can produce thin walls, depending on the binder and powder properties.
Details	Offers high-detail resolution.	Offers high-detail resolution.	Lower detail resolution compared to powder bed fusion methods.	Lower detail resolution, dependent on nozzle diameter.	Offers moderate detail resolution.
Holes and tubes	It can produce small holes down to 0.5 mm.	It can produce small holes down to 0.5 mm.	The ability to produce holes/tubes depends on nozzle diameter and layer height.	Dependent on nozzle diameter.	It can produce small holes, depending on the binder and powder properties.
Machining stock	Not typically required.	Not typically required.	It may be necessary for high-precision components.	It may be needed for high-precision components.	Not typically required.
Clearance	It can produce small clearances, down to 0.2 mm.	It can produce small clearances, down to 0.2 mm.	Clearance is dependent on nozzle diameter and layer height.	Dependent on nozzle diameter.	It can produce small clearances, depending on the binder and powder properties.
Hollowing	It can produce hollow structures.	It can produce hollow structures.	It can produce hollow structures.	It can produce hollow structures.	It can produce hollow structures.
Screw threads	It can produce fine threads down to M2.	It can produce fine threads down to M2.	Ability to produce screw threads dependent on nozzle diameter and layer height.	Dependent on nozzle diameter.	It can produce threads, depending on the binder and powder properties.

Feature type	L-PBF	EBM	FFF	DED	BJT
Surface finish	It can achieve smooth surfaces but may require post-processing.	It can achieve smooth surfaces but may require post-processing.	Surface roughness can be high and depends on layer height.	Surface roughness can be high and depends on the deposition rate.	Surface roughness can be high and depends on powder size.
Infill	It has a solid infill but can be used to design internal lattice structures.	It has a solid infill but can be used to design internal lattice structures.	It can vary in infill density and pattern.	It has a solid infill but can be used to design internal lattice structures.	It has a solid infill but can be used to design internal lattice structures.
Overhangs	Supports needed for overhangs >45°.	Supports needed for overhangs >45°.	Supports needed for overhangs >45°.	Supports are not typically required due to high deposition rates.	Supports unnecessary as the unbound powder acts as a support.
Bridging	It can bridge small gaps, depending on material and process parameters.	It can bridge small gaps, depending on material and process parameters.	It can bridge gaps, but it is dependent on material and cooling.	It can bridge gaps, but it is dependent on the deposition rate.	It can bridge small gaps but depends on binder and powder properties.

Table 2.
Comparative analysis of feature types across L-PBF, EBM, FFF, DED, and BJT [1, 2, 18, 38, 40, 54, 59–61, 80–85].

functionality [41, 42, 76, 86]. Optimization methodologies are commonly used as design tools to improve performance and fulfill the functional requirements of components. Several aspects need to be analyzed, which might vary with the application of components.

Several areas can benefit from a design and/or redesign strategy for improved performance and the use of additive manufacturability for component fabrication [42]. Areas like health, components, molding, and tooling can benefit from DfAM since all components have margins that need to be improved. In addition to cost reduction, which has been mentioned several times, there is the possibility of creating components that are 100% movable, eliminating assembly steps and production operations [41, 87].

For example, in health-related applications (medical implants and prosthetics, for instance, like the examples in the previous picture), a combination of the material's biocompatibility and the body is required to get shapes that can accommodate it. The first step is to evaluate the patient's surrounding tissue through Computer Tomography (CT) scans, data that will be used for reverse engineering (RE), enabling time-saving in development and modeling [41].

Comparing this AM strategy to traditional manufacturing, where components with standard dimensions usually exist, patients will benefit from having a 100% dedicated component. In traditional manufacturing, a 100% adapted component is expensive since that component includes development and manufacturing costs; AM enables a single production for one component only, enabling high-cost savings.

4.3 Exploiting geometric freedom

Utilizing the geometric freedom of AM technologies is a fundamental aspect of DfAM. This freedom enables the creation of intricate and optimized structures previously unfeasible with traditional manufacturing methods. TO software allows the generation of designs based on specific performance criteria such as stress distribution, weight reduction, and material efficiency. This approach helps create efficient and optimized structures for their intended function. TO can lead to innovative designs that maximize performance while minimizing material usage, as the suggestion of robotics arm gripper clamps shown in **Figure 7(a)** and **(b)** [81].

AM's ability to create shapes impossible with traditional manufacturing, such as intricate lattice structures, internal cooling channels, conformal cooling features, and complex surface textures, enhances the final product's functionality, esthetics, and performance [88, 89]. These features significantly improve thermal management, structural integrity, and overall product performance without adding unnecessary weight. By designing components to consolidate multiple components into a single structure, the need for assembly is reduced, decreasing the number of potential failure points and simplifying the supply chain. Component consolidation can also lead to significant cost savings and improved reliability [44].

Moreover, by designing self-supporting geometries, reducing or eliminating the need for support structures during printing is possible. This reduces material waste and minimizes post-processing efforts, leading to faster production times and lower costs [90]. Exploiting the geometric freedom of AM technologies allows designers to create highly optimized and functional products. By leveraging TO, complex geometries, components consolidation, and minimizing support structures, DfAM can significantly advance product design and manufacturing efficiency [81].

4.4 Material selection

Selecting appropriate metals is a crucial aspect of DfAM. The chosen metals must be compatible with the specific AM process, as different AM technologies have unique material requirements. Understanding these metals' properties and suitability for the intended application is critical to ensuring successful outcomes [45].

For instance, L-PBF typically uses metals like titanium, aluminum, and stainless steel, which are prized for their strength-to-weight ratios and are commonly used in

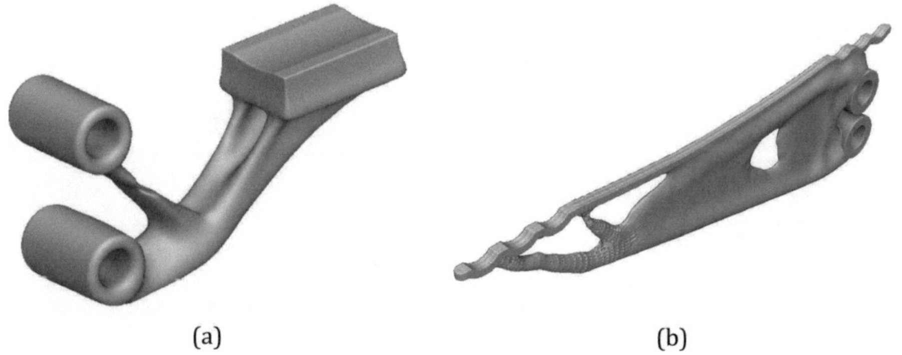

(a) (b)

Figure 7.
Robotics arm gripper optimized [77, 80].

aerospace and automotive industries. Similarly, EBM is effective with materials such as titanium alloys and cobalt-chromium, which are known for their high strength and biocompatibility, making them suitable for medical implants and aerospace components [4]. Custom metal development is another critical consideration. By developing or customizing metal alloys, designers can enhance mechanical properties or achieve characteristics unique to AM, such as improved thermal stability, increased toughness, or enhanced corrosion resistance. This is particularly valuable in high-performance applications where standard metals may only meet some design requirements [23]. Focusing on creating lightweight structures without compromising strength and durability is essential. Techniques such as lattice structures and hollowing can significantly reduce material usage while maintaining the necessary mechanical properties. These strategies are crucial in industries like aerospace and automotive, where weight reduction is critical for performance and fuel efficiency [91].

By carefully selecting and possibly developing metal alloys tailored to the specific needs of the design and AM process, it is possible to achieve optimal performance, cost-efficiency, and innovative solutions that leverage the full potential of AM.

4.5 Process selection and manufacturability

Each additive manufacturing (AM) technology has its own strengths and limitations, so choosing the appropriate AM process that aligns with your design requirements is crucial, considering factors such as resolution, material compatibility, and build volume. Additionally, design with post-processing requirements in mind, planning for necessary finishing processes like heat treatment, support removal, surface smoothing, or machining to achieve the final desired properties [92].

When functional requirements are selected and the manufacturing process is chosen, it is required to confirm if the design needs to be modified to improve manufacturability. For AM, several software analyses are made, with specific software that enables the simulation of manufacturing strategy, which will allow cost reduction and quality increase and is manufacturing method dependent [41, 42, 93]. The production method and orientation of how the component is built in the AM equipment chamber influence the result of the component. An "optimal" build direction will affect how building supports are created during construction and who must be removed during post-processing, as shown in **Figure 8(a)**. Some components will be expensive or scraped if supports are in difficult or inaccessible places, as shown in **Figure 8(b)** [94].

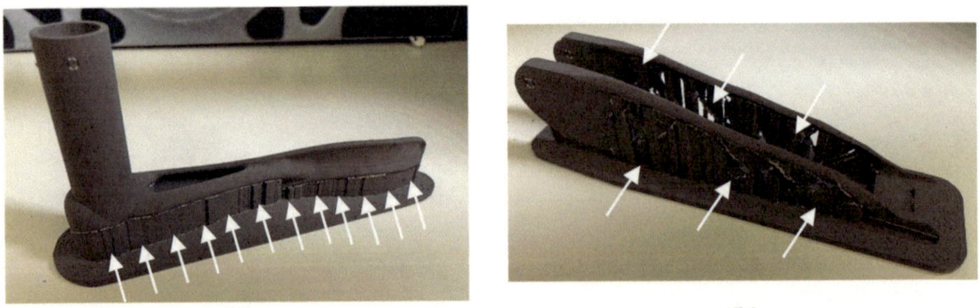

(a) (b)

Figure 8.
Building orientation and supports required for components topologically optimized and produced through the FFF process (a) door handle [40] and (b) office stapler [38].

It is required to check different parameters, even component orientation in the building tray, to confirm that the component will not have overhangs, materials displacements, and non-conformities between what was designed and the outcome from manufacturing. Sometimes, geometry, size, and shape in specific areas are even required to improve the manufacturability. It is expected to simultaneously iterate design for function, performance, and manufacturability to improve outcomes [41, 42, 76]. When considering manufacturability, it is also required to define the post-processing operation, like trimming, blasting, heat treatment, and polishing, among others, which will increase the cost and lead time of the component; thus, this operation will increase the finishing of component, as also—many times—reduce the effective cost since it will minimize operation time in AM equipment [95].

Manufacturability also requires knowledge about the materials (availability and metallurgical properties) and equipment (different technologies, with other suppliers) since both will directly influence components' dimensional accuracy, surface roughness, and minimal building wall thickness and size [96]. It can now be realized that components that were developed through proper methods of DfAM, with a structural optimization technique that improves the material layout within a specified design space for a defined set of loads and boundary conditions, have high specific stiffness and strength through reducing (or eliminating) material wastage, which happens mainly due to AM's synergy with TO [64].

5. Benefits

As seen, DfAM is revolutionizing product development by offering unparalleled design freedom, customization capabilities, and efficiency gains. By harnessing the potential of DfAM, practitioners can unlock new avenues for innovation, streamline production processes, and achieve superior product performance. The main benefits are:

1. Design flexibility: DfAM is a cornerstone of design flexibility, enabling the creation of intricate, complex geometries unachievable by traditional manufacturing methods. This includes fabricating organic shapes, internal structures, and consolidated assemblies, empowering designers to explore unprecedented design possibilities [17, 56].

2. Lightweighting and material efficiency: DfAM allows for topology-optimized designs where the material is placed only where necessary, reducing weight without compromising strength [97, 98]. Using lattice structures and lightweight designs significantly reduces material usage, contributing to sustainability goals and cost-effectiveness [24, 99, 100].

3. Customization and personalization: AM enhances customization and personalization, allowing the production of custom, individualized components tailored to specific needs [2, 101]. This capability is particularly transformative in sectors like healthcare, where personalized medical implants can be manufactured based on a patient's unique anatomy [26, 92].

4. Rapid prototyping and iteration: DfAM accelerates rapid prototyping and iteration processes, enabling quick turnaround times for design iterations and

producing prototypes or small batches without the need for expensive tooling or setup changes [66, 86].

5. Simplification of assemblies: DfAM facilitates the integration of multiple components into a single piece, reducing the need for assembly, potentially improving product reliability, and reducing maintenance requirements. This approach is particularly advantageous in industries characterized by complex assemblies, such as aerospace and automotive [10, 102].

6. Waste minimization: DfAM minimizes material waste by using only the necessary amount of material required for the design, contrasting with subtractive manufacturing methods that often produce significant waste. This efficiency contributes to environmental sustainability by lowering energy consumption and minimizing environmental impact [37, 47].

7. Agile and on-demand manufacturing: DfAM supports agile and on-demand manufacturing, enabling production closer to the point of use. This capability is advantageous for producing spare components, reducing inventory costs, and promoting more agile supply chains [103–105].

8. Innovation and design optimization: Innovation and design optimization are intrinsic to DfAM, encouraging designers to push the boundaries of traditional design constraints and create new, improved, and more efficient products [39, 106].

9. Sustainability: DfAM's reduced waste, lighter designs, and on-demand production lower energy consumption and minimize environmental impact compared to traditional manufacturing processes [74, 107].

While DfAM offers numerous advantages, challenges such as certification, material properties, post-processing requirements, and cost considerations should be carefully evaluated during implementation to maximize its benefits.

6. Challenges and limitations

Practitioners can overcome the inherent limitations of DfAM by addressing issues such as design complexity, material selection constraints, post-processing requirements, cost considerations, and compliance with industry standards. Overcoming these challenges is crucial for unlocking the full potential of AM, enabling the creation of optimized, functional, and efficient designs that meet the evolving needs of diverse industries. The main challenges and limitations are:

1. Geometric constraints: One of the primary challenges in DfAM involves managing geometric constraints, particularly with overhangs and support structures. AM processes often require support for overhanging features, impacting surface finish, and necessitating additional post-processing efforts. Designers must strategically plan for self-supporting geometries to minimize or eliminate the need for support, optimizing production efficiency and final component quality [102, 106].

2. Surface finish: Achieving smooth, high-quality surface finishes remains a significant challenge in AM due to its layer-by-layer nature. Visible layer lines and resolution limitations can affect surface esthetics and functional performance. Effective post-processing techniques are essential to mitigate these issues. However, they add complexity to the manufacturing process and require careful integration into the initial design phase to ensure the desired outcomes [15, 108, 109].

3. Post-processing requirements: Post-processing in AM involves support removal, surface finishing, and heat treatment to achieve desired mechanical properties and surface finish. Integrating these requirements into the design phase is critical for streamlining production workflows and minimizing manual labor and equipment usage costs [15, 106].

4. Material constraints: The availability of materials suitable for AM processes varies, and not all materials are compatible with every AM technology. Designers may encounter limitations in material options or may need to compromise on material properties to align with the capabilities of a specific AM process. Balancing material selection with design requirements and performance expectations is essential for optimizing components' functionality and ensuring long-term durability and market availability [59].

5. Cost implications: While AM offers unique advantages, including design flexibility and rapid prototyping, its implementation can be cost-prohibitive compared to traditional manufacturing methods. Factors contributing to higher costs include equipment investment, material expenses, and skilled labor requirements. Managing these cost implications effectively is crucial for determining the feasibility and economic viability of adopting AM for specific applications [26, 110].

6. Complexity and design validation: Overly complex designs enabled by AM technologies often require rigorous validation processes to ensure structural integrity and performance reliability. Simulation tools and physical testing are essential for verifying design feasibility and identifying potential failure points. However, these validation processes add complexity and time to the design iteration cycle, requiring careful management of resources and expertise [11, 52, 102].

7. Regulatory and standards compliance: In aerospace, automotive, and healthcare industries, meeting stringent regulatory requirements and industry standards is imperative for certifying AM-produced components. Ensuring compliance with quality assurance protocols and safety standards poses significant challenges, influencing design considerations and manufacturing practices to achieve regulatory approval and market acceptance [49, 111].

8. Knowledge and expertise: Effectively utilizing AM technologies demands specialized knowledge and expertise across various disciplines, including materials science, engineering design, and manufacturing processes. The need for more qualified personnel and educational resources in DfAM presents a barrier to widespread adoption and implementation. Investing in training and development initiatives is essential for building a skilled workforce capable of leveraging AM's full potential in industrial applications [13, 48].

The challenges and limitations of DfAM require a systematic approach that integrates technical expertise, strategic planning, and continuous innovation. By addressing these obstacles proactively, stakeholders can capitalize on the transformative capabilities of AM, driving advancements in product design, manufacturing efficiency, and overall industry competitiveness.

7. DfAM in industrial applications

DfAM in industry offers a comprehensive exploration of the practical implementations and transformative potential of DfAM across diverse sectors. This section delves into detailed case studies and real-world applications, highlighting how DfAM principles have effectively addressed industry-specific challenges and achieved tangible outcomes. By examining these examples, valuable insights and lessons learned are elucidated, providing a deeper understanding of the impact and versatility of AM technologies in industrial settings. These case studies highlight the innovative solutions enabled by DfAM and inspire further advancements in integrating AM into mainstream industrial practices:

1. Automotive sector: DfAM has seen widespread adoption in the automotive industry, leveraging techniques such as L-PBF and EBM to streamline production processes. AM technologies have proven crucial by reducing component count, weight, and material waste while enhancing or maintaining performance standards. Key insights underscore the need for tailored metallic materials, optimized designs, and refined post-processing methods to exploit AM's benefits in mass production fully [92, 112, 113].

2. Medical sector: DfAM has revolutionized the production of patient-specific implants like cranial and orthopedic devices in healthcare. Metal AM methods such as EBM and DED enable precise customization to individual anatomies, significantly improving fit and patient outcomes. Critical considerations include precision engineering, biocompatible material selection, and strict adherence to regulatory standards, ensuring the success of personalized medical applications [6, 104, 114, 115].

3. Aeronautical industry: DfAM has made significant strides in aerospace by transforming the design and manufacturing of complex components such as fuel nozzles. Utilizing metal AM technologies like L-PBF, manufacturers consolidate multiple components into durable single-piece designs, optimizing internal geometries for enhanced efficiency. This highlights the pivotal role of design optimization in AM, reducing component counts and maximizing the creative freedom inherent in AM [98, 101, 112].

4. Aerospace industry: In space exploration, DfAM has enabled the production of lightweight, intricate structures that are impossible to achieve with traditional manufacturing methods. AM facilitates the creation of components with complex geometry and internal features, reducing overall weight without compromising strength or functionality, thereby advancing space missions' efficiency and reliability [48, 52, 53, 102].

5. Tooling/mold industry: DfAM has revolutionized tooling and mold production by enabling the creation of highly intricate, high-performance tools with shortened lead times. Techniques like FFF and BJT in metal AM technologies emphasize designing for functional performance, meticulous material selection, and effective post-processing strategies. This evolution supports the development of tools with advanced features, such as conformal cooling channels, enhancing molding processes' efficiency and product quality [39, 59, 108, 116].

6. Energy sector: In the energy industry, DfAM plays a pivotal role in optimizing the performance of components for renewable energy systems like wind turbines and solar panels. By fabricating complex geometries that improve efficiency and durability, AM technologies contribute to advancing sustainable energy solutions [50, 92, 112].

8. Materials innovation in DfAM

Materials play a crucial role in metal additive manufacturing (AM), where this technology revolutionizes material development and application by enabling the creation of components with exceptional properties such as high strength, heat resistance, and corrosion resistance. Innovations in various alloys cater to diverse industrial needs, redefining material innovation in metal manufacturing. Titanium alloys, such as Ti-6Al-4V, offer outstanding strength-to-weight ratios and are commonly used in aerospace and medical applications. Aluminum alloys like AlSi10Mg provide excellent thermal properties and are favored in the automotive and aerospace industries. Nickel-based superalloys, including Inconel 718, are known for their high-temperature performance and are essential in the aerospace and energy sectors. Specialized steels, such as maraging steel, provide high strength and toughness, making them suitable for tooling and high-performance engineering applications [49, 117, 118]. This evolution empowers practitioners to pioneer new material science frontiers using AM technologies' diverse capabilities, developing advanced alloys and composites that significantly enhance performance, functionality, and sustainability across various industries [119, 120].

Hybrid metal composites integrate different metals or blend with non-metallic materials, offering enhanced properties such as improved strength-to-weight ratios and specific thermal or electrical characteristics [121, 122]. These advancements expand applications from aerospace to electronics, underscoring AM's transformative impact in material science by enabling innovative components to meet stringent performance requirements across diverse industries.

Advancements in powder metallurgy techniques optimize metal powders' particle size distribution, flowability, and oxygen content, enhancing material performance and suitability for AM processes like L-PBF, EBM, DED, FFF, and BJT. These refinements are critical in achieving high-quality, reliable components through AM, supporting versatile applications across industries [118, 123–127].

AM's quick prototyping accelerates material development cycles, enabling rapid iteration and facilitating faster testing and validation of new alloys and composites. AM's design freedom pushes material composition and structure boundaries, fostering innovation to develop advanced materials optimized for specific processes and applications. Furthermore, AM facilitates multi-material printing, integrating multiple metals or alloys within a single print. Graded metal structures allow for tailored material properties through gradual transitions between different metals, while functional integration incorporates dissimilar metals to achieve complex geometries and multifunctional capabilities [122, 128–130].

AM materials exhibit specific properties crucial for achieving desired outcomes throughout the design process. Variations in layer adhesion and porosity impact printed components' strength and reliability. Understanding unique mechanical characteristics such as tensile strength, elasticity, and fatigue resistance ensures meeting performance requirements [92, 131, 132]. Critical thermal properties like heat resistance, thermal expansion, and conductivity are essential for high-temperature applications. Specific post-processing methods are necessary to achieve the desired surface finish and texture, impacting the final product's visual esthetics and functional integrity [67, 71, 133, 134]. These considerations underscore the complexity and importance of material selection and design optimization in AM applications.

9. DfAM software and tools

DfAM is a rapidly evolving field driven by advancements in software solutions tailored to AM workflows. These tools represent a significant progression in enabling designers to create more optimized and efficient designs suited for AM. Integrating advanced CAD software, precise simulation tools, and powerful TO techniques accelerates the evolution of DfAM [135, 136]. This integration facilitates the development of complex, high-performance, and sustainable products across various industries [52, 137].

The landscape of CAD software for DfAM has seen significant advancements. There is a noticeable shift toward solutions that cater specifically to the intricacies of AM. Software like Autodesk Fusion 360, SolidWorks, and Siemens NX offer specific modules for AM, including features like automated lattice generation and topology optimization. Specialized CAD tools facilitate more intuitive design processes that align with AM technologies' unique requirements and capabilities, integrating features with complex geometries and support from multiple materials. Specialized CAD tools facilitate more intuitive design processes that align with AM technologies' unique requirements and capabilities [39, 45].

Simulation tools have also advanced considerably, offering detailed simulations of the printing process. These tools allow designers to anticipate potential issues, optimize designs, and ensure the manufacturability of complex structures. By simulating the layer-by-layer building process, designers can identify and mitigate problems such as warping, residual stress, and support structure placement, leading to more reliable and high-quality AM components. For instance, software like ANSYS Additive Suite and Hexagon Simufact Additive provide comprehensive simulation environments that include thermal, mechanical, and microstructural analyses [59, 74, 138].

TO techniques have gained prominence in the DfAM toolkit, empowering designers to create intricate, lightweight, and highly functional designs. These techniques algorithmically determine the optimal distribution of material within a given design space. TO leverages the design freedom provided by AM to reduce material usage while maintaining or enhancing the structural integrity and performance of the component. This results in innovative designs that would be challenging or impossible to produce using traditional manufacturing methods [98, 139, 140].

DOI: http://dx.doi.org/10.5772/intechopen.1007309

By exploring advancements in CAD software, simulation tools, and TO techniques, we gain insights into the technological innovations driving the evolution of DfAM practices. Through critical analysis and evaluation of the capabilities and functionalities of these software tools, we can highlight their role in facilitating the design, optimization, and simulation of AM processes. This exploration underscores the importance of these tools in the ongoing evolution of DfAM and their potential to shape the future of manufacturing [141, 142].

10. Conclusions, future trends, and directions

DfAM empowers designers and engineers to harness the transformative potential of AM technology. Its ability to optimize designs for AM's inherent capabilities, such as geometric freedom, material efficiency, and functional integration, challenges the limitations of traditional manufacturing practices. DfAM drives innovation across diverse industries, enabling performance gains and novel product applications.

Future advancements in AM materials, software tools, design methodologies, and process control will further accelerate DfAM adoption. It is poised to become a cornerstone of modern, high-performance manufacturing, fundamentally reshaping how products are designed, produced, and utilized. This comprehensive approach to design and manufacturing represents the future of production, promising a world of products limited only by the bounds of our imagination.

Forecasts suggest a broader adoption of AM techniques across various industries, including aerospace, automotive, healthcare, and consumer goods, as companies recognize the potential for cost-effective, customized production. The trend toward on-demand manufacturing, reducing inventory, and enabling more localized and agile production will continue as AM capabilities improve.

Continued development of new materials tailored for AM, such as high-performance polymers, metal alloys, and composite materials, will expand the range of applications and improve material properties. Further advancements in multi-material and multi-color printing technologies will enable more complex and functional components, allowing graded structures and diverse functionalities within a single print.

Post-processing techniques and technologies will improve surface finish and structural integrity and integrate secondary processes within the AM workflow. Ongoing research will concentrate on developing sophisticated design software tailored explicitly for AM, focusing on optimizing and automating design processes to leverage the full potential of AM.

Integrating machine learning and AI in DfAM will assist in design optimization, material selection, and process control, further enhancing efficiency and innovation. Research will continue to advance in bioprinting for tissue engineering and regenerative medicine, aiming for complex, functional biological structures for medical use.

Research will also focus on sustainable material development, recycling techniques, and reducing waste, aligning with the principles of the circular economy. These future trends and directions highlight the potential of DfAM in reshaping the future of manufacturing, promising a world of products limited only by the bounds of our imagination.

Acknowledgements

This work is a result of Agenda "Hi-Rev – Recuperação do Setor de Componentes Automóveis", nr. C644864375-00000002, investment project nr. 64, financed by the Recovery and Resilience Plan (PRR) and by the European Union - NextGeneration EU.

Conflict of interest

The authors declare no conflicts of interest.

Thanks

The authors express their gratitude to the former students and alums of the University of Porto's Faculty of Engineering Master's in Materials Engineering: Beatriz S. Monteiro, Carolina S. Oliveira, Francisca A. Rocha, E. Sofia Alves, Francisca L. Nunes, Margarida P. Mata, Mariana C. Maia, Mariana M. Trindade, Mariana S. Cunha, and Mateus F. Pinto.

Abbreviations

AI	artificial intelligence
AM	additive manufacturing
BJT	binder jetting
CAD	Computer-Aided Design
CAE	Computer-Aided Engineering
CAM	Computer-Aided Manufacturing
CT	Computer Tomography
DED	Direct Energy Deposition
DfAM	Design for Additive Manufacturing
EBM	Electron Beam Melting
FEA	Finite Element Analysis
FFF	Fused Filament Fabrication
L-PBF	Laser Powder Bed Fusion
OEM	Original Equipment Manufacturing
RE	reverse engineering
TO	topology optimization

Author details

Jose M. Costa[1,2]*, Elsa W. Sequeiros[1,2], Daniel Figueiredo[3], Ana R. Reis[1,2] and Manuel F. Vieira[1,2]

1 Faculty of Engineering, University of Porto, Porto, Portugal

2 LAETA, Institute of Science and Innovation in Mechanical and Industrial Engineering, Porto, Portugal

3 Palbit, Branca, Portugal

*Address all correspondence to: jose.costa@fe.up.pt

References

[1] Costa J, Sequeiros E, Vieira MT, Vieira M. Additive manufacturing: Material extrusion of metallic parts. U Porto Journal of Engineering. 2021;7:53-69. DOI: 10.24840/2183-6493_007.003_0005

[2] Costa JM, Sequeiros EW, Vieira MF. Fused filament fabrication for metallic materials: A brief review. Materials. 2023;**16**:7505. DOI: 10.3390/ma16247505

[3] Costa JM, Sequeiros EW, Santos RF, Vieira MF. Benchmarking L-PBF systems for die production: Powder, dimensional, surface, microstructural and mechanical characterisation. Metals-Basel. 2024;**14**:520. DOI: 10.3390/met14050520

[4] Chowdhury S, Yadaiah N, Prakash C, Ramakrishna S, Dixit S, Gupta LR, et al. Laser powder bed fusion: A state-of-the-art review of the technology, materials, properties & defects, and numerical modelling. Journal of Materials Research and Technology. 2022;**20**:2109-2172. DOI: 10.1016/j.jmrt.2022.07.121

[5] Mostafaei A, Ghiaasiaan R, Ho IT, Strayer S, Chang K-C, Shamsaei N, et al. Additive manufacturing of nickel-based superalloys: A state-of-the-art review on process-structure-defect-property relationship. Progress in Materials Science. 2023;**101108**:14-99. DOI: 10.1016/j.pmatsci.2023.101108

[6] Gebhardt A, Hötter JS. Additive Manufacturing: 3D Printing for Prototyping and Manufacturing. Hanser Publications; 2016. pp. 33-70, 101-193. DOI: 10.3139/9781569907030.007

[7] Acquesta A, Monetta T. As-built EBM and DMLS Ti-6Al-4V parts: Topography–corrosion resistance relationship in a simulated body fluid. Metals-Basel. 2020;**10**:1015. DOI: 10.3390/met10081015

[8] Ataee A, Li Y, Fraser D, Song G, Wen C. Anisotropic Ti-6Al-4V gyroid scaffolds manufactured by electron beam melting (EBM) for bone implant applications. Mater Design. 2018;**137**:345-354. DOI: 10.1016/j.matdes.2017.10.040

[9] Diegel O, Nordin A, Motte D. Additive Manufacturing Technologies. Springer Singapore; 2019. pp. 19-39, 41-77. DOI: 10.1007/978-981-13-8281-9_2

[10] Pei E, Monzón M, Bernard A. Additive Manufacturing-Developments in Training and Education. Springer; 2019. pp. 53-83. DOI: 10.1007/978-3-319-76084-1

[11] Pei E. Springer Handbook of Additive Manufacturing. Springer; 2023. pp. 31-71. DOI: 10.1007/978-3-031-20752-5

[12] Sam M, Jojith R, Radhika N. Progression in manufacturing of functionally graded materials and impact of thermal treatment—A critical review. Journal of Manufacturing Processes. 2021;**68**:1339-1377. DOI: 10.1016/j.jmapro.2021.06.062

[13] Dutta B, Froes FH. Additive Manufacturing of Titanium Alloys: State of the Art, Challenges and Opportunities. Elsevier Science; 2016. pp. 25-40, 61-73. DOI: 10.1016/C2015-0-02470-4

[14] Dutta B, Palaniswamy S, Choi J, Song LJ, Mazumder J, Fasm. Additive manufacturing by direct metal deposition. Advanced Materials & Processes. 2011;**169**:33-36

[15] Ge J, Pillay S, Ning H. Post-process treatments for additive-manufactured metallic structures: A comprehensive review. Journal of Materials Engineering and Performance. 2023;**32**:7073-7122. DOI: 10.1007/s11665-023-08051-9

[16] Rahman M, Islam KS, Dip TM, Chowdhury MFM, Debnath SR, Hasan SMM, et al. A review on nanomaterial-based additive manufacturing: Dynamics in properties, prospects, and challenges. Progress in Additive Manufacturing. 2023;**9**:1197-1224. DOI: 10.1007/s40964-023-00514-8

[17] Leary M. Design for Additive Manufacturing. Elsevier Science; 2019. pp. 7-31, 91-122, 165-268, 295-334. DOI: 10.1016/C2017-0-04238-6

[18] Mostafaei A, Elliott AM, Barnes JE, Li F, Tan W, Cramer CL, et al. Binder jet 3D printing—Process parameters, materials, properties, modeling, and challenges. Progress in Materials Science. 2021;**119**:100707. DOI: 10.1016/j.pmatsci.2020.100707

[19] Lores A, Azurmendi N, Agote I, Zuza E. A review on recent developments in binder jetting metal additive manufacturing: Materials and process characteristics. Powder Metallurgy. 2019;**62**:267-296. DOI: 10.1080/00325899.2019.1669299

[20] Sames WJ, List FA, Pannala S, Dehoff RR, Babu SS. The metallurgy and processing science of metal additive manufacturing. International Materials Review. 2016;**61**:315-360. DOI: 10.1080/09506608.2015.1116649

[21] Costa JM. Metal-based additive manufacturing: Evaluation of metallic parts produced with additive manufacturing technology at YAZAKI Europe Limited [master]. Universidade do Porto, Faculdade de Engenharia da Universidade do Porto; 2016. pp. 3-28. Available from: https://hdl.handle.net/10216/87466

[22] DebRoy T, Mukherjee T, Wei HL, Elmer JW, Milewski JO. Metallurgy, mechanistic models and machine learning in metal printing. Nature Reviews Materials. 2021;**6**:48-68. DOI: 10.1038/s41578-020-00236-1

[23] Cooke S, Ahmadi K, Willerth S, Herring R. Metal additive manufacturing: Technology, metallurgy and modelling. Journal of Manufacturing Processes. 2020;**57**:978-1003. DOI: 10.1016/j.jmapro.2020.07.025

[24] Oliveira DF, Vieira JS, Duarte I, Vincze G, Oliveira JM, Miranda G. 18Ni300 maraging steel lattice structures fabricated via laser powder bed fusion—mechanical behavior and gas permeability. Metals-Basel. 1982;**2023**:13

[25] Mukherjee T, Elmer JW, Wei HL, Lienert TJ, Zhang W, Kou S, et al. Control of grain structure, phases, and defects in additive manufacturing of high-performance metallic components. Progress in Materials Science. 2023;**138**:101153. DOI: 10.1016/j.pmatsci.2023.101153

[26] Diegel O, Nordin A, Motte D. A Practical Guide to Design for Additive Manufacturing. Springer; 2019. pp. 1-77. DOI: 10.1007/978-981-13-8281-9

[27] Nagarajan B, Hu Z, Song X, Zhai W, Wei J. Development of micro selective laser melting: The state of the art and future perspectives. PRC Engineering. 2019;**5**:702-720. DOI: 10.1016/j.eng.2019.07.002

[28] Guo N, Leu MC. Additive manufacturing: Technology, applications and research needs. Frontiers of

Mechanical Engineering. 2013;**8**:215-243. DOI: 10.1007/s11465-013-0248-8

[29] Tebianian M, Aghaie S, Razavi Jafari N, Elmi Hosseini S, Pereira A, Fernandes F, et al. A review of the metal additive manufacturing processes. Materials. 2023;**16**:7514. DOI: 10.3390/ma16247514

[30] Kovalenko E, Krasanov I, Valdaytseva E, Klimova-Korsmik O, Gushchina M. Influence of laser direct energy deposition process parameters on the structure and phase composition of a high-entropy alloy FeCoNiCrMn. Metals-Basel. 2023;**13**:534. DOI: 10.3390/met13030534

[31] Zhang X, Liu S, Liu Y, Guo H, Shi W. Titanium alloy fabricated by additive manufacturing for medical applications: Obtaining, characterization and application—Review. Metals-Basel. 2023;**13**:462. DOI: 10.3390/met13030462

[32] Dávila JL, Neto PI, Noritomi PY, Coelho RT, Da Silva JVL. Hybrid manufacturing: A review of the synergy between directed energy deposition and subtractive processes. The International Journal of Advanced Manufacturing Technology. 2020;**110**:3377-3390. DOI: 10.1007/s00170-020-06062-7

[33] Feenstra DR, Banerjee R, Fraser HL, Huang A, Molotnikov A, Birbilis N. Critical review of the state of the art in multi-material fabrication via directed energy deposition. Current Opinion in Solid State and Materials Science. 2021;**25**:100924. DOI: 10.1016/j.cossms.2021.100924

[34] Srinivasan D, Ananth K. Recent advances in alloy development for metal additive manufacturing in gas turbine/aerospace applications: A review. Journal of the Indian Institute of Science. 2022;**102**:311-349. DOI: 10.1007/s41745-022-00290-4

[35] Rahman KM, Wei A, Miyanaji H, Williams CB. Impact of binder on part densification: Enhancing binder jetting part properties through the fabrication of shelled geometries. Additive Manufacturing. 2023;**62**:103377. DOI: 10.1016/j.addma.2022.103377

[36] Careri F, Khan RHU, Todd C, Attallah MM. Additive manufacturing of heat exchangers in aerospace applications: A review. Applied Thermal Engineering. 2023;**235**:121387. DOI: 10.1016/j.applthermaleng.2023.121387

[37] Alfaify A, Saleh M, Abdullah FM, Al-Ahmari AM. Design for additive manufacturing: A systematic review. Sustainability. 2020;**12**:7936. DOI: 10.3390/su12197936

[38] Oliveira C, Maia M, Costa J. Production of an office stapler by material extrusion process, using DfAM as optimization strategy. U Porto Journal of Engineering. 2023;**9**:28-41. DOI: 10.24840/2183-6493_009-001_001635

[39] Egan PF. Design for Additive Manufacturing: Recent innovations and future directions. Designs. 2023;**7**:83. DOI: 10.3390/designs7040083

[40] Mata M, Pinto M, Costa J. Topological optimization of a metal extruded Doorhandle using nTopology. U Porto Journal of Engineering. 2023;**9**:42-54. DOI: 10.24840/2183-6493_009-001_001620

[41] Hällgren S. Some Aspects on Designing for Metal Powder Bed Fusion. Örebro University; 2017. pp. 1-63. URN: urn:nbn:se:oru:diva-62947

[42] Hällgren S, Pejryd L, Ekengren J. (Re)design for additive manufacturing. Procedia CIRP. 2016;**50**:246-251. DOI: 10.1016/j.procir.2016.04.150

[43] Thompson MK, Moroni G, Vaneker T, Fadel G, Campbell RI, Gibson I, et al. Design for additive manufacturing: Trends, opportunities, considerations, and constraints. Cirp Annals-Manufacturing Technology. 2016;**65**:737-760. DOI: 10.1016/j.cirp.2016.05.004

[44] Sossou G, Demoly F, Montavon G, Gomes S. An additive manufacturing oriented design approach to mechanical assemblies. Journal of Computational Design and Engineering. 2017;**5**:3-18. DOI: 10.1016/j.jcde.2017.11.005

[45] Vaneker T, Bernard A, Moroni G, Gibson I, Zhang Y. Design for additive manufacturing: Framework and methodology. CIRP Annals. 2020;**69**:578-599. DOI: 10.1016/j.cirp.2020.05.006

[46] Gao W, Zhang YB, Ramanujan D, Ramani K, Chen Y, Williams CB, et al. The status, challenges, and future of additive manufacturing in engineering. Computer-aided Design. 2015;**69**:65-89. DOI: 10.1016/j.cad.2015.04.001

[47] Vafadar A, Guzzomi F, Rassau A, Hayward K. Advances in metal additive manufacturing: A review of common processes, industrial applications, and current challenges. Applied Sciences. 2021;**11**:1213. DOI: 10.3390/app11031213

[48] Mohd Yusuf S, Cutler S, Gao N. Review: The impact of metal additive manufacturing on the aerospace industry. Metals-Basel. 2019;**9**:1286. DOI: 10.3390/met9121286

[49] Srivastava M, Rathee S, Maheshwari S, Kundra T. Additive Manufacturing: Fundamentals and Advancements. CRC Press; 2019. pp. 1-58. ISBN: 9781351049382

[50] Jayakrishna M, Vijay M, Khan B. An overview of extensive analysis of 3D printing applications in the manufacturing sector. Journal of Engineering. 2023;**2023**:1-6

[51] Trojanowska J, Ciszak O, Machado JM, Pavlenko I. Advances in Manufacturing II: Volume 1 - Solutions for Industry 4.0. Springer International Publishing; 2019. pp. 150-164. DOI: 10.1007/978-3-030-18715-6

[52] Badiru AB, Valencia VV, Liu D. Additive Manufacturing Handbook: Product Development for the Defense Industry. CRC Press; 2017. pp. 149-208. ISBN 1482264099

[53] Sacco E, Moon SK. Additive manufacturing for space: Status and promises. The International Journal of Advanced Manufacturing Technology. 2019;**105**:4123-4146. DOI: 10.1007/s00170-019-03786-z

[54] Nunes F, Trindade M, Costa J. Production and nTopology optimization of an additive manufactured support for smart glasses. U Porto Journal of Engineering. 2024;**10**:20-33. DOI: 10.24840/2183-6493_0010-001_002057

[55] Software M, Nastran MSC. Design Sensitivity and Optimization User's Guide. 2018:2017. Available from: https://simcompanion.hexagon.com/customers/s/article/msc-nastran-2017-design-sensitivity-and-optimization-user-s-guide-doc11148

[56] Mata M. nTopology Optimization Applied to Additive Manufactured Hydrofoil Wing Components. Universidade do Porto - Faculdade de Engenharia, Repositório Aberto da

Universidade do Porto; 2022. pp. 6-21. Available from: https://hdl.handle.net/10216/142981

[57] Esfarjani SM, Dadashi A, Azadi M. Topology optimization of additive-manufactured metamaterial structures: A review focused on multi-material types. Forces in Mechanics. 2022;**7**:100100. DOI: 10.1016/j.finmec.2022.100100

[58] Wu J, Sigmund O, Groen JP. Topology optimization of multi-scale structures: A review. Structural and Multidisciplinary Optimization. 2021;**63**:1455-1480. DOI: 10.1007/s00158-021-02881-8

[59] Gibson I, Rosen D, Stucker B, Khorasani M. Design for Additive Manufacturing. Springer International Publishing; 2021. pp. 555-607. DOI: 10.1007/978-3-030-56127-7_19

[60] Gebisa AW, Lemu HG. Design for manufacturing to design for additive manufacturing: Analysis of implications for design optimality and product sustainability. Procedia Manufacturing. 2017;**13**:724-731. DOI: 10.1016/j.promfg.2017.09.120

[61] Yang S, Zhao YF. Additive manufacturing-enabled design theory and methodology: A critical review. The International Journal of Advanced Manufacturing Technology. 2015;**80**:327-342. DOI: 10.1007/s00170-015-6994-5

[62] McMillan M, Leary M, Brandt M. Computationally efficient finite difference method for metal additive manufacturing: A reduced-order DFAM tool applied to SLM. Materials & Design. 2017;**132**:226-243. DOI: 10.1016/j.matdes.2017.06.058

[63] Brackett D, Ashcroft I, Hague R. Topology optimization for additive manufacturing. In: Proceedings of the 22nd Annual International Solid Freeform Fabrication Symposium - An Additive Manufacturing Conference, SFF 2011, Austin, TX. Austin: University of Texas; 2011. pp. 348-362

[64] Panesar A, Abdi M, Hickman D, Ashcroft I. Strategies for functionally graded lattice structures derived using topology optimisation for additive manufacturing. Additive Manufacturing. 2018;**19**:81-94. DOI: 10.1016/j.addma.2017.11.008

[65] Refai K, Montemurro M, Brugger C, Saintier N. Determination of the effective elastic properties of titanium lattice structures. Mechanics of Advanced Materials and Structures. 2020;**27**:1966-1982. DOI: 10.1080/15376494.2018.1536816

[66] Diegel O, Schutte J, Ferreira A, Chan YL. Design for additive manufacturing process for a lightweight hydraulic manifold. Additive Manufacturing. 2020;**36**:101446. DOI: 10.1016/j.addma.2020.101446

[67] Khoda B, Ahsan AMMN, Shovon AN, Alam AI. 3D metal lattice structure manufacturing with continuous rods. Scientific Reports. 2021;**11**. DOI: 10.1038/s41598-020-79826-6

[68] Hoang V-N, Tran P, Vu V-T, Nguyen-Xuan H. Design of lattice structures with direct multiscale topology optimization. Composite Structures. 2020;**252**:112718. DOI: 10.1016/j.compstruct.2020.112718

[69] Klahn C, Leutenecker B, Meboldt M. Design strategies for the process of additive manufacturing. Procedia CIRP. 2015;**36**:230-235. DOI: 10.1016/j.procir.2015.01.082

[70] Maskery I, Aremu AO, Parry L, Wildman RD, Tuck CJ, Ashcroft IA.

Effective design and simulation of surface-based lattice structures featuring volume fraction and cell type grading. Mater Design. 2018;**155**:220-232. DOI: 10.1016/j.matdes.2018.05.058

[71] Babamiri BB, Barnes B, Soltani-Tehrani A, Shamsaei N, Hazeli K. Designing additively manufactured lattice structures based on deformation mechanisms. Additive Manufacturing. 2021;**46**:102143. DOI: 10.1016/j.addma.2021.102143

[72] Krish S. A practical generative design method. Computer Aided Design. 2011;**43**:88-100. DOI: 10.1016/j.cad.2010.09.009

[73] Zhang Y, Wang Z, Zhang Y, Gomes S, Bernard A. Bio-inspired generative design for support structure generation and optimization in Additive Manufacturing (AM). CIRP Annals. 2020;**69**:117-120. DOI: 10.1016/j.cirp.2020.04.091

[74] Mandolini M, Pradel P, Cicconi P. Design for additive manufacturing: Methods and tools. Applied Sciences. 2022;**12**:6548. DOI: 10.3390/app12136548

[75] Klahn C, Singer D, Meboldt M. Design guidelines for additive manufactured snap-fit joints. Procedia CIRP. 2016;**50**:264-269. DOI: 10.1016/j.procir.2016.04.130

[76] Manfredi D, Calignano F, Krishnan M, Canali R, Ambrosio EP, Atzeni E. From powders to dense metal parts: Characterization of a commercial AlSiMg alloy processed through direct metal laser sintering. Materials (1996-1944). 2013;**6**:856

[77] Cunha M, Alves S, Costa J. Design and optimization of a gripper clamp. U Porto Journal of Engineering. 2024;**10**:34-44. DOI: 10.24840/2183-6493_0010-001_002059

[78] Matilainen V, Piili H, Salminen A, Syvänen T, Nyrhilä O. Characterization of process efficiency improvement in laser additive manufacturing. Physics Procedia. 2014;**56**:317-326. DOI: 10.1016/j.phpro.2014.08.177

[79] Ritt S. EOS and SLM Solutions agree a cross-licence for Laser Sintering/Selective Laser Melting. Available from: http://www.eos.info/Press/eos_slm_solutions_agree_cross_licence_for_laser_sintering [Accessed: June 27]

[80] Monteiro B, Rocha F, Costa J. Topology optimization of a robot gripper with nTopology. U Porto Journal of Engineering. 2024;**10**:11-19. DOI: 10.24840/2183-6493_010-001_002051

[81] Gibson I, Rosen DW, Stucker B. Additive Manufacturing Technologies: Rapid Prototyping to Direct Digital Manufacturing. New York: Springer US; 2010

[82] Khorasani M, Gibson I, Ghasemi AH, Hadavi E, Rolfe B. Laser subtractive and laser powder bed fusion of metals: Review of process and production features. Rapid Prototyping Journal. 2023;**29**:935-958. DOI: 10.1108/rpj-03-2021-0055

[83] Leary M, Downing D, Lozanovski B, Harris J. 5 - Design principles. In: Yadroitsev I, Yadroitsava I, du Plessis A, MacDonald E, editors. Fundamentals of Laser Powder Bed Fusion of Metals. Elsevier; 2021. pp. 91-119, 165-184. DOI: 10.1016/C2017-0-04238-6

[84] Mata M, Bencatel R, Sequeiros EW, Vieira MF, Costa JM. Topology optimization applied to additive-manufactured hydrofoil wing components. Academia Materials

Science. 2024;**1**(2):1-12. DOI: 10.20935/AcadMatSci6213

[85] Nunes FLdS. Anisotropy in Sintering of 316L Parts Produced by Binder Jetting 3D Printing. University of Porto, Faculty of Engineering; 2023. pp. 3-18. Available from: https://hdl.handle.net/10216/153600

[86] Atzeni E, Salmi A. Economics of additive manufacturing for end-usable metal parts. International Journal of Advanced Manufacturing Technology. 2012;**62**:1147-1155. DOI: 10.1007/s00170-011-3878-1

[87] Hällgren S, Pejryd L, Ekengren J. Additive manufacturing and high speed machining-cost comparison of short lead time manufacturing methods. Procedia CIRP. 2016;**50**:384-389. DOI: 10.1016/j.procir.2016.05.049

[88] King WE, Anderson AT, Ferencz RM, Hodge NE, Kamath C, Khairallah SA, et al. Laser powder bed fusion additive manufacturing of metals; physics, computational, and materials challenges. Applied Physics Reviews. 2015;**2**:041304. DOI: 10.1063/1.4937809

[89] Clausen A, Aage N, Sigmund O. Exploiting additive manufacturing infill in topology optimization for improved buckling load. PRC Engineering. 2016;**2**:250-257. DOI: 10.1016/J.Eng.2016.02.006

[90] Ye J, Guo Q, Lu H, Kyvelou P, Zhao Y, Gardner L, et al. Topology optimisation of self-supporting structures based on the multi-directional additive manufacturing technique. Virtual and Physical Prototyping. 2023;**18**:1-19. DOI: 10.1080/17452759.2023.2271458

[91] Zhu L, Li N, Childs PRN. Light-weighting in aerospace component and system design. Propulsion and Power Research. 2018;7:103-119. DOI: 10.1016/j.jppr.2018.04.001

[92] Kanishka K, Acherjee B. Revolutionizing manufacturing: A comprehensive overview of additive manufacturing processes, materials, developments, and challenges. Journal of Manufacturing Processes. 2023;**107**:574-619. DOI: 10.1016/j.jmapro.2023.10.024

[93] Kruth J-P, Badrossamay M, Yasa E, Deckers J, Thijs L, Van Humbeeck J. Part and material properties in selective laser melting of metals. In: Proceedings of the Proceedings of the 16th International Symposium on Electromachining (ISEM XVI); Shanghai, China. 2010

[94] Das P, Chandran R, Samant R, Anand S. Optimum part build orientation in additive manufacturing for minimizing part errors and support structures. Procedia Manufacturing. 2015;**1**:343-354. DOI: 10.1016/j.promfg.2015.09.041

[95] Smith J, Butler D. Post-processing for Additive Manufactured Metal Parts: A Brief Introduction. Springer International Publishing; 2023. pp. 821-832. DOI: 10.1007/978-3-031-20752-5_48

[96] Du Plessis A. 6 - Porosity in laser powder bed fusion. In: Yadroitsev I, Yadroitsava I, du Plessis A, MacDonald E, editors. Fundamentals of Laser Powder Bed Fusion of Metals. Elsevier; 2021. pp. 155-178. DOI: 10.1016/B978-0-12-824090-8.00007-X

[97] Schmelzle J, Kline EV, Dickman CJ, Reutzel EW, Jones G, Simpson TW. (Re)designing for part consolidation: Understanding the challenges of metal additive manufacturing. Journal of Mechanical Design. 2015;**137**:111404-111412. DOI: 10.1115/1.4031156

[98] Plocher J, Panesar A. Review on design and structural optimisation in additive manufacturing: Towards next-generation lightweight structures. Mater Design. 2019;**183**:108164. DOI: 10.1016/j.matdes.2019.108164

[99] Yang L, Hsu K, Baughman B, Godfrey D, Medina F, Menon M, et al. Additive Manufacturing of Metals: The Technology, Materials, Design and Production. Springer; 2017. pp. 1-15. DOI: 10.1016/j.matdes.2019.108164

[100] Veloso F, Gomes-Fonseca J, Morais P, Correia-Pinto J, Pinho ACM, Vilaça JL. Overview of methods and software for the design of functionally graded lattice structures. Advanced Engineering Materials. 2022;**24**:2200483. DOI: 10.1002/adem.202200483

[101] Thomas DS, Gilbert SW. Costs and Cost Effectiveness of Additive Manufacturing. NIST Special Publication; 2014. pp. 5-27. DOI: 10.6028/NIST.SP.1176 ER

[102] Blakey-Milner B, Gradl P, Snedden G, Brooks M, Pitot J, Lopez E, et al. Metal additive manufacturing in aerospace: A review. Mater Design. 2021;**209**:110008. DOI: 10.1016/j.matdes.2021.110008

[103] Montero J, Weber S, Bleckmann M, Paetzold K. A methodology for the decentralised design and production of additive manufactured spare parts. Production & Manufacturing Research. 2020;**8**:313-334. DOI: 10.1080/21693277.2020.1790437

[104] Lopez Taborda LL, Maury H, Pacheco J. Design for additive manufacturing: A comprehensive review of the tendencies and limitations of methodologies. Rapid Prototyping Journal. 2021;**27**:918-966. DOI: 10.1108/rpj-11-2019-0296

[105] Thomas D. Costs, benefits, and adoption of additive manufacturing: A supply chain perspective. International Journal of Advanced Manufacturing Technology. 2016;**85**:1857-1876. DOI: 10.1007/s00170-015-7973-6

[106] Wiberg A, Persson J, Ölvander J. Design for additive manufacturing – A review of available design methods and software. Rapid Prototyping Journal. 2019;**25**:1080-1094. DOI: 10.1108/rpj-10-2018-0262

[107] Armstrong M, Mehrabi H, Naveed N. An overview of modern metal additive manufacturing technology. Journal of Manufacturing Processes. 2022;**84**:1001-1029. DOI: 10.1016/j.jmapro.2022.10.060

[108] Nazir A, Gokcekaya O, Md Masum Billah K, Ertugrul O, Jiang J, Sun J, et al. Multi-material additive manufacturing: A systematic review of design, properties, applications, challenges, and 3D printing of materials and cellular metamaterials. Mater Design. 2023;**226**:111661. DOI: 10.1016/j.matdes.2023.111661

[109] Leary M. Surface roughness optimisation for selective laser melting (SLM). In: Laser Additive Manufacturing. Melbourne, VIC, Australia: RMIT University, Centre for Additive Manufacturing; 2017. pp. 99-118

[110] Groover MP. Fundamentals of Modern Manufacturing: Materials, Processes, and Systems. Wiley; 2020. pp. W1-W19. ISBN 1119722012

[111] Muthu SS, Savalani MM. Handbook of Sustainability in Additive Manufacturing. Vol. 1. Springer Singapore; 2016. pp. 110-126. ISBN 9811005478

[112] Nyamekye P, Rahimpour Golroudbary S, Piili H, Luukka P,

Kraslawski A. Impact of additive manufacturing on titanium supply chain: Case of titanium alloys in automotive and aerospace industries. Advances in Industrial and Manufacturing Engineering. 2023;**6**:100112. DOI: 10.1016/j.aime.2023.100112

[113] Sampaio ÁM, Pontes AJ. Product Design and Development by Additive Manufacturing for Automotive Systems. Springer International Publishing; 2021. pp. 61-66. DOI: 10.1007/978-3-030-80462-6_8

[114] Zadpoor AA. Design for additive bio-Manufacturing: From patient-specific medical devices to rationally designed meta-biomaterials. International Journal of Molecular Sciences. 2017;**18**:1-18. DOI: 10.3390/ijms18081607

[115] Wang Z, Wang C, Li C, Qin Y, Zhong L, Chen B, et al. Analysis of factors influencing bone ingrowth into three-dimensional printed porous metal scaffolds: A review. Journal of Alloys and Compounds. 2017;**717**:271-285. DOI: 10.1016/j.jallcom.2017.05.079

[116] Asnafi N. Tool and die making, surface treatment, and repair by laser-based additive processes. BHM Berg- und Huttenmannische Monatshefte. 2021;**166**:225-236. DOI: 10.1007/s00501-021-01113-2

[117] Bar-Cohen Y. Advances in Manufacturing and Processing of Materials and Structures. CRC Press; 2018. pp. 51-94. ISBN 1351867962

[118] Schmidt M, Merklein M, Bourell D, Dimitrov D, Hausotte T, Wegener K, et al. Laser based additive manufacturing in industry and academia. CIRP Annals - Manufacturing Technology. 2017;**66**:561-583. DOI: 10.1016/j.cirp.2017.05.011

[119] Silberschmidt VV, Matveenko VP. Mechanics of Advanced Materials: Analysis of Properties and Performance. Springer International Publishing; 2015. pp. 43-78. DOI: 10.1007/978-3-319-17118-0

[120] Bar-Cohen Y. High Temperature Materials and Mechanisms. Taylor & Francis; 2014. pp. 17-68. ISBN 1466566450

[121] Chawla N, Chawla KK. Metal Matrix Composites. New York: Springer; 2013

[122] Hayat MD, Singh H, He Z, Cao P. Titanium metal matrix composites: An overview. Composites Part A: Applied Science and Manufacturing. 2019;**121**:418-438. DOI: 10.1016/j.compositesa.2019.04.005

[123] Dey A, Roan Eagle IN, Yodo N. A review on filament materials for fused filament fabrication. Journal of Manufacturing and Materials Processing. 2021;**5**:69. DOI: 10.3390/jmmp5030069

[124] Mirzababaei S, Pasebani S. A review on binder jet additive manufacturing of 316L stainless steel. Journal of Manufacturing and Materials Processing. 2019;**3**:82. DOI: 10.3390/jmmp3030082

[125] Gaytan SM, Cadena MA, Karim H, Delfin D, Lin Y, Espalin D, et al. Fabrication of barium titanate by binder jetting additive manufacturing technology. Ceramics International. 2015;**41**:6610-6619. DOI: 10.1016/j.ceramint.2015.01.108

[126] Murr LE, Gaytan SM, Ramirez DA, Martinez E, Hernandez J, Amato KN, et al. Metal fabrication by additive manufacturing using laser and electron beam melting technologies. Journal of Materials Science and

Technology. 2012;**28**:1-14. DOI: 10.1016/S1005-0302(12)60016-4

[127] Gibson BT, Bandari YK, Richardson BS, Henry WC, Vetland EJ, Sundermann TW, et al. Melt pool size control through multiple closed-loop modalities in laser-wire directed energy deposition of Ti-6Al-4V. Additive Manufacturing. 2020;**32**:1-12. DOI: 10.1016/j.addma.2019.100993

[128] Qian M, Froes FH. Titanium Powder Metallurgy: Science, Technology and Applications. Elsevier Science; 2015. pp. 1-6. DOI: 10.1016/B978-0-12-814062-8.00001-7

[129] Tofail SAM, Koumoulos EP, Bandyopadhyay A, Bose S, O'Donoghue L, Charitidis C. Additive manufacturing: Scientific and technological challenges, market uptake and opportunities. Materials Today. 2018;**21**:22-37. DOI: 10.1016/j.mattod.2017.07.001

[130] Yan L, Chen Y, Liou F. Additive manufacturing of functionally graded metallic materials using laser metal deposition. Additive Manufacturing. 2020;**31**:100901. DOI: 10.1016/j.addma.2019.100901

[131] du Plessis A, Yadroitsava I, Yadroitsev I. Effects of defects on mechanical properties in metal additive manufacturing: A review focusing on X-ray tomography insights. Mater Design. 2020;**187**:108385. DOI: 10.1016/j.matdes.2019.108385

[132] Bidare P, Jiménez A, Hassanin H, Essa K. Porosity, cracks, and mechanical properties of additively manufactured tooling alloys: A review. Advances in Manufacturing. 2022;**10**:175-204. DOI: 10.1007/s40436-021-00365-y

[133] Al-Maharma AY, Patil SP, Markert B. Effects of porosity on the mechanical properties of additively manufactured components: A critical review. Materials Research Express. 2020;**7**:122001. DOI: 10.1088/2053-1591/abcc5d

[134] Wu Y, Fang J, Wu C, Li C, Sun G, Li Q. Additively manufactured materials and structures: A state-of-the-art review on their mechanical characteristics and energy absorption. International Journal of Mechanical Sciences. 2023;**246**:108102. DOI: 10.1016/j.ijmecsci.2023.108102

[135] Peng T, Zhu Y, Leu M, Bourell D. Additive manufacturing-enabled design, manufacturing, and lifecycle performance. Additive Manufacturing. 2020;**36**:101646. DOI: 10.1016/j.addma.2020.101646

[136] Zerbst U, Bruno G, Buffière J-Y, Wegener T, Niendorf T, Wu T, et al. Damage tolerant design of additively manufactured metallic components subjected to cyclic loading: State of the art and challenges. Progress in Materials Science. 2021;**121**:100786. DOI: 10.1016/j.pmatsci.2021.100786

[137] Newell A, George A, Papakostas N, Lhachemi H, Malik A, Shorten R. On Design for Additive Manufacturing: Review of Challenges and Opportunities Utilising Visualisation Technologies. Valbonne Sophia-Antipolis, France: IEEE; 2019

[138] Nguyen-Van V, Li S, Liu J, Nguyen K, Tran P. Modelling of 3D concrete printing process: A perspective on material and structural simulations. Additive Manufacturing. 2023;**61**:103333. DOI: 10.1016/j.addma.2022.103333

[139] Ibhadode O, Zhang Z, Sixt J, Nsiempba KM, Orakwe J, Martinez-Marchese A, et al. Topology optimization for metal additive manufacturing: Current trends,

challenges, and future outlook. Virtual and Physical Prototyping. 2023;**18**:1-35. DOI: 10.1080/17452759.2023.2181192

[140] Zhu J, Zhou H, Wang C, Zhou L, Yuan S, Zhang W. A review of topology optimization for additive manufacturing: Status and challenges. Chinese Journal of Aeronautics. 2021;**34**:91-110. DOI: 10.1016/j.cja.2020.09.020

[141] Gerhard D, Köring T, Neges M. Generative Engineering and Design – A Comparison of Different Approaches to Utilize Artificial Intelligence in CAD Software Tools. Springer Nature Switzerland; 2023. pp. 206-215. DOI: 10.1007/978-3-031-25182-5_21

[142] Regassa Hunde B, Debebe Woldeyohannes A. Future prospects of computer-aided design (CAD) – A review from the perspective of artificial intelligence (AI), extended reality, and 3D printing. Results in Engineering. 2022;**14**:100478. DOI: 10.1016/j.rineng.2022.100478

Chapter 10

Post-Process Treatments for Additive Manufactured Titanium Alloys

Hao Deng, Wenhao Liu, Lvjun Zhou, Sheng Cao and Longqing Chen

Abstract

This chapter summarized the influences of post-process treatments on the microstructural evolution and mechanical properties of titanium (Ti) alloys prepared by powder bed additive manufacturing fabrication (AM) including laser powder bed melting (L-PBF) and electron beam powder bed melting (EB-PBF) technologies. Described post-process treatments included stress-relieving, annealing, $\alpha + \beta$ solution treating and aging, β solution treating and aging, and hot isostatic pressing. The effects of post-process treatment parameters such as temperature, time, cooling rates and pressure on stress relief, phase transformation, precipitation, and defect mitigation were discussed. Furthermore, the influences of microstructure characteristics on mechanical properties of tensile, fatigue, and fracture toughness were described. This chapter provided post-process guidelines for customizing microstructure, optimizing the performance, and enhancing reliability of AM titanium components.

Keywords: post-process treatments, additive manufacturing, titanium alloys, L-PBF/EB-PBF, mechanical property

1. Introduction

In general, the conventional subtractive manufactured titanium (Ti) alloys are expensive due to the significant material wastage, high energy input, and long lead times [1–4]. In contrast, the emerging additive manufacturing (AM) technology has attracted significant attention and has been implemented in various industries over the last two decades. AM is known for fabricating nearly fully dense and complex shaped components at a high material utilization with limited machining (material wastage), and shortened lead times [5–7]. As a result, the AM technology has been widely used in the industries of aerospace, medical, energy, and automotive. In addition, it has become a key technology in designing and producing high-performance components.

ISO/ASTM 52900 standard classifies AM processes into seven categories: binder ejection (BJP), direct energy deposition (DED), material extrusion 3D printing (extrusion 3D printing), material jetting, powder bed fusion (PBF), laminated object manufacturing (LOM), and light curing. The PBF and DED processes are widely used direct additive manufacturing routes with better integrity, dimensional accuracy, and high application potential than other techniques. Two typical metal additive manufacturing processes—powder bed fusion and directed energy deposition—are shown in **Figure 1** [8].

In powder bed fusion processes, a layer of powder is spread on the forming substrate before the high energy beam melting process. In comparison, powder or wire material is fed into the focused high energy beam in the deposition process. In addition, powder bed fusion processes have better surface quality and higher printing accuracy than directed energy deposition processes [9]. These advantages make powder bed fusion particularly valuable in advanced fields like aerospace, energy, and chemical industries, where precision is paramount [6, 7]. Consequently, this section will primarily focus on the powder bed process. The powder bed fusion process can further be classified into laser powder bed fusion (L-PBF) and electron beam melting (EB-PBM) depending on the heat source used.

L-PBF, also known as selective laser melting (SLM) or direct metal laser melting (DMLM), was first developed by the Fraunhofer Institute ILT in Germany in 1995 [10]. This innovative technique utilizes a focused laser beam to selectively melt metal powder based on cross-sectional slices of a three-dimensional (3D) CAD model. The laser scanning vectors precisely fill the contours of each slice, melting the powder to create a molten pool that rapidly solidifies in protective atmosphere of inert gas, forming a solid layer. After scanning and solidifying one layer, the build platform lowers by a predefined layer thickness; a re-coater then spreads a new layer of powder over the surface. The laser then scans and melts the powder layer according to the

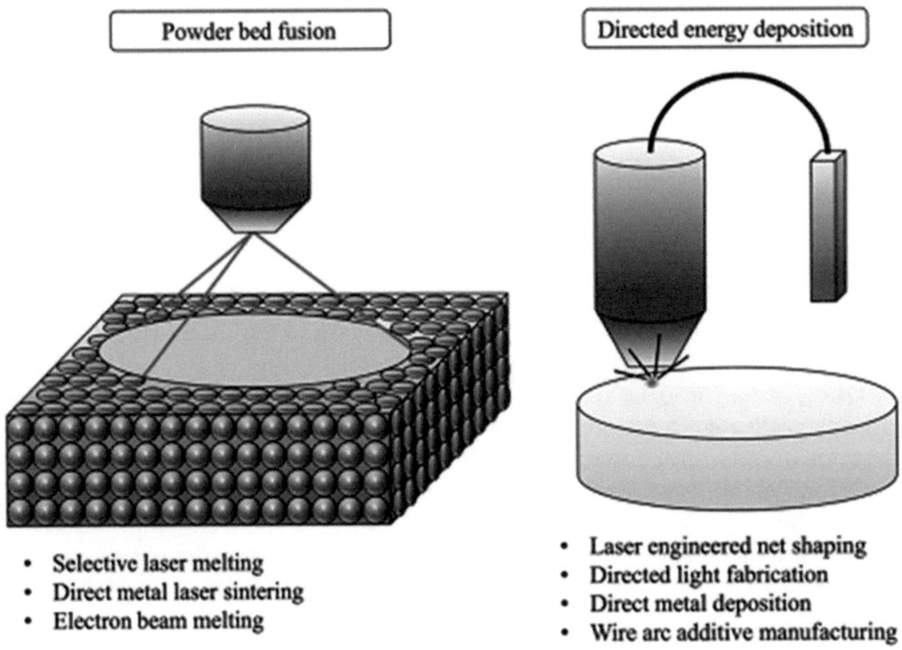

Figure 1.
Categorization of metal AM process by the raw material feeding condition [8].

cross-sectional CAD slice of the next layer. Such a layer-by-layer fabrication repeats until the whole CAD files built [11].

Figure 2 illustrates a typical L-PBF system, which includes several key components: a processing laser, an automatic powder feed and spill system, and an inert gas protection system [12]. The processing laser is responsible for the precision melting of the powder, while the automatic powder feed system ensures a consistent supply of powder to the build area. The inert gas protection system maintains the controlled atmosphere necessary for high-quality production. Upon completion of the manufacturing process, any unused powder can be collected from spills and powder beds for recycling. This not only reduces material waste but also enhances the cost-effectiveness of the process. The ability to recycle unused powder is a significant advantage, particularly when working with expensive materials like titanium or specialized alloys [13, 14].

The main difference between EB-PBF and L-PBF production methods is in the input energy source: EB-PBF utilizes electron beam, while L-PBF employs laser beam. This fundamental distinction leads to variations in process parameters, material compatibility, and final part characteristics [16, 17]. In the EB-PBF process, the build chamber is first evacuated to create a vacuum atmosphere to the required level. This controlled environment minimizes the risk of oxidation and contamination, which is particularly crucial for reactive metals like titanium. After the chamber condition is stabilized, the substrate is preheated to reduce thermal gradients and residual stresses in turn. Like L-PBF, EB-PBF is also a layer-by-layer process. The actual fabrication involves directing the electron beam to selectively sinter the powder layer in pre-scan and melt the powder in the melting process. **Figure 3** shows a diagram of the EB-PBF process, with a diagram of the EB-PBF device on the left and a diagram of building one layer on the right [15].

The electron beam's high energy density allows efficient melting, and the process is highly controlled to ensure precision. After each layer is completed, the build

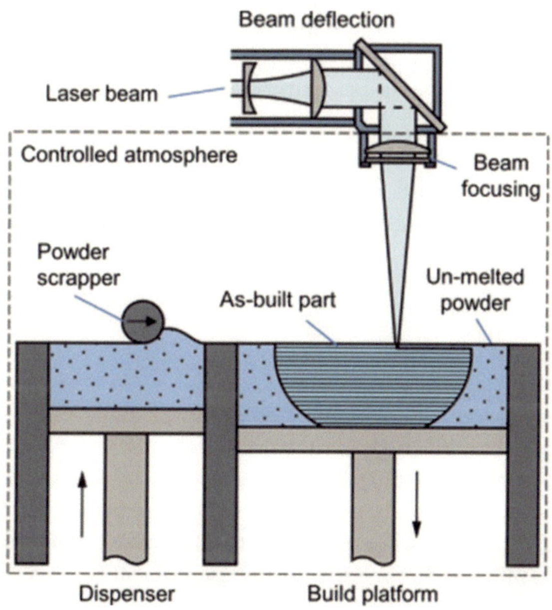

Figure 2.
A schematic diagram of L-PBF system [12].

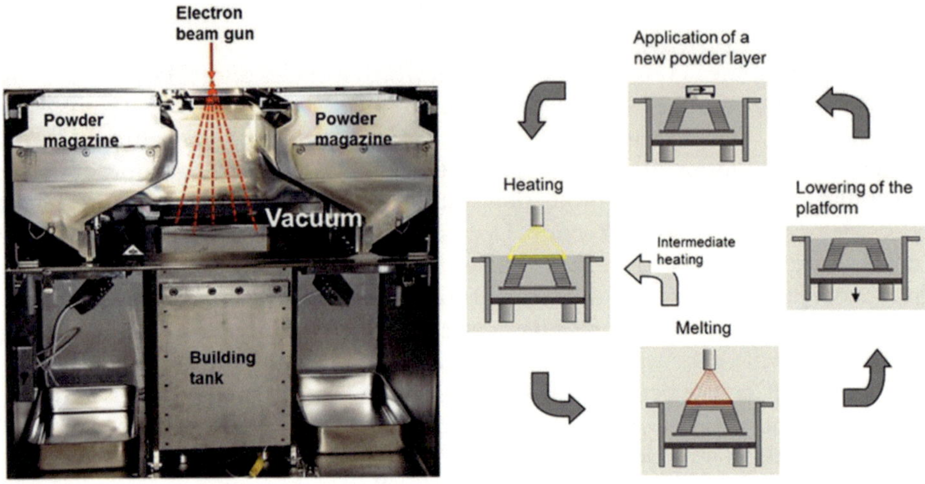

Figure 3.
EB-PBF process. Left: Process chamber. Right: 4-step process for building one layer [15].

platform lowers by a specified layer thickness, and a new layer of powder is spread across the build area. This cycle of preheating, melting, and rapid cooling is repeated until the entire part is formed. The vacuum atmosphere not only prevents oxidation but also aids in achieving good dimensional stability and high-density components. EB-PBF offers several advantages, especially for producing alloys that are prone to reacting with oxygen, such as titanium and its alloys.

However, EB-PBF also has its challenges. The surface finish of EB-PBF-produced parts is typically rougher compared to those produced by L-PBF, which may necessitate additional post-processing steps [18]. The equipment and operational costs for EB-PBF are also generally higher due to the need for vacuum systems and high-energy electron beams. Despite these challenges, the advantages of EB-PBF make it a valuable technology in fields such as aerospace, medical implants, and high-performance engineering.

To date, most research in the field of PBF has focused on commercial titanium alloys, such as CP-Ti, Ti-6Al-4 V, and Ti-5Al-5Mo-5 V-3Cr. Among these, PBF fabricated $\alpha + \beta$ alloy of Ti-6Al-4 V components have been employed in large-scale industrial applications [19]. Therefore, in this chapter, we will focus on the extensively studied $\alpha + \beta$ titanium alloys, such as Ti-6Al-4 V, Ti-6Al-7Nb and Ti-6Al-2Sn-4Zr-6Mo (Ti-6246).

2. Process and microstructure

Metal components produced by PBF often exhibit intrinsic defects, residual stress, metastable phases, and microstructure anisotropy and heterogeneity. These issues contribute to poor mechanical properties with large variations and property anisotropy. To address these issues, post-process treatments are employed to tailor the microstructure. However, selecting appropriate treatments and parameters is crucial in achieving desired microstructure and property. Therefore, this section covers evolutions of defects, residual stress, and microstructure during the post-process treatments of stress-relieving, annealing, $\alpha + \beta$ solution treating and aging, β solution treating and aging, and hot isostatic pressing.

2.1 AM process and defect

In general, inappropriate AM parameters and poor powder quality can introduce defects of gas pores, lack of fusion (LOF), keyhole, and balling (beading-up). These defects can significantly affect the mechanical properties. Defects formed in L-PBF can generally be categorized into LOF, keyhole, and balling [19, 22]. Luo et al. developed a processing map for L-PBF manufactured Ti-6Al-4 V (**Figure 4**). Keyhole tends to occur at high laser powers, low scan speeds, and high volumetric energy densities (VED). Meanwhile, lack of fusion (LOF) is more likely to happen at low laser powers, high scan speeds, and low VED. In addition, LOF is frequently observed at large hatch distance or layer thickness conditions. Between the VED range of LOF and keyhole, a suitable process VED range can be obtained to achieve high relative densities. However, balling occurs at suitable VED ranges when both laser power and scan speed are high.

For LOF, these occur when the beam power is insufficient or the scanning speed is too high (corresponding to the lower-right area of **Figure 4** [20]), resulting in inadequate energy density. Improper layer thickness or scanning spacing can also cause the powder to not fully melt, leading to unfused defects. These defects manifest as weak bonds between layers, reducing the structural integrity of the part [23, 24]. **Figure 5** highlights the LOF defect in the process of L-PBF printing Ti-6Al-4 V. For keyhole, excessive laser power combined with too low scanning speed can overheat the molten pool (corresponding to the top-left area of **Figure 4** [20]), creating deep and narrow holes (keyhole phenomenon). High energy input can also cause evaporation within the molten pool, forming evaporation holes. These holes compromise the density and strength of the part, making it susceptible to failure under stress [25]. **Figure 6** highlights a keyhole pore in a Ti-6Al-4 V L-PBF-printed part. For balling or beading-up defects, high laser power and high scanning speed can cause the molten metal to break under surface tension, forming spherical particles due to Marangoni flow and Plateau-Rayleigh instability. This phenomenon leads to a rough surface finish and uneven layer formation, affecting the part's esthetic and functional properties [26–28]. **Figure 7** shows that in the process of printing AlSi10Mg

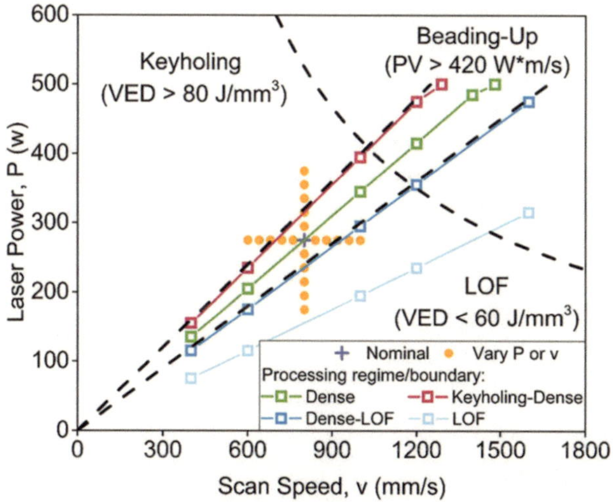

Figure 4.
A typical process map for L-PBF-built Ti–6Al–4 V alloy [20].

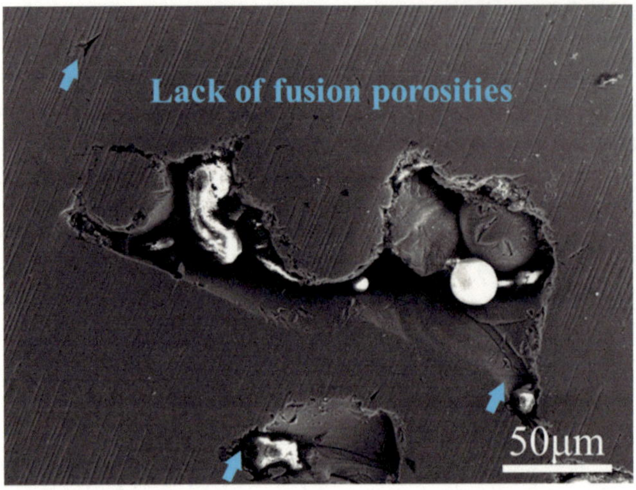

Figure 5.
A SEM image showing LOF defects in L-PBF processes.

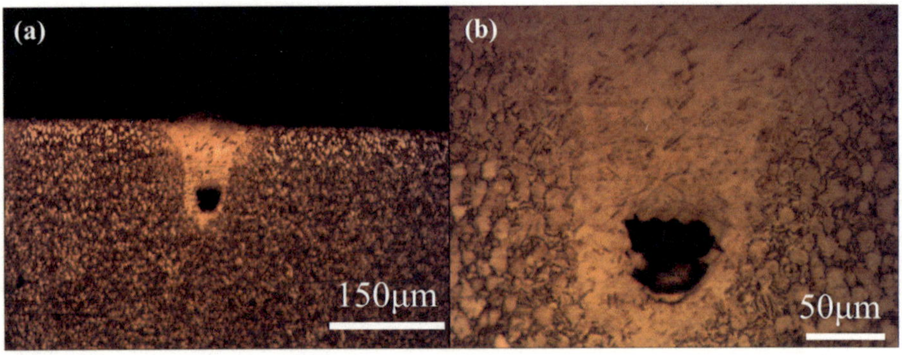

Figure 6.
Optical microscope (OM) images highlighting the location and shape of a keyhole defect in L-PBF processes.

alloy by L-PBF technology, the degree of balling of the surface increases with the increase of scanning speed [21]. To minimize these defects, it is crucial to optimize process parameters such as laser power, scanning speed, layer thickness, and scanning spacing. Appropriate combination of processing parameters leads to suitable energy inputs, which can provide dense L-PBF parts. At the same time, the quality of the powder is crucial in reducing defect formation. Factors such as particle size distribution, sphericity, recycle condition, and chemical variations between powder batches/recycle times directly influence the printed relative density. Spherical powders with a particle size distribution of 15–53 μm enhance powder layering uniformity than coarse and fine powders, thereby reducing porosity and unfused defects. Additionally, the powder sphericity and recycle are vital; spherical particles enhance fluidity, and powders used multiple times can develop defects due to morphological changes or hygroscopicity.

In addition, defects can also form in the EB-PBF fabrication. Two of the most common defects in EB-PBF manufacturing include spherical gas pores and LOF defects. **Figure 8** shows two types of typical defects found in EB-PBF fabricated Ti-6Al-4 V.

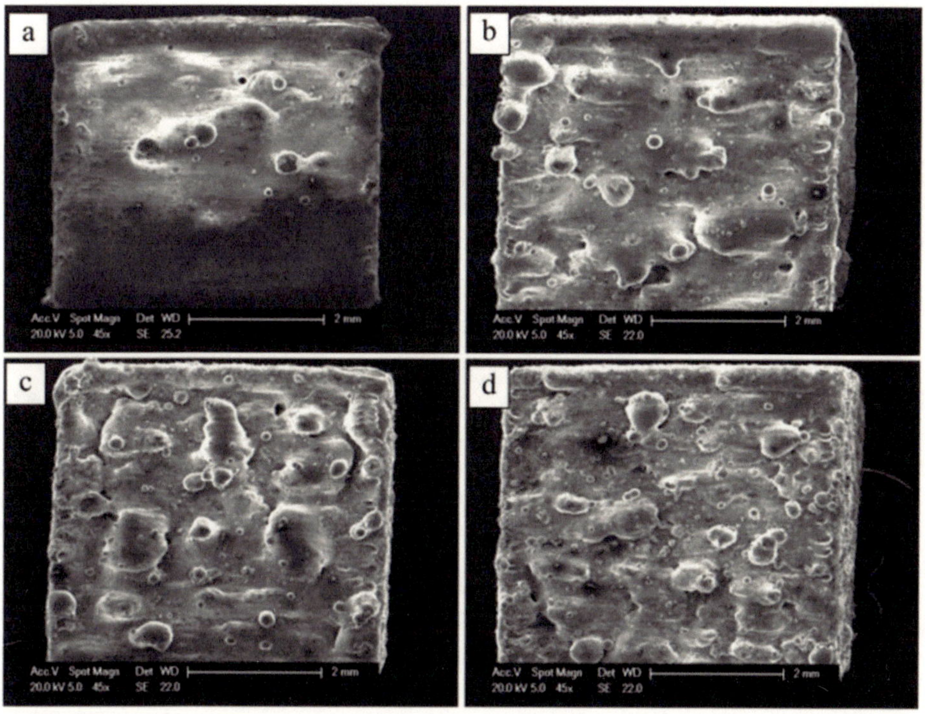

Figure 7.
The severity of the balling (beading-up) phenomena increases as the scan speed increases, as shown by the L-PBF printed surface profiles with scan speeds of (a) 250 mm/s, (b) 500 mm/s, (c) 750mm/s and (d) 1000 mm/s [21].

Spherical pores (**Figure 8a**) are gas pores formed when gas trapped in the molten material, often due to the residual gas in the powder or improper melting conditions. Gas pores weaken the material by acting as stress concentrators, particularly affecting fatigue resistance [29]. In contrast, unfused defects (**Figure 8b**) are irregular shaped voids caused by insufficient fusion between layers, often due to low electron beam energy. This lack of energy prevents complete melting, leaving gaps or unbonded regions that compromise the material's structural integrity, leading to potential failure under mechanical loads. Both types of defects (spherical pores and unfused voids)

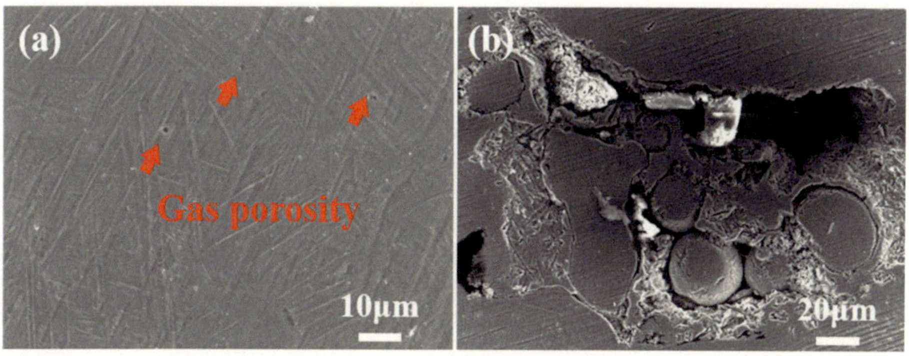

Figure 8.
OM images of gas porosity (a) and a lack-of-fusion void (b) in EB-PBF processes.

diminish the mechanical properties of EB-PBF-built parts, reducing tensile strength, toughness, and fatigue life. However, these issues can be mitigated by optimizing parameters such as beam energy, scan speed, and powder quality. The lack-of-fusion defects are typically oriented perpendicular to the build direction and are located at the interface between layers [30]. While LOF defects and gas porosity are localized to the interior of the part and can be mitigated by post-process heat treatments, other defects such as melt balling or layer delamination result in build failure and are therefore omitted from this analysis.

2.2 Residual stress

In the LPFB process, the high cooling rate and large thermal gradient result in high residual stress, which can lead to warping, cracking, or delamination of the AM fabricated object. Comparing to the L-PBF process, the preheated powder bed and uniform distributed electron beam energy in EB-PBF result in a relatively slower cooling rate and a smaller thermal gradient comparing to those in L-PBF. As a result, EB-PBF fabricated parts have lower residual stress than L-PBF fabricated parts, which also reduces the risk of deformation and cracking.

Currently, numerous scholars have investigated the residual stress in titanium alloy produced by L-PBF printing. Yadroitsev et al. [31] conducted X-ray diffraction residual stress measurements on small-sized Ti6Al4V samples processed by L-PBF. Their findings revealed that residual stress in the scanning direction is more tensile compared to the perpendicular direction, with the maximum stress observed at the build-substrate interface. Bilal Ahmad et al. [32] utilized the contour method to measure residual stress in various directions of Ti-6Al-4 V alloy fabricated through L-PBF, identifying the general distribution pattern of residual stress in this material. Typically, tensile residual stress is present in the first row of printing, with the maximum tensile stress occurring at the start and end of this row. For each layer, compressive residual stress is found at the center, while tensile stress is present at the edges. Along the building direction, the first layer and the sample's surface exhibit tensile residual stress, whereas the sample's center has compressive residual stress. These observations have also been corroborated by other studies on L-PBF printed alloys, both through measurement and simulation (**Figure 9**). The impact of residual stress can be notably mitigated by adjusting various printing parameters, such as the residence time for each layer and the rotation angle of each layer's scanning. However, to completely eliminate residual stress, stress-relief annealing is usually required.

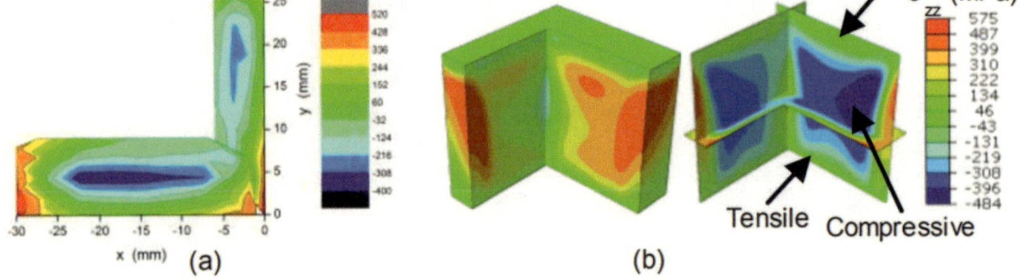

Figure 9.
Typical residual stress in build direction of as-L-PBF part: (a) cross-sectional residual stress contours measured using neutron diffraction [32], (b) predicted residual stress [33].

2.3 As-fabricated microstructure

2.3.1 L-PBF fabricated microstructure

As the L-PBF process involves directional solidification, rapid cooling (10^4–10^6 K/s), and thermal cycling, the as-built Ti-6Al-4 V has a nonequilibrium microstructure (**Figure 10**) including coarse prior β grains and intergranular fine α′ martensite. The formation of these columnar β grains is due to the limited grain refinement effects of alloying elements like Al and V. Due to the directional solidification, columnar prior β grains generally have a strong <100>β fiber texture along the build direction. During molten pool solidification, the solidification front shifts from planar to dendritic, with <100>β dendrites preferentially nucleating. These dendrites, driven by a significant thermal gradient along the build direction, grow across layers, merge into coarse columnar β grains, and form a <100>β texture [34]. The presence of columnar β grains in α + β titanium alloys can lead to anisotropic mechanical properties. To address this, researchers have explored methods to transform columnar β grains into equiaxed β grains during the printing process, including adjusting process parameters, adding grain refiners, or using auxiliary energy fields to prevent epitaxial growth.

The thermal cycle of the L-PBF process leads to the martensitic decomposition (α′ → α + β) and the coexistence of α′ and β phases at certain temperatures. As the thermal cycling effects are different at different locations within the same layer and different at different build height, L-PBF prepared Ti-6Al-4 V alloys display not only

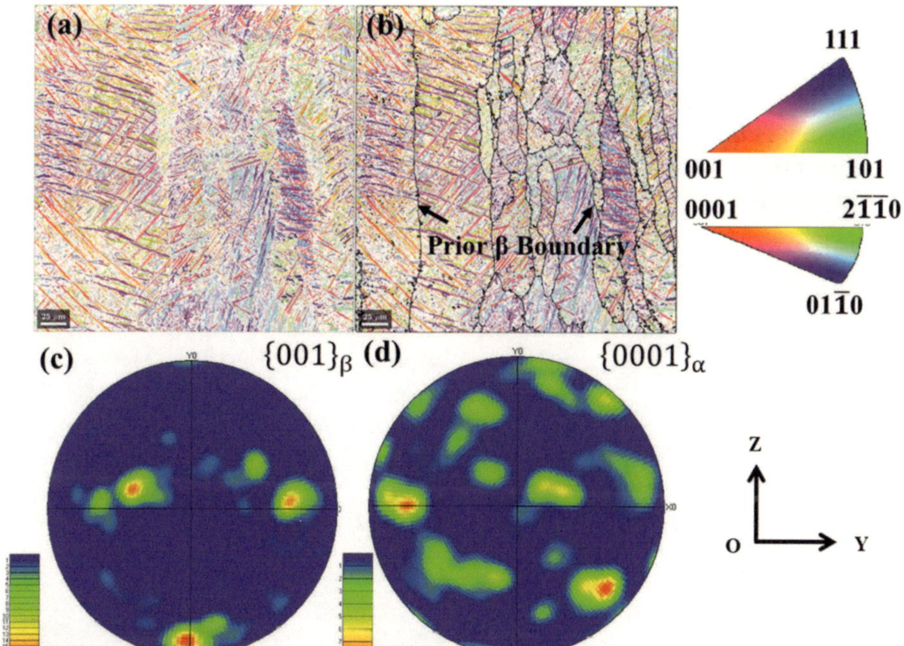

Figure 10.
(a) EBSD IPF map of L-PBF fabricated Ti6Al4V cube on YZ (Vertical) plane, (b) corresponding numerically reconstructed prior β IPF map, (c) corresponding $(001)_\beta$ PF, and (d) corresponding $(0001)_\alpha$ PF [35].

a nonequilibrium microstructure but also various nonuniform microstructures at different locations within the alloy, including α + β lamellar regions, α + β phase areas, and α′-martensite needle regions. Therefore, post-heat treatment is essential for optimizing the industrial application of L-PBF-prepared Ti-6Al-4 V alloys.

Ti-6Al-7Nb is designed for orthopedic applications such as hip prosthetics because of its excellent corrosion resistance and biocompatibility. As shown in **Figure 11(a)**, the L-PBF fabricated Ti-6Al-7Nb microstructure consists of martensitic α' phase embedded within columnar prior β grains, which is similar to the microstructure observed in Ti-6Al-4 V produced by L-PBF [36, 38, 39]. For another α + β Ti alloy, Ti-6Al-2Sn-4Zr-6Mo (Ti-6246) offers superior high-temperature properties and excellent corrosion resistance. This heat-treatable alloy, known for its good hardenability, is an ideal substitute for Ti-6Al-4 V alloys where higher strength is required. Due to its relatively high content of β-stabilizers, which enhances resistance to martensitic phase transitions, Ti-6Al-2Sn-4Zr-6Mo produced by L-PBF typically features a rhombic α′ phase as shown in **Figure 11(b)** [37, 40].

2.3.2 EB-PBF fabricated microstructure

Although the main working principles of EB-PBF (Electron Beam Melting) and L-PBF (Selective Laser Melting) processes are similar, both featuring large temperature gradients and rapid cooling rates, they differ significantly in build temperatures and cooling rates. As a result, the same alloy processed by these two methods will exhibit different microstructures. The EB-PBF build temperature ranges between 500 and 800°C, with the cooling rates between 10^3 and 10^5°C/s. Therefore, the α + β titanium alloy produced by EB-PBF cannot maintain its solidified structure as martensite decomposition usually occurs at the high build temperatures, resulting in a microstructure composed of α and β phases. However, a small portion of hexagonal α′ martensite can still be found in larger EB-PBF fabricated Ti-6Al-4 V parts, particularly in sections very close to the top surface where high cooling rates occur. Studies (**Figure 12**) have also shown that as the deposition height increases, the cooling rate slows down, leading to the coarsening of the microstructure. Consequently, the as-built specimens exhibit a typical gradient structure: equiaxed β

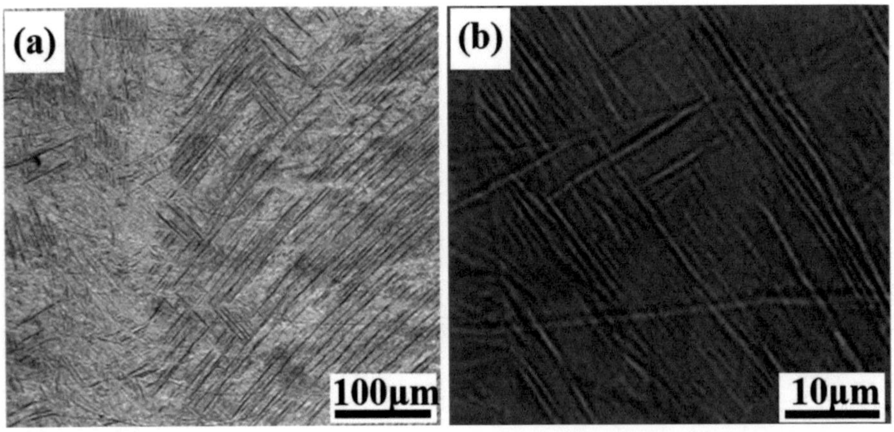

Figure 11.
Microstructure of different Ti alloys prepared by L-PBF: (a) Ti-6Al-7Nb, (b) Ti-6Al-2Sn-4Zr-6Mo [36, 37].

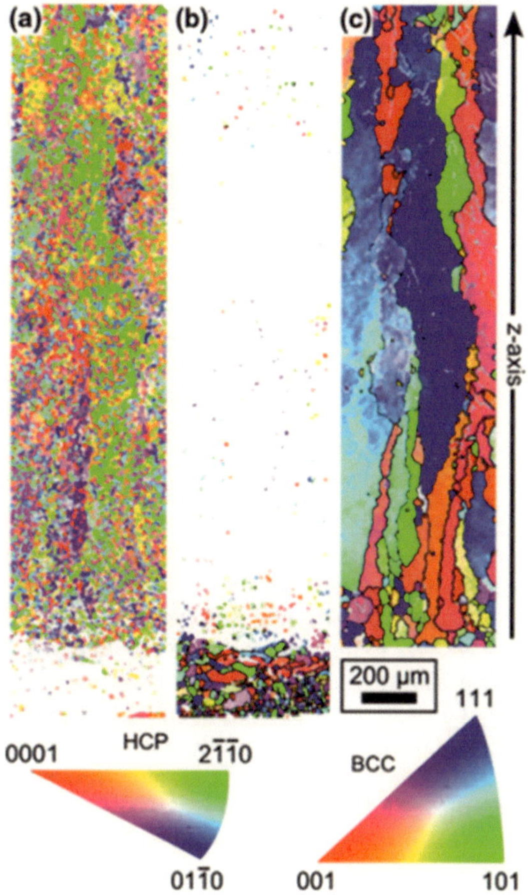

Figure 12.
Grain orientations and textures of EB-PBF fabricated Ti-6Al-4 V [41].

grains near the substrate transitioning to columnar β grains with increasing deposition height. The width of the columnar crystals and the number of cross-layer growth layers increase, resulting in a coarser lamellar α phase [41].

2.4 Post-process treatment

2.4.1 Stress-relieving

Parts produced through PBF often exhibit significant residual stress, which can adversely affect their performance. The primary goal of stress relieving is to reduce or eliminate these stresses. Typically, this process is carried out within a temperature range of 450–670°C. Lower temperatures are generally used, with both temperature and time playing crucial roles in stress relief efficiency. For instance, holding at 595°C for 1 hour is equivalent to 5 hours at 540°C, as illustrated in **Figure 13** [42]. When stress relieving α + β titanium alloys fabricated by PBF at relatively low temperatures, there may be a slight coarsening of the microstructure, though the size and variant of the primary and secondary α/α′ laths remain unchanged. Annealing PBF titanium alloys at 680°C for 3 hours can lead to a 90% reduction in tensile residual stresses [43],

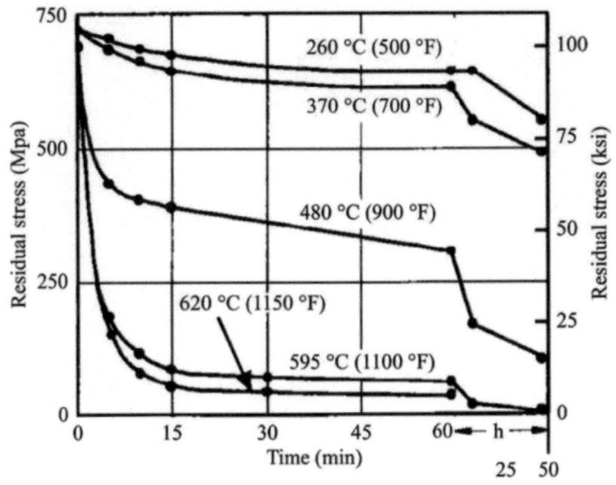

Figure 13.
Effect of stress-relieving temperatures on residual stress in Ti-6Al-4 V [42].

and this method also enhances fatigue crack growth resistance. In addition, Cecchel et al. documented the reduction of residual stresses in as-received samples after various heat treatments. According to this study [44], the most effective way to reduce the stress from −428 to 14 MPa was to perform 0.5 hours of solution at 1015°C, followed by tempering at 730°C for 2 hours. In this study, the stresses were reduced from −428 to 14 MPa in the case of the −428 to 14 MPa. At a temperature of 550°C, the Al and O content can influence the material and potentially cause the precipitation of hardened α_2 particles from the α phase [45]. Selection of a suitable temperature is particularly important for residual stress relief.

2.4.2 Annealing

Comparing to stress-relieving, annealing is carried out at a higher temperature range between 650and 950°C. The as-built acicular martensite α′ microstructure can be partially decomposed to α′ + α + β microstructure or fully decomposed to α + β lamellar microstructure depending on the annealing temperature and time. In general, a higher annealing temperature or a longer annealing time would promote the martensite decomposition. Annealing of L-PBF α + β Ti alloys at temperatures below 600°C results in microstructure changes that are minimal compared to the as-built. Heating to 850°C (**Figure 14**) generates an α + β phase with a fine α phase. In addition, the study claims that [46, 47] higher temperatures increase the driving force for the decomposition of the α-phase and its transformation into the β-phase. According to the ASTM 2924-14 [48] specification for Ti-6Al-4 V fabricated by AM, an elongation requirement of 10% is not met by this low ductility. Hence, the research [49] was focused on promoting the nonequilibrium α' martensite decomposition in the prepared microstructures to equilibrium α + β microstructures by annealing. Previous studies have shown that annealing in the 750–950°C range can effectively promote the decomposition of α′ martensite and improve plasticity.

2.4.3 α + β solution treating and aging

In addition to the one-stage stress-relieving/annealing treatment, solution treating and aging (STA) treatment is a two-stage treatment including a high temperature

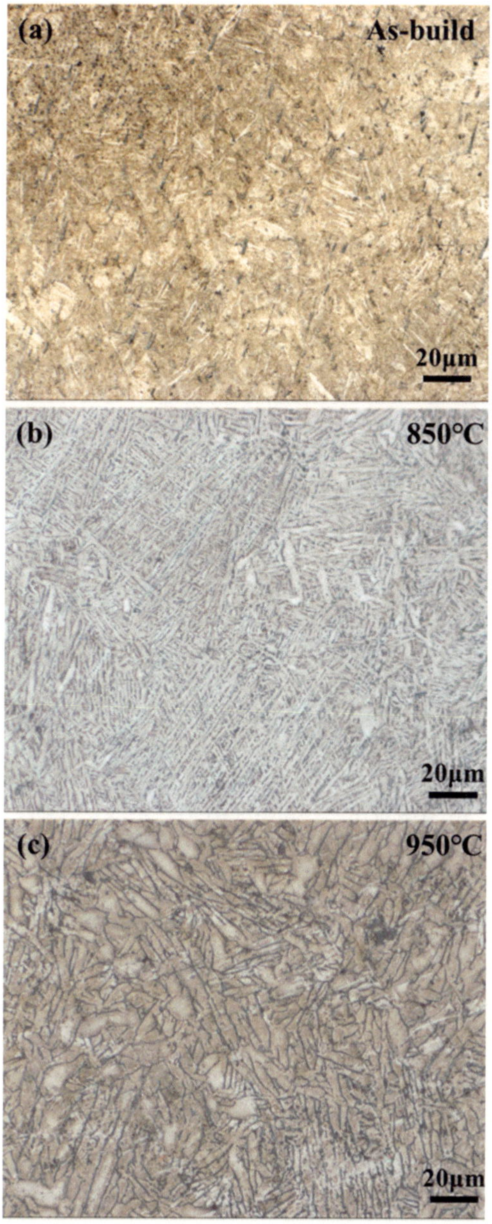

Figure 14.
Ti-6Al-4 V microstructure evolution at different annealing temperatures [50].

solution and a low temperature aging. STA treatments can be further categorized into α + β STA and β STA based on the solution temperatures. α + β STA has the solution temperature below the β transus. It can lead to the formation of bi-lamellar or bi-modal microstructure in PBF fabricated α + β Ti alloys. This process allows for precise control over the size, volume fraction, and morphology of each phase, enabling the development of various microstructures. The impact of the solution stage on the microstructure is complex and plays a crucial role in determining the alloy's characteristics.

Table 1 [51] summarizes a study by Haar and Becker on the heat treatment of Ti-6Al-4 V alloy produced by L-PBF. Treating the alloy at 750°C for 8 hours, followed by air cooling (AC), promotes the decomposition of martensite, resulting in a lamellar α + β structure. As the solution temperature increases, the size of the primary α phase (α_p) also increases significantly (**Figure 15(a)** and **(c)**). Specifically, when the solution temperature is raised from 750 to 870°C, the α_p size increases from 1.5 to 2.5 μm. At temperatures near the β transus, grain boundary α fractures, leading to the spheroidization of α_p into an equiaxed α phase. However, when the heat treatment is conducted below the β transus, the precipitation of grain boundary α impedes element diffusion and does not alter the morphology of columnar β grains (**Figure 16**), resulting in minimal impact of solution temperature on the size of β grains.

In addition to solution temperature, the cooling rate after solution treatment significantly affects the growth of α_p, with furnace cooling (FC) being particularly effective in promoting this growth (**Figure 15(c)** and **(d)**). Compared to air-cooled samples, FC samples exhibit smaller β grain size, a higher α_p phase fraction, and reduced secondary α (α_s) precipitation during aging, which is associated with a higher proportion of α_p. The holding time during the solution stage also greatly influences the microstructure. Comparing the volume fraction of α_p at different holding times at the same solution temperature reveals that a 30-minute hold is insufficient to achieve steady-state equilibrium. The volume fraction of α_p increases with longer holding times, stabilizing at 41%, indicating equilibrium is reached between 30 minutes and 2 hours (**Figure 17(a)–(c)**). Although the size of α_p does not change significantly with holding time, a noticeable increase occurs when holding time is extended to 8 hours (**Figure 17(d)**). In contrast, the aging stage primarily influences the microstructure by affecting α_s. Lower aging temperatures promote the precipitation of fine α_s, significantly enhancing alloy strength, while extending the aging time ensures complete α_s precipitation.

2.4.4 β solution treating and aging

Compared to α + β STA treatments, β STA involves performing the solution treatment at temperatures above the β transus. The β STA process typically serves two main purposes: converting columnar β grains into equiaxed grains and achieving a

Heat treatment	α_p phase fraction (%)	α_p grain width (μm)
750°C/8 h + AC	87	1.5
800°C/2 h + FC	87	2
870°C/4 h + AC	65	2.5
870°C/2 h + FC	88	3.5
910°C/2 h + WQ	59	1.5
910°C/8 h + WQ	59	2.8
945°C/4 h + WQ	80	3.5
960°C/30mi + FC	42	1.5
960°C/4 h + FC	92	1.5
945°C/4 h + FC to 930°C	70	8
960°C/4 h + FC to 900°C	52	11

Table 1.
Effects of heat treatment on microstructure [51].

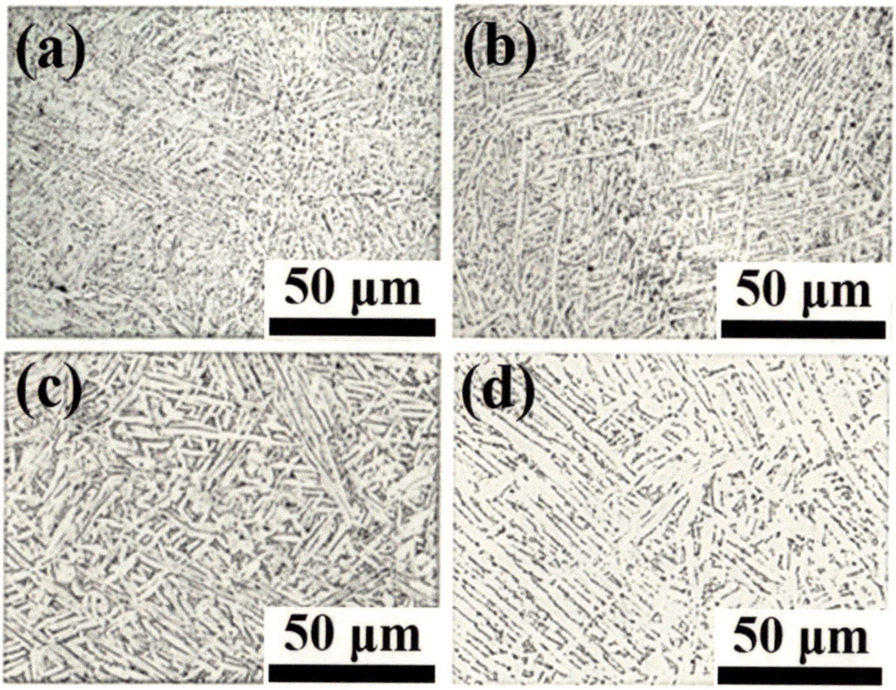

Figure 15.
Microstructure of L-PBF fabricated Ti-6Al-4 V alloys at different heat treatment: (a) 750°C, 8 h AC, (b) 800°C, 2 h FC, (c) 870°C, 4 h AC, (d) 870°C, 2 h FC [51].

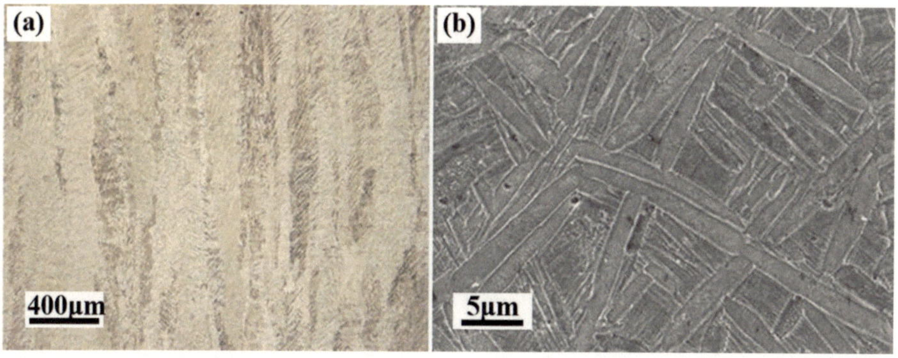

Figure 16.
Microstructures of the subtransus solution treatment 1223 K (950°C/1 h WQ): (a) optical micrograph showing the grain morphology and (b) SEM micrograph in the SE mode showing the α, α′, and β phases [52].

uniform α + β microstructure. As illustrated in **Figure 18**, during solution treatment above the β transus, all α′/α phases transform into a β phase solid solution. The columnar β grains develop into equiaxed grains due to surface tension while retaining the <100>β texture. Following water quenching (WQ), the body-centered cubic phase (β) undergoes a displacive transformation into the hexagonal phase (α′) through a shearing mechanism, accompanied by nonthermal nucleation without energy release.

However, because the heat treatment occurs above the β transus and lacks the barrier of grain boundary α, the initial growth of prior β grains is unimpeded, leading to

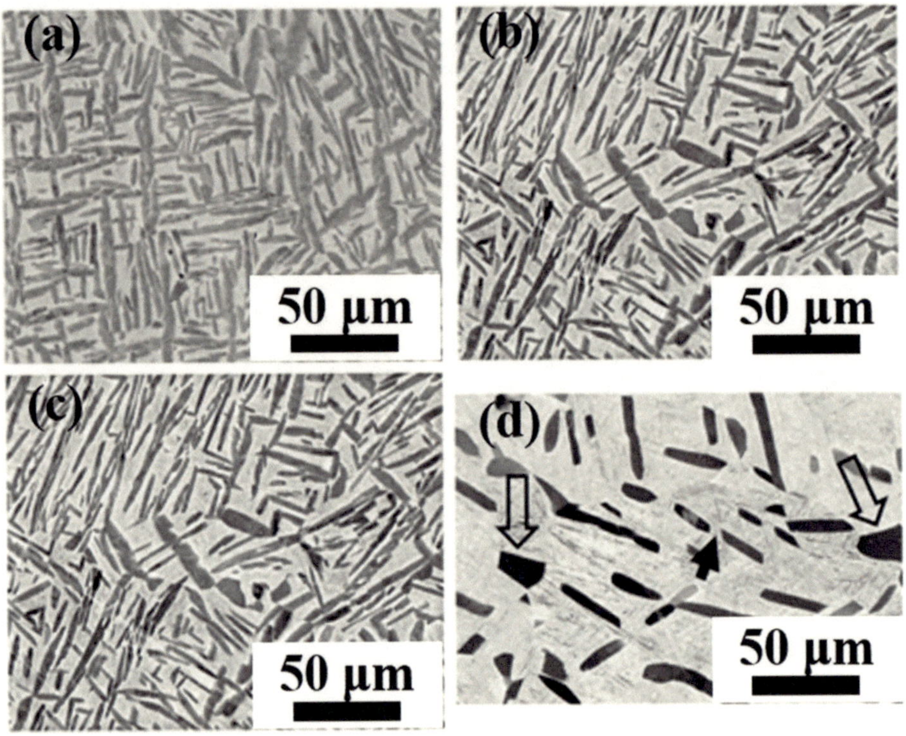

Figure 17.
Microstructure of L-PBF fabricated Ti-6Al-4 V alloys at different heat treatment: (a) 910°C, 30 min, (b) 910°C, 2 h, (c) 910°C, 8 h, (d) 945°C, 4 h [51].

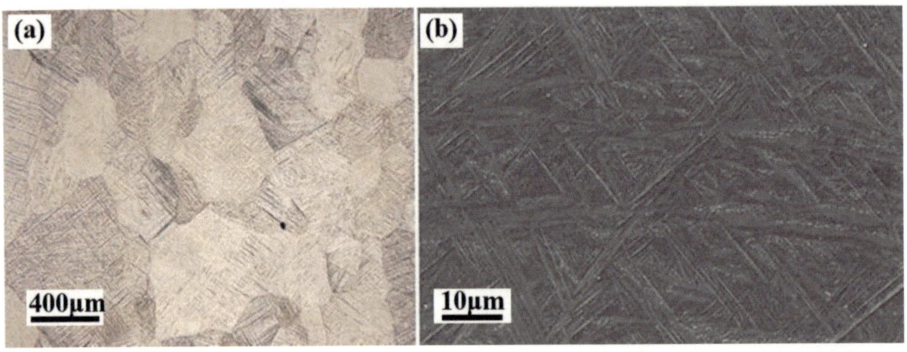

Figure 18.
Microstructures of the supertransus solution treatment 1323 K (1050°C/1 h WQ): (a) optical micrograph showing the grain morphology and (b) SEM micrograph in the SE mode showing the α' needles [52].

a rapid growth rate. This makes it challenging to control the size of the prior β grains during heat treatment above the β transus, often resulting in coarse prior β grains and a subsequent deterioration in material performance. As a result, this heat treatment method is seldom used in practice.

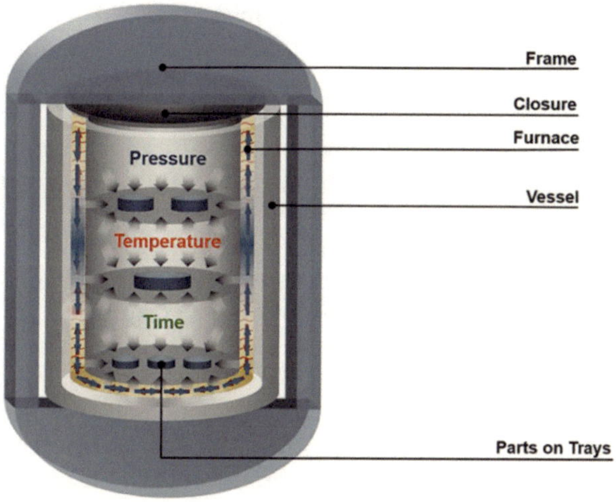

Figure 19.
A simplified schematic picture of the hot isostatic pressing equipment and process.

2.4.5 HIP

Hot isostatic pressing (HIP) is a process in which an inert gas is used as a pressure-transfer medium, and the samples are placed in a closed vessel, where high temperatures and high pressures are combined to exert an isotropic and equal pressure on the samples. **Figure 19** shows a simplified schematic picture of the hot isostatic pressing equipment and process. For AM (L-PBF and EB-PBF) α + β titanium alloy, hot isostatic pressing heat treatments can significantly reduce or eliminate defects within the component, thereby improving the mechanical and fatigue properties of the component. During HIP treatments, internal porosity can be eliminated if the compressive stress generated by the HIP pressure is greater than the yield strength of the material at the HIP temperature. All [48] standard HIP treatments for Ti-6Al-4 V are below the β-phase transition point (T_{β}) and typically include a slow cooling rate. During these standard HIP treatments below the phase transition point, high temperatures and pressures promote atomic diffusion, leading to grain structure refinement. Larger grains can be broken down into smaller, more homogeneous grains, thereby increasing the strength and toughness of the material. Residual stresses due to rapid heating and cooling cycles in the AM process can also be reduced by HIP.

3. Microstructure and mechanical properties

3.1 Tensile property

The microstructure of PBF fabricated α + β titanium alloys plays a crucial role in determining yield strength, primarily influenced by variations in α′/α thickness. The rapid cooling in L-PBF promotes the formation of a twinned α′ martensitic structure, characterized by high hardness and strength but lower ductility. To achieve optimal mechanical properties, choosing proper heat treatment is essential. **Table 2** highlights the impact of several common heat treatments on the microstructure and mechanical properties. As indicated in **Table 2**, stress-relieving treatments at temperatures up to 700°C effectively reduce internal and residual stresses in L-PBF α + β titanium alloys, resulting in minimal strength changes while enhancing plasticity. Initially, the microstructure is fully martensitic; however, low-temperature annealing transforms it into a mixed α cluster and α′ basketweave structure, while at the β-transus temperature (T_β), it develops into a coarse α/β lamellar structure. Solution treatment at the martensitic phase transition temperature (T_{α}') induces a phase change from α′ to α + β. The significant formation of α during this transition transforms the high-strength, low-plasticity characteristics of the original samples into a balance of relatively high strength and improved plasticity. When considering the structural evolution of L-PBF α + β titanium alloys in relation to T_β, heat treatment above, T_β alters the prior β microstructure, whereas solution treatment below the β-phase transition point regulates phase content and lath size without affecting the prior β grains. Overall, tensile strength decreases as the α′ martensite content decreases and the α′/α thickness increases.

3.2 Fatigue property

The fatigue cracking mechanism of additively manufactured (AM) α + β titanium alloys is significantly influenced by the microstructure and the presence of defects. Unfused regions, pores, and other printing-related defects, along with the low-plasticity fully martensitic structure and coarse prior β microstructure within the laths, negatively impact fatigue properties. Therefore, the primary strategies for optimizing fatigue resistance focus on enhancing the microstructure and eliminating defects. Post-treatment with HIP is highly effective in removing internal defects, thereby improving the fatigue performance of machined specimens and significantly reducing the variability in fatigue data. It has been shown that HIP post-treatment can alter the fatigue damage mechanism, shifting the initiation of fatigue cracks from surface and build defects to larger microstructural features. Furthermore, the fatigue performance of PBF specimens treated in this way is comparable to that of conventionally processed materials.

Conditions	Heat treatments	phase	YS (MPa)	UTS (MPa)	El. (%)
As-built	—	α′	1065 (H)	1250 (H)	4 (H) [53]
			1137 ± 20 (V)	1206 ± 8 (V)	7.6 ± 2 (V) [54]
Stress-relieving	670°C/5 h	α′/α	1015 (V)	1090 (V)	10 (V) [55]
Annealing	730°C/2 h/FC	α′/α + β	958 ± 6 (H)	1057 ± 8 (H)	12.4 ± 0.7 (H)
			937 ± 9 (V)	1052 ± 11 (V)	9.6 ± 0.9 (V) [56]
Annealing	800°C/2 h/FC	α′/α + β	937 ± 4 (H)	1041 ± 5 (H)	19 ± 1 (H) [49]
Annealing	920°C/2 h/FC	α + β	850 (V)	933 (V)	15 (V) [57]
β solution	1050°C/2 h/FC	α + β	—	986 ± 45 (V)	13.8 ± 0.8 (V) [58]

Table 2.
Microstructural and mechanical properties of Ti-6Al-4 V prepared by L-PBF under printing and heat treatment conditions.

Microstructural characteristics play a crucial role in dislocation motion, crack initiation, and crack propagation during fatigue deformation, which in turn influences the fatigue performance and load-bearing capacity of Ti6Al4V alloys. Generally, a finer microstructure correlates with higher fatigue strength. The lath martensitic structure formed by L-PBF, though lacking βr between the α′ laths, consists of differently oriented neighboring α′ laths, which restrict dislocation motion to within the laths, making each lath an individual deformation unit. The fine martensitic structure in L-PBF-formed α + β titanium alloys enhances fatigue performance. In terms of fatigue resistance, the hierarchy is as follows: martensitic structure > aged bimodal structure (STA) > other solid solution bimodal structures (ST, AC + ST, FC + ST). Within the lamellar structure, significant differences exist between the external and internal conditions (such as phase compositions and lath dimensions), leading to a load retention factor ranging from 1.18 to 4.46, which is lower than that of equiaxed microstructures. In the lamellar structure, the size of the deformation unit during fatigue corresponds to the effective dislocation slip distance. As this slip distance increases, the dislocation pinning force at interfaces also increases, leading to stress concentration, crack initiation, and accelerated fatigue failure.

3.3 Fracture toughness

Fracture mechanics, which examines the strength and crack propagation behavior of components containing cracks, is closely related to the field of fatigue science. The primary focus of fracture mechanics is to study the fracture toughness of various materials in order to determine whether a cracked component will fail under certain external loading conditions and to analyze the crack propagation behavior under applied loads. Research has demonstrated that altering the chemical composition, microstructure, and texture of alloys can significantly enhance the fracture toughness of titanium alloys. However, improving fracture toughness often requires a trade-off with other desirable properties. The plane strain fracture toughness, K_{IC}, is particularly important because the critical crack size for unstable crack growth is proportional to K_{IC}. In titanium alloys, it is sometimes necessary to slightly reduce K_{IC} in order to improve strength.

Differences in the chemical composition of titanium alloys lead to variations in their mechanical properties. The presence of an α′-transformed microstructure significantly increases toughness while only slightly decreasing the alloy's strength. For PBF Ti6Al4V alloys with an α′ + β microstructure, dynamic fracture toughness is influenced by the interaction between the α-phase content and microstructural features, such as secondary α-phase morphology. The microstructure of the L-PBF α + β titanium alloy is formed by a brief solid solution treatment around the point of β-phase transformation, followed by aging, which can lead to the development of a coarse lamellar microstructure. This has two advantages for improving fracture toughness: (1) the α + β microstructure is more ductile than the α′-martensitic microstructure, and (2) a coarser lamellar microstructure creates a rougher crack front profile, increasing crack curvature and thereby enhancing fracture toughness.

4. Conclusion

In conclusion, this chapter has thoroughly explored the impact of various post-process treatments on titanium alloys produced *via* additive manufacturing techniques, such as laser powder bed fusion (L-PBF) and electron beam powder bed fusion (EB-PBF). The discussion has highlighted how parameters like temperature, cooling rate, and pressure influence key mechanical properties including tensile strength, fatigue resistance, and fracture toughness. These findings emphasize the importance of optimizing post-process treatments to reduce defects, manage residual stress, and tailor the microstructure for improved performance and reliability. As additive manufacturing technologies continue to evolve, these insights will play a critical role in advancing the application of titanium alloys across industries, ensuring that components not only meet but also exceed performance expectations.

Author details

Hao Deng[1,2*], Wenhao Liu[3], Lvjun Zhou[4], Sheng Cao[5] and Longqing Chen[4]

1 Institute of Metal Research, Chinese Academy of Sciences, Shenyang, China

2 China National Erzhong Group Co., Deyang Wanhang Die Forging Co., Ltd., Deyang, China

3 College of Physics, Sichuan University, Chengdu, China

4 Key Laboratory of Radiation Physics and Technology of Ministry of Education, Institute of Nuclear Science and Technology, Sichuan University, Chengdu, China

5 Institute of Advanced Wear & Corrosion Resistant and Functional Materials, College of Chemistry and Materials Science, Jinan University, Guangzhou, China

*Address all correspondence to: denghaoscu@126.com

References

[1] E. Technologies, U.S.D.o.E.O.o.E. Efficiency, R. Energy, O.R.N. Laboratory, H.V.P.M. Program, Summary of Emerging Titanium Cost Reduction Technologies: A Study Performed for US Department of Energy and Oak Ridge National Laboratory, EHK Technologies. 2003

[2] Bambach M, Sizova I, Szyndler J, Bennett J, Hyatt G, Cao J, et al. On the hot deformation behavior of Ti-6Al-4V made by additive manufacturing. Journal of Materials Processing Technology. 2021;**288**:116840

[3] Dutta B, Froes FS. The additive manufacturing (AM) of titanium alloys. Metal Powder Report. 2017;**72**(2):96-106

[4] Bambach M, Sizova I, Sydow B, Hemes S, Meiners F. Hybrid manufacturing of components from Ti-6Al-4V by metal forming and wire-arc additive manufacturing. Journal of Materials Processing Technology. 2020;**282**:116689

[5] Li H-G, Che P-C, Yang X-K, Huang Y-J, Ning Z-L, Sun J-F, et al. Enhanced tensile properties and wear resistance of additively manufactured CoCrFeMnNi high-entropy alloy at cryogenic temperature. Rare Metals. 2022;**41**(4):1210-1216

[6] Tan C, Weng F, Sui S, Chew Y, Bi G. Progress and perspectives in laser additive manufacturing of key aeroengine materials. International Journal of Machine Tools and Manufacture. 2021;**170**:103804

[7] Sun S-B, Zheng L-J, Liu J-H, Zhang H. Microstructure, cracking behavior and control of Al–Fe–V–Si alloy produced by selective laser melting. Rare Metals. 2023;**42**:1353-1362

[8] Wang H, Lee YJ, Bai Y, Zhang J. Post-Processing Techniques for Metal-Based Additive Manufacturing: Towards Precision Fabrication. Boca Raton: CRC Press; 2023

[9] Ramachandiran N, Asgari H, Dibia F, Eybel R, Gerlich A, Toyserkani E. Effect of non-lamellar α precipitate morphology on the mechanical properties of Ti5553 parts made by laser powder-bed fusion at high laser scan speeds. Materials Science and Engineering: A. 2022;**841**:143039

[10] Brandt M. The role of lasers in additive manufacturing. Laser Additive Manufacturing. Cambridge: Woodhead Publishing; 2016

[11] Gu D, Shi X, Poprawe R, Bourell DL, Setchi R, Zhu J. Material-structure-performance integrated laser-metal additive manufacturing. Science. 2021;**372**(6545):eabg1487

[12] Pragana JPM, Sampaio RFV, Bragança IMF, Silva CMA, Martins PAF. Engineering, hybrid metal additive manufacturing: A state–of–the-art review. Advances in Industrial and Manufacturing Engineering. 2021;**2**:100032

[13] Strondl A, Lyckfeldt O, Brodin HK, Ackelid U. Characterization and control of powder properties for additive manufacturing. JOM. 2015;**67**:549-554

[14] Shipley H, McDonnell D, Culleton M, Coull R, Lupoi R, O'Donnell G, et al. Optimisation of process parameters to address fundamental challenges during selective laser melting of Ti-6Al-4V: A review. International Journal of Machine Tools and Manufacture. 2018;**128**:1-20

[15] Körner C. Additive manufacturing of metallic components by selective electron

beam melting—A review. International Materials Reviews. 2016;**61**(5):361-377

[16] Gong X, Anderson T, Chou K. Review on powder-based electron beam additive manufacturing technology. In: International Symposium on Flexible Automation, American Society of Mechanical Engineers. 2012;**45110**:507-515

[17] Murr LE, Martinez E, Amato KN, Gaytan SM, Hernandez J, Ramirez DA, et al. Fabrication of metal and alloy components by additive manufacturing: Examples of 3D materials science. Journal of Materials Research and Technology. 2012;**1**(1):42-54

[18] Ramsperger M, Mújica Roncery L, Lopez-Galilea I, Singer RF, Theisen W, Körner C. Solution heat treatment of the single crystal nickel-base superalloy CMSX-4 fabricated by selective electron beam melting. Advanced Engineering Materials. 2015;**17**(10):1486-1493

[19] Su J-L, Jiang F-L, Teng J, Chen L-Q, Requena G, Yan M, et al. Laser additive manufacturing of titanium alloys: Process, materials and post-processing. Rare Metals. 2024;**43**:6288-6328

[20] Luo Q, Yin L, Simpson TW, Beese AM. Effect of processing parameters on pore structures, grain features, and mechanical properties in Ti-6Al-4V by laser powder bed fusion. Additive Manufacturing. 2022;**56**:102915

[21] Aboulkhair NT, Everitt NM, Ashcroft I, Tuck CJAM. Reducing porosity in AlSi10Mg parts processed by selective laser melting. Additive manufacturing. 2014;**1**:77-86

[22] Guillen D, Wahlquist S, Ali A. Critical review of LPBF metal print defects detection: Roles of selective sensing technology. Applied Sciences. 2024;**14**(15):6718

[23] Eliasu A, Czekanski A, Boakye-Yiadom S. Effect of laser powder bed fusion parameters on the microstructural evolution and hardness of 316L stainless steel. The International Journal of Advanced Manufacturing Technology. 2021;**113**:2651-2669

[24] Heiden MJ, Deibler LA, Rodelas JM, Koepke JR, Tung DJ, Saiz DJ, et al. Evolution of 316L stainless steel feedstock due to laser powder bed fusion process. Additive Manufacturing. 2019;**25**:84-103

[25] Liu Y, Li S, Wang H, Hou W, Hao Y, Yang R, et al. Microstructure, defects and mechanical behavior of beta-type titanium porous structures manufactured by electron beam melting and selective laser melting. Acta Materialia. 2016;**113**:56-67

[26] Shen YF, Gu D, Pan Y. Balling process in selective laser sintering 316 stainless steel powder. Key Engineering Materials. 2006;**315**:357-360

[27] Yakout M, Elbestawi M, Veldhuis SC. A study of thermal expansion coefficients and microstructure during selective laser melting of invar 36 and stainless steel 316L. Additive Manufacturing. 2018;**24**:405-418

[28] Olatunde Olakanmi E, Dalgarno KW, Cochrane RF. Laser sintering of blended Al-Si powders. Rapid Prototyping Journal. 2012;**18**(2):109-119

[29] Guo C, Ge W, Lin F. Effects of scanning parameters on material deposition during electron beam selective melting of Ti-6Al-4V powder. Journal of Materials Processing Technology. 2015;**217**:148-157

[30] Hrabe N, Gnäupel-Herold T, Quinn T. Fatigue properties of a titanium alloy (Ti–6Al–4V) fabricated via electron beam melting (EBM): Effects of internal

defects and residual stress. International Journal of Fatigue. 2017;**94**:202-210

[31] Yadroitsev I, Yadroitsava I. Evaluation of residual stress in stainless steel 316L and Ti6Al4V samples produced by selective laser melting. Virtual and Physical Prototyping. 2015;**10**(2):67-76

[32] Ahmad B, van der Veen SO, Fitzpatrick ME, Guo H. Residual stress evaluation in selective-laser-melting additively manufactured titanium (Ti-6Al-4V) and inconel 718 using the contour method and numerical simulation. Additive Manufacturing. 2018;**22**:571-582

[33] Li C, Liu Z, Fang X, Guo Y. On the simulation scalability of predicting residual stress and distortion in selective laser melting. Journal of Manufacturing Science and Engineering. 2018;**140**(4):041013

[34] Simonelli M, Tse YY, Tuck C. On the texture formation of selective laser melted Ti-6Al-4V. Metallurgical and Materials Transactions A. 2014;**45**:2863-2872

[35] Pantawane MV, Yang T, Jin Y, Joshi SS, Dasari S, Sharma A, et al. Crystallographic texture dependent bulk anisotropic elastic response of additively manufactured Ti6Al4V. Scientific Reports. 2021;**11**(1):633

[36] Chlebus E, Kuźnicka B, Kurzynowski T, Dybała BJMC. Microstructure and mechanical behaviour of Ti—6Al—7Nb alloy produced by selective laser melting. 2011;**62**(5):488-495

[37] Carrozza A, Aversa A, Fino P, Lombardi M. Towards customized heat treatments and mechanical properties in the LPBF-processed Ti-6Al-2Sn-4Zr-6Mo alloy. Materials & Design. 2022;**215**:110512

[38] Mondal B, Mukherjee T, DebRoy T. Crack free metal printing using physics informed machine learning. Acta Materialia. 2022;**226**:117612

[39] Hojjatzadeh SMH, Parab ND, Yan W, Guo Q, Xiong L, Zhao C, et al. Pore elimination mechanisms during 3D printing of metals. Nature Communications. 2019;**10**(1):3088

[40] Yang M, Wang L, Yan W. Phase-field modeling of grain evolution in additive manufacturing with addition of reinforcing particles. Additive Manufacturing. 2021;**47**:102286

[41] Al-Bermani S, Blackmore M, Zhang W, Todd I. The origin of microstructural diversity, texture, and mechanical properties in electron beam melted Ti-6Al-4V. Metallurgical and Materials Transactions A. 2010;**41**:3422-3434

[42] Brunette DM, Tengvall P, Textor M, Thomsen P, Schenk R. The corrosion properties of titanium and titanium alloys, titanium in medicine: Material science. Surface Science, Engineering, Biological Responses and Medical Applications. Berlin: Springer; 2001

[43] Syed AK, Ahmad B, Guo H, Machry T, Eatock D, Meyer J, et al. An experimental study of residual stress and direction-dependence of fatigue crack growth behaviour in as-built and stress-relieved selective-laser-melted Ti6Al4V. Materials Science and Engineering: A. 2019;**755**:246-257

[44] Cecchel S, Ferrario D, Cornacchia G, Gelfi M. Development of heat treatments for selective laser melting Ti6Al4V alloy: Effect on microstructure, mechanical properties, and corrosion resistance. Advanced Engineering Materials. 2020;**22**(8):2000359

[45] Haubrich J, Gussone J, Barriobero-Vila P, Kürnsteiner P, Jägle EA, Raabe D, et al. The role of lattice defects, element partitioning and intrinsic heat effects on the microstructure in selective laser melted Ti-6Al-4V. Acta Materialia. 2019;**167**:136-148

[46] He J, Li D, Jiang W, Ke L, Qin G, Ye Y, et al. The martensitic transformation and mechanical properties of Ti6Al4V prepared via selective laser melting. Materials. 2019;**12**(2):321

[47] Zhao Z-Y, Li L, Bai P-K, Jin Y, Wu L-Y, Li J, et al. The heat treatment influence on the microstructure and hardness of TC4 titanium alloy manufactured via selective laser melting. Materials. 2018;**11**(8):1318

[48] A. ASTM, F2924-14. Standard Specification for Additive Manufacturing Titanium-6 Aluminum-4 Vanadium with Powder Bed Fusion. West Conshohocken, PA, USA: ASTM; 2014

[49] Cao S, Chu R, Zhou X, Yang K, Jia Q, Lim CVS, et al. Role of martensite decomposition in tensile properties of selective laser melted Ti-6Al-4V. Journal of Alloys and Compounds. 2018;**744**:357-363

[50] Yu G, Li Z, Hua Y, Liu H, Zhao X, Li W, et al. The effects of post heat treatment on the microstructural and mechanical properties of an additive-manufactured porous titanium alloy. Materials. 2020;**13**(3):593

[51] Ter Haar GM, Becker TH. Selective laser melting produced Ti-6Al-4V: Post-process heat treatments to achieve superior tensile properties. Materials. 2018;**11**(1):146

[52] Vilaro T, Colin C, Bartout J-D. As-fabricated and heat-treated microstructures of the Ti-6Al-4V alloy processed by selective laser melting. Metallurgical and Materials Transactions A. 2011;**42**(10):3190-3199

[53] Cao S, Chen Z, Lim CVS, Yang K, Jia Q, Jarvis T, et al. Defect, microstructure, and mechanical property of Ti-6Al-4V alloy fabricated by high-power selective laser melting. JOM. 2017;**69**(12):2684-2692

[54] Yang Y, Chen J, Wang H, Zhang Z, Lu Y, Wu S, et al. Crystallographic features of α variants and β phase for Ti-6Al-4V alloy fabricated by selective laser melting. Materials Science and Engineering: A. 2017;**707**:548-558

[55] Benedetti M, Torresani E, Leoni M, Fontanari V, Bandini M, Pederzolli C, et al. The effect of post-sintering treatments on the fatigue and biological behavior of Ti-6Al-4V ELI parts made by selective laser melting. Journal of the Mechanical Behavior of Biomedical Materials. 2017;**71**:295-306

[56] Simonelli M, Tse YY, Tuck C. Effect of the build orientation on the mechanical properties and fracture modes of SLM Ti–6Al–4V. Materials Science and Engineering: A. 2014;**616**:1-11

[57] Yan X, Yin S, Chen C, Huang C, Bolot R, Lupoi R, et al. Effect of heat treatment on the phase transformation and mechanical properties of Ti6Al4V fabricated by selective laser melting. Journal of Alloys and Compounds. 2018;**764**:1056-1071

[58] Leuders S, Lieneke T, Lammers S, Tröster T, Niendorf T. On the fatigue properties of metals manufactured by selective laser melting–the role of ductility. Journal of Materials Research. 2014;**29**(17):1911-1919

Chapter 11

Material Extrusion Additive Technologies: Benefits, Challenges, and Niche Applications

Gemma Herranz, Cristina Berges and Javier Hidalgo

Abstract

Material extrusion (MEX) additive manufacturing (AM) technologies for metals and ceramics offer unparalleled versatility and cost-effectiveness, making them accessible to a broad range of users and enabling the democratization of three-dimensional (3D) printing. This chapter provides a comprehensive overview of MEX, focusing on the key components and conditions of each process stage. The powder selection considers the particle size, distribution, and morphology, while the binder needs to meet the rheological requirements for pellets or filaments extrusion. The component design and its position during printing, debinding, and sintering are also discussed, providing recommendations in terms of shrinkage, wall thickness, overhangs and bridges, corners, edges, etc. The chapter highlights the opportunities of MEX in producing components with tailored microstructures, combining dense and porous regions, and integrating complex geometries. Besides, the benefits and disadvantages of this AM technology over others are discussed, emphasizing its competitive cost, isotropic microstructure with mitigated internal stresses and finer grain size, leading to mechanical properties in the range of powder injection molding (PIM) specimens. The versatility of MEX is further explored through its compatibility with PIM technology and its ability to adapt to novel material systems. Despite current challenges, such as achieving high-precision and mitigating shrinkage effects, the ongoing advancements in feedstock development, process optimization, and material innovation position MEX as a transformative force in additive manufacturing. The chapter concludes with specific emphasis placed on its emerging capabilities, like multi-material printing and reinforced composites, showcasing their potential for advanced applications in aerospace, automotive, and medical industries.

Keywords: material extrusion, additive manufacturing, 3D printing, advanced materials, sintering

1. Introduction

Material extrusion (MEX) is an additive manufacturing alternative that stands out for the versatile, cost-effective, and efficient fabrication of high-complexity components from a wide variety of materials, including polymers, metals, ceramics, and composites. This chapter will focus on the MEX variants that are a part of powder technology for the fabrication metal and ceramics involving a final sintering step to consolidate the component. More precisely, it will address MEX variants using thermoplastic pellets and filaments as feedstocks. The main particularities and benefits of these technologies to enable breakthrough applications and leveraging niche industrial sectors will not only be highlighted, but also their main limitations and current challenges. It will also cover, with a dedicated section, the convenience of MEX for the manufacturing of emerging reinforced metal and ceramic composites and multi-material printing. The potential use of MEX to boost other compatible advanced technologies, such as powder injection molding (PIM), is also discussed.

Defined by the ASTM (American Society for Testing and Materials) standard F2792, MEX additive technologies refer to processes by which material is dispensed through a nozzle or orifice to be selectively deposited layer by layer to create a specified three-dimensional (3D) geometry. Over the years, this technology has received various names and identities. The variant MEAM (acronym of material extrusion additive manufacturing) is recently becoming popular to distinguish it from other manufacturing technologies that are also based on extrusion. MEX was initially referred to as Fused Deposition Modeling (FDM), a term coined by Stratasys, the pioneering company that patented and commercialized the technology in 1992 [1]. As more players entered the market, they coined their term, Fused Filament Fabrication (FFF), ensuring they sidestepped any potential legal entanglements while embracing the same fundamental process. Both terms, FDM and FFF, have been widely adopted by academics and have appeared in seminal works, cementing their place in the literature of additive manufacturing. However, they have their limitations, ignoring the broader spectrum of material extrusion technologies currently available. FFF narrowly defines the technology as filament-based and shares with FDM the convention that material must be melted. Material extrusion encompasses more than just filament-based processes, and material melting is not a prerequisite. Indeed, robocasting, also referred to as Direct Ink Writing (DIW), can be considered a type of MEX that uses a paste consisting of a dispersion of a metal or ceramic powder with a binder that is directly deposited without the need of melting.

As depicted in **Figure 1**, MEX technologies use different raw materials, heat requirements, and extrusion mechanisms, but share the core principle of layer-by-layer material deposition. A crucial aspect of processing materials other than polymers by MEX is the use of a binder to agglutinate fine particle powders. During the extrusion process, the binder holds the powder particles together, providing the necessary cohesion and workability. After shaping, the binder is typically removed through a debinding process, followed by sintering to consolidate the powder particles into a solid structure. However, in reinforced polymer composites, the binder remains as a matrix. In these composites, whiskers or long fibers can also be combined with the binder to enhance the polymeric material's properties. Depending on the material family and application, different extrusion methods and feedstock are adequate. Understanding these variations is essential for matching the appropriate technology to specific applications based on their capabilities and limitations.

1.1 Plunger-based technologies

Plunger-based mechanisms are linked to DIW and used to push a semi-liquid or paste form slurry through the nozzle, avoiding the need for heating [2]. The absence of heating simplifies the process, making it suitable for temperature-sensitive

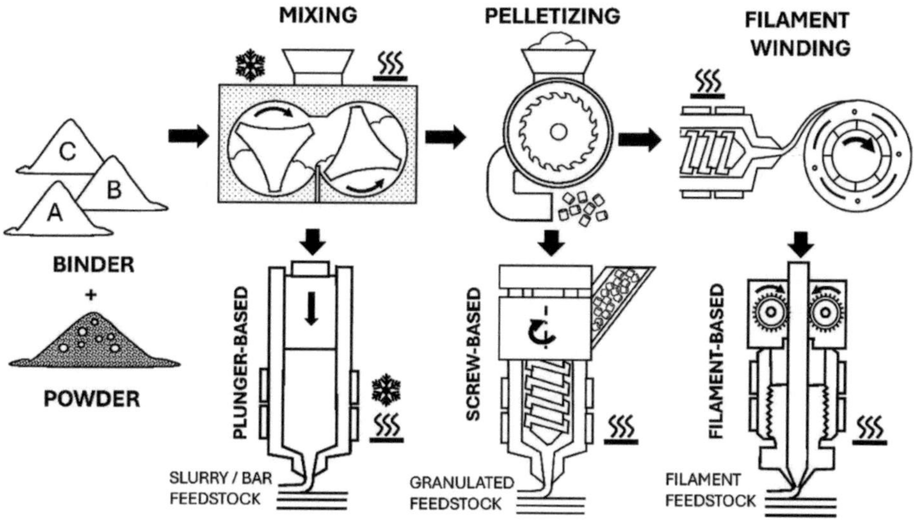

Figure 1.
Schematic of the different MEX technologies and the corresponding feedstock forms.

materials and reducing energy consumption. Feedstock preparation only requires a mixing step to combine all the components, though adjusting the feedstock formulation to achieve adequate slurry properties can be complex as will be discussed in subsequent sections. As there is no transition between the melt and solid state, the printed parts may lack consistency, and rapid debinding and freezing might be required to maintain part integrity. Additionally, the plunger mechanism may not be as precise, potentially resulting in lower-resolution layers and consistency compared to other MEX technologies.

The plunger-based method can also be applied to granulated materials or solid bars that are melted to enable them to be pushed, flowed, and deposited by the nozzle. They are then cooled to form a solid part. This approach can address the disadvantages of slurries, such as inconsistent parts and lower-resolution layers, by ensuring a controlled transition from melt to solid state. However, the use of solid bars limits the printable volume without interrupting the printing process, as the bar needs to be replenished once it is exhausted. This limitation also applies to slurries.

1.2 Screw-based technologies

Screw-based printers are fed with granulated feedstock, typically in the form of pellets, which are advanced by a rotating screw. The feedstock progresses through a chamber where it is heated to its glass transition or melting temperature. The screw also homogenizes the softened material and removes air bubbles. Eventually, the material is pushed through the nozzle and deposited on a flatbed. The main advantage of screw-based systems is that they enable continuous printing, eliminating the need to interrupt the printing for replenishment. This capability enables high production rates compared to the other variants.

Screw-based MEX, sometimes referred to as composite extrusion modeling (CEM), stands out as the most adaptable material extrusion system regarding material selection [3]. This technology benefits from compatibility with feedstocks developed for mature processes like powder injection molding, ensuring high readiness and availability of materials while leveraging consolidated knowledge for further developments. Pellets offer several advantages over slurries and filaments. The standards for feedstocks are less restrictive, and they can commonly accommodate higher solid particle loads (60–70 vol.%) while maintaining printability. Additionally, granulates are more straightforward to process, requiring fewer steps, and thus are typically more economical than their filament counterparts [4]. Furthermore, any defective or flawed printed single-material green parts would be easily reusable by crushing and sieving them before reintroducing the material into the printer hopper. The properties of these recycled leftovers might not be as suitable as fresh pellets for meeting high-quality demands, which would depend on the feedstock and binder type. It is common to establish a number of cycles for reutilizing the feedstocks and to mix reused feedstock with virgin raw feedstock to maintain the desired properties. Screw-based systems also enable continuous printing without the need for interruptions, resulting in higher production rates compared to other variants. All these advantages make pellet-based systems a popular choice.

Despite the advantages, screw-based systems come with certain drawbacks. Unlike filaments, where the feedstock properties are more consistent, pellets can vary in size and shape, making extrusion more challenging. Therefore, the size of the granulated feedstock must be carefully controlled to minimize defects caused by air entrapment and variable material flow rates during printing. Variations in material flow can lead to nonuniform deposition, resulting in material faults and poor layer adhesion. Studies have shown that granules sized between 3 and 5 mm significantly contribute to maximizing relative density after sintering by ensuring reliable performance throughout the printing process and minimizing defects [5]. Achieving precise temperature control is also essential, with chambers often incorporating different heating zones and cooling fans. The pellets should be evenly melted at the precise moment to avoid issues. The temperature near the nozzle must be high enough to ensure proper fluidity, avoiding nozzle clogging or degrading the feedstock, while farther from the nozzle, the temperature must not rise to a point where pellets begin clumping, resulting in blockages. Failure to manage these factors can result in printing jams, poor shape retention, and deficient surface finish. To address the challenges of using pellets, screw-based systems tend to be more complex, which increases equipment costs compared to simpler plunger- or filament-based systems. Despite these challenges, advancements in pellet manufacturing and printer design continue to face these issues, aiming to improve the precision and reliability of pellet-based 3D printing technologies.

1.3 Filament-based technologies

Filament-based technology offers the highest precision among MEX technologies, ensuring consistent material flow and uniform layer deposition [4]. This method utilizes filament feedstock, typically composed of thermoplastic materials wound on spools. The filament is fed into the extruder through a gear-driven mechanism, which pushes and directs it to the nozzle. At the nozzle, the filament is heated to its melting point and then deposited to build layers. Filament-based MEX is widely adopted for polymer printing due to its simplicity and low equipment costs. This method has limitations when applied to materials like metals and ceramics. Highly powder-loaded filaments have been successfully validated in open-source 3D printers [6], increasing accessibility for a wider range of users. However, these users still require additional debinding and sintering systems, which are often not readily available and may need to be subcontracted to specialized companies. Besides, the demands on filaments in terms of flexibility, stiffness, viscosity in the melted state, and debinding can be compromised and challenging to achieve with highly loaded metal and ceramic powdered materials. Thus, filament-based MEX is restricted to materials that can be formed into functional filaments. Optimal solid loading attainable can be notably lower than that for the pellet systems, which has implications in geometry control. Additionally, processing filaments requires additional steps compared to pelletizing, which raises the price of the feedstock. Overall, the precision achievable by filament still has no rival among MEX technologies, making it the preferred choice for demanding applications.

2. Selection of powder and binder

2.1 Powder

The MEX technology can be applied to the manufacture of virtually any particulate material, though the limit is set by the availability of powders with features suitable for this technology. One significant advantage of MEX among other AM technologies is that it shares with PIM many of the demands qualifying the powder suitable for the process. This compatibility means that MEX can leverage the extensive range of existing PIM powders, which have been optimized for properties like particle size distribution, shape, and composition [7]. This not only broadens the material options available for MEX but also reduces the development time and cost for new materials, as manufacturers can use well-established knowledge.

MEX can utilize diverse types of powders such as prealloyed, mixed-alloyed, or masteralloy powders. Prealloyed powders contain uniformly distributed alloying elements within each particle, ensuring consistent composition and properties throughout the part. This consistency simplifies the manufacturing process by allowing standardized process parameters, and it enhances microstructural uniformity, improving overall performance. However, prealloyed powders are typically more expensive and offer limited flexibility in tailoring material properties once produced. Mixed-alloyed powders, which are blends of different elemental or alloyed powders, provide significant flexibility in customizing the final composition and properties of the part. This approach can be cost-effective compared to prealloyed powders, but it presents challenges in achieving uniform distribution of alloying elements. Ensuring a consistent blend during the extrusion process requires precise control, and there is a risk of segregation during handling, leading to compositional variations. Masteralloy powders are highly concentrated alloy powders that, when mixed with base metal powders, enable efficient and flexible alloying. This method allows for the creation of a wide range of compositions by adjusting the amount of masteralloy added. It often enhances process efficiency and improves shape retention during sintering, which is critical for parts with intricate geometries. This compensates for the higher cost of masteralloy powders, and the challenges associated with blending them homogeneously. Although these powders are expensive, the method can offer overall cost savings in new alloy developments and reducing energy consumption through more efficient sintering.

Like PIM, various powder morphologies are accepted in MEX, each impacting the entire process. Powder characteristics significantly influence the feedstock's rheology and sinter ability [8], which are critical parameters in MEX. For instance, spherical or rounded powders typically offer better flowability and packing density, which are beneficial during both the extrusion and sintering processes. The improved flowability aids in the consistent extrusion of the feedstock, while the higher packing density enhances densification during sintering, leading to parts with fewer defects and higher final densities. In contrast, irregular or angular powders may present challenges in flowability and uniformity due to their poorer packing capacity. Nevertheless, these powders can improve mechanical interlocking between particles, potentially enhancing the strength of the green and brown parts before sintering, making them easier to handle during processing. Nevertheless, the reduced packing efficiency of irregular powders can lead to increased porosity and defects in the final sintered part, potentially compromising its mechanical properties.

Particle size is another important parameter. Fine particles provide a high surface area-to-volume ratio, which can enhance sinter ability and lead to denser final parts. This is particularly interesting in ceramic materials to trigger diffusion mechanisms and attain parts with higher mechanical strength and lower porosity. However, the increased surface area of fine powders leads to higher interparticle friction and stronger Van der Waals forces, reducing their flowability. Besides, fine particles are prone to agglomerate and can form large clusters that obstruct the narrow pathways within the extrusion system, leading to clogs. In this respect, too large powders can also result in clogging. If the powder particles are too large relative to the nozzle diameter, they can physically block the nozzle, preventing the flow of the feedstock. A common guideline is that the powder particle size should be no more than one-tenth to one-fifth of the nozzle diameter to ensure smooth flow and minimize clogging risks. A nozzle with a larger diameter can accommodate larger particles, but this must be balanced against the desired resolution and precision of the printed part.

The powder size distribution (PSD) significantly influences the performance and quality of the final product in MEX technologies. PSD is often quantified by the S_w parameter, where low S_w values indicate a broad PSD. The ideal S_w should be close to 2, whereas S_w higher than 7 might result in inadequate for the process [8]. A broad PSD allows for better packing density (PD) of the powders (**Figure 2**). This contributes to higher part strength before sintering, essential for handling and transportation between various stages of the manufacturing process. Particularly, more densely packed powder structure reduces the risk of deformation or collapse of the brown part when the binder is removed. During sintering, a broader PSD facilitates more efficient particle rearrangement and densification, resulting

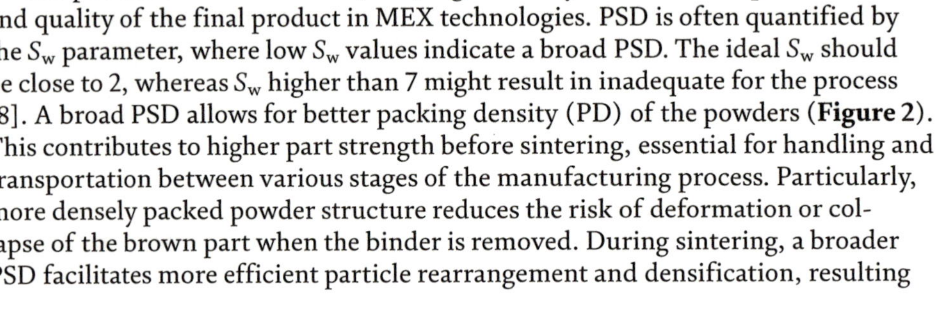

Figure 2.
(a) Comparison of two powders with the same average particle size (D50), but different particle size distributions (S_w). (b) Effect of particle size and particle distribution on feedstock viscosity. (c) Effect of solid loading (Φ) and particle shape on feestock viscosity. PD = packing density of the powder.

in parts with fewer defects and higher final densities. Additionally, a broad PSD enables a higher powder fraction, i.e., solid loading (Φ), while maintaining adequate processing conditions. Increasing the powder fraction is beneficial for reducing the pronounced shrinkage that occurs during the sintering process as powder particles join and densify. This reduction in shrinkage helps attain better dimensional control of the final product.

Abrasion and wear of the nozzle in MEX technologies are significant concerns that directly affect the service life of the nozzle and the quality of the printed parts [4]. The extent of nozzle wear is closely related to the characteristics of the powders used in the process, including their hardness, size, shape, and composition. Powders made from materials with high hardness, such as ceramics and certain metals (e.g., tungsten, carbide), are more likely to cause significant abrasion and wear on the nozzle. These hard particles can erode the nozzle material over time, leading to increased nozzle diameter and compromised precision. Irregular or angular particles with sharp edges are more abrasive than spherical particles. These shapes can cause localized stress points on the nozzle surface, leading to quicker degradation. Chemically reactive powders can interact with the nozzle material, potentially leading to chemical erosion, in addition to mechanical wear. To mitigate the effects of abrasion and wear, various materials are used for nozzle construction, each offering distinct levels of resistance and durability. Hardened steel nozzles are robust and offer good resistance to wear from abrasive powders. They are cost-effective and widely used in general applications. Other variants exist offering high wear resistance to hard powders. Ceramic materials like tungsten carbide and ruby-tipped nozzles provide excellent wear resistance due to their extreme hardness and durability. These nozzles are more expensive and may be brittle, making them more prone to cracking or breaking under mechanical stress. Nozzles coated with wear-resistant materials, such as diamond-like carbon (DLC) or other hard coatings, offer a balance between cost and durability.

2.2 Binder

The material extrusion technologies directly depend on the binder [9]. Its composition is determined by the printing system employed and material and should confer the feedstock with adequate properties to succeed in all the stages. In fact, the binder influences the shaping of green parts, as well as the debinding and sintering steps. Although the type of powder is also a determinant factor in the process stages, a suitable selection of binders can make possible the use of complicated powders, powder mixtures, or powders as reinforcement, maximizing the attainable powder fraction in the feedstock. When discussing MEX variants such as screw-based and plunger-based systems that utilize pellets, it is common to consider the same feedstocks used in PIM. This approach presents a significant advantage because the binder development in PIM is backed by extensive and well-established knowledge, which can be leveraged for new material applications in MEX. Even with the broad knowledge acquired in PIM over the years, adjusting the binder formulation remains a complex task, as binders consist of multiple compounds, typically a mixture of different thermoplastic polymers and other organic substances. These components serve different functions and can be broadly classified into four main families: backbone polymers, plasticizers, fluidizers, and surfactants. Plasticizers increase the flexibility and workability of the feedstock, facilitating processing and shaping. Fluidizers reduce the viscosity of the feedstock, improving flow properties during extrusion. Finally, surfactants improve the dispersion and compatibility of different components within the binder, ensuring a homogeneous mixture and avoiding segregation of the different compounds during printing.

However, despite the similarities, MEX and PIM processes exhibit significant technical differences that impact the suitability of feedstocks between the two technologies. In both PIM and MEX processes, the feedstock needs to exhibit pseudoplastic behavior, where viscosity decreases with increasing shear rate. This characteristic ensures proper flow under the varying shear conditions encountered in each process. One critical difference between the processes lies in the range of shear rates experienced by the molten feedstock. PIM typically operates at high shear rates during the injection phase (100–10,000 s^{-1}), facilitating the flow of feedstock with lower viscosity under these conditions. In contrast, MEX operates at much lower shear rates during the extrusion process (10–500 s^{-1}), which can adversely affect the flow characteristics of the same feedstock. Another key factor is the sensitivity of the feedstock's viscosity to temperature variations, characterized by the activation energy parameter. For PIM, a low activation energy (typically below 40 kJ/mol) is essential to maintain flow as the feedstock encounters the cool mold walls. In MEX, however, the feedstock must maintain an optimal temperature to ensure good flow and layer deposition while rapidly increasing viscosity upon exiting the nozzle to ensure sufficient shape retention. A higher activation energy is preferable for MEX feedstocks to facilitate this quick transition [10].

The ratio of each component in the binder should be optimized for each powder nature. For example, to avoid the powder loading reduction in the feedstock when designing the binder and maximize the sintering capacity of the particles, an increase in the fluidizer content with respect to the backbone should be attained, so the viscosity is controlled for the printing process.

In filament-based Material Extrusion (MEX), selecting the right binder is even more challenging due to the complex balance it must achieve between various performance attributes [3, 10]. The binder must facilitate the filament's ability to be spooled and unwound smoothly while ensuring it maintains adequate fluidity for the extrusion process. Additionally, the binder must provide enough mechanical resistance to withstand the forces exerted during printing and the dynamic movements of the printing head. Marks or indentations left by the gear teeth can affect the filament's diameter consistency, leading to issues like inconsistent extrusion or clogging. Moreover, wear caused by the feeding gear can introduce debris or create weak points in the filament, further exacerbating problems with filament flow and print quality. Achieving a good filament's properties balance often requires a careful selection of elastomers and additives to enhance the flexibility and reduce its viscosity, coupled with a robust polymer backbone that imparts stiffness and strength to the green part. Ideally, the filament should contain a high solid loading, typically around 60–65 vol.%, to match the performance of pellet-based MEX. However, this high solid loading can increase the brittleness of the filament, making it more prone to breakage, so in practice, a slightly lower solid loading is often used to improve handling and durability. Filament properties are sensitive to storage conditions, including humidity and the passage of time. Moisture absorption can lead to swelling, inconsistent extrusion, and poor print quality [11], while long-term exposure can result in material degradation and increased brittleness. These factors must be carefully considered when selecting the binder, as they influence the filament's stability and performance. Proper storage and handling practices are essential to maintain filament quality and ensure reliable printing outcomes.

In the case of filaments, the use of additional polymers in a significant concentration providing flexibility into the feedstock can make the debinding stage trickier. Normally, thermoplastic elastomers show swelling or low solubility in many organic solvents, as well as in water. However, avoiding the solvent step in the debinding stage might lead to highly slow thermal cycles, which also remains challenging. Since the goal of debinding is to remove the binder in the shortest time with the least impact on the compact, this strategy could be insufficient to keep brown parts without bubbles, cracks, or deformation.

Another requirement of the feedstock is that it must remain sufficiently tacky to promote strong adhesion between subsequent layers, which is common for both MEX technologies. This requires careful control of the extrusion temperature and print speed to balance flowability and cooling and emphasizes the crucial role of feedstock rheology, tailorable by the binder composition. There is limited detailed literature on the specific requirements for MEX feedstocks, particularly regarding their activation energy and interlayer adhesion characteristics. This gap in knowledge highlights the need for further investigation and development of MEX-specific binder formulations. Incorporating adhesion promoters as tackifiers to enhance interlayer bonding is one potential area for research, aiming to achieve high-quality prints and robust mechanical properties.

The nature of the binder dictates the type of debinding process required, impacting the efficiency, cost, and accessibility of the debinding process. Debinding is time-intensive and critical for ensuring the quality of the final product. The two most commonly used industrial debinding methods are catalytic debinding and solvent-thermal debinding, a methodology that has been transferred from PIM. Catalytic debinding uses a catalytic reaction to rapidly break down the binder at low temperatures, making it one of the most efficient and reliable debinding methods. The most common example is the use of an acid catalyst to decompose a polyacetal-based binder system. The process demands precise control of temperature, catalyst concentration, and reaction time to avoid incomplete debinding or overreaction, which can damage the part and thus, the process relies on highly automated systems with sophisticated control mechanisms to ensure consistency and repeatability. Hence, catalytic debinding requires high investments and specific handling and safety protocols, making it less accessible to smaller manufacturers or those with limited resources. In thermal debinding, organic components are degraded and vaporized by a controlled heating process in a furnace, which requires significantly less investment than the catalytic route. The process can take several days, especially for thicker parts, as careful control is required to prevent defects caused by the build-up of vapors that cannot escape from the part. Therefore, in the initial phase, the green parts are commonly immersed in a solvent that dissolves the primary binder component, typically a wax or low-molecular-weight polymer. This step removes a significant portion of the binder, reducing the risk of defects such as cracking or deformation during the subsequent thermal debinding stage. This enables a reduction in the debinding time, though it introduces other challenges like controlling swelling for proper shape retention. Additionally, handling and disposing of solvents that pose environmental and safety risks requires proper management and regulatory compliance. Other less commonly used methods, such as wick debinding [12] and supercritical debinding [13], are also possible. Wick debinding can be beneficial for thick parts to avoid defects, while supercritical debinding can significantly reduce debinding times.

Another promising approach that could potentially be applied to the PIM industry is the use of hybrid binders [14]. These materials are designed to meet the rheological properties required for both PIM and MEX technologies, allowing them to be either injected or printed. One significant advantage of this approach is its potential to streamline prototype development. In PIM, the mold design and manufacturing phase often creates a bottleneck, limiting production speed due to its cost and time-intensive nature. By using MEX to produce prototypes from the same feedstock intended for injection molding, manufacturers can significantly reduce mold design time and lead times. This capability accelerates product development cycles, lowers costs, and enables rapid iterations before finalizing the mold design. Developing feedstocks based on rheological prediction criteria will help manufacturers expand the potential of this family of additive manufacturing technologies.

3. Component design considerations in MEX

To be successful in MEX technology, component design is critical, as in other powder-based technologies. Since the MEX process requires post-printing debinding and sintering steps, design factors, such as cross-sectional thickness and geometric features, must be carefully considered. Additionally, printing time and cost are important concerns, and part design can be optimized to minimize material usage and reduce printing time (**Figure 3**).

Shrinkage: Shrinkage occurs during feedstock cooling in printing, and subsequently during debinding and sintering. Precisely accounting for shrinkage is essential to ensure dimensional accuracy of the final component. Shrinkage is not a universal constant and determining accurate shrinkage factors requires empirical data, typically obtained through 2–3 iterations for each material and geometry combination. Scaling factors should be applied in the computer-aided design (CAD) model to compensate for this shrinkage.

During printing, binder solidification often results in shrinkage that depends on the feedstock composition and printing conditions, typically ranging from 0.5% to

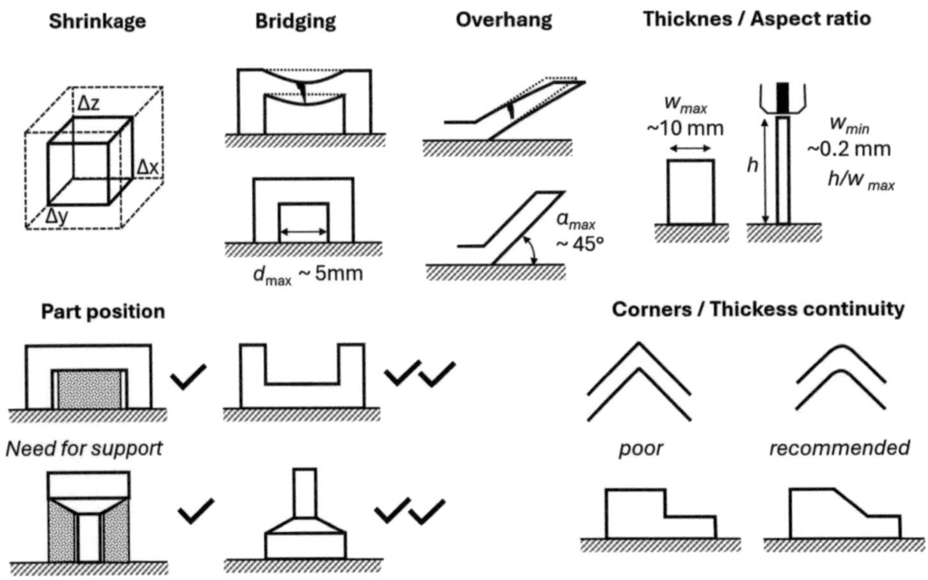

Figure 3.
Key considerations in part design and positioning for MEX technologies.

2% of the part dimensions. Significant shrinkage occurs during the debinding and sintering steps. Metal and ceramic parts can shrink by 15–25% during sintering, depending on the material, the powder fraction in the feedstock, and the specific sintering conditions. It is worth noting that shrinkage is commonly anisotropic, with the z-axis (printing head direction) often shrinking more than the x- and y axes due to layer-by-layer deposition and the effect of gravity. Part orientation during debinding and sintering also affects shrinkage and deformation. Placing the largest flat surface on the sintering tray minimizes warping and ensures uniform shrinkage. Maintaining the same orientation from printing through sintering is critical to avoid introducing new stresses and deformations.

Wall thickness: Wall thickness is a critical parameter in the design of components for MEX. It is constrained by both the printing process and the subsequent debinding step, making it essential to ensure adequate wall thickness for achieving high-quality, functional parts. The minimum wall thickness is primarily determined by the nozzle diameter and the precision of the extrusion system. Nozzle diameters typically range from 0.2 to 1 mm. Thin, tall walls are particularly prone to deformation due to stresses imposed during the printing process and the movement of the printing head, especially if the material's resistance is insufficient. Additionally, feedstock properties, including stiffness and strength at variable temperatures, significantly influence the lower limits of wall thickness.

The debinding step sets the upper threshold for wall thickness. Thick walls can impede effective binder removal, either by solvents, thermal or catalytic degradation, leading to incomplete debinding and potential defects. The maximum allowable wall thickness is feedstock-dependent but should not exceed 15 mm. Ideally, wall thickness should be below 10 mm to ensure effective debinding minimizing debinding times. Adjusting the infill percentage can facilitate debinding by creating internal channels within the part, allowing for thicker walls [13, 15]. However, this approach reduces the mechanical performance of the part and must be balanced against the need for structural integrity. Besides, one should consider that closed cavities pose significant challenges during debinding. Solvents may not adequately penetrate them, and even if they do, removed polymer might be trapped in the cavity and deposited on its bottom when the solvent is evaporated. Besides, thermal degradation fumes might be trapped in the cavity preventing effective binder removal, deforming the part, or even creating cracks by the accumulated pressure. Hence, design strategies considering hollow spaces should incorporate channels to ensure solvent access and binder escape.

Lastly, maintaining uniform wall thickness is crucial to prevent warping and dimensional variability during processing. Warping can result from differences in cooling rates, binder removal rates, material shrinkage, and residual stresses during printing, debinding, and sintering, which are influenced by wall thickness. If design constraints require changes in wall thickness, a gradual transition is essential. A recommended guideline is to implement a thickness change over a distance at least three times the thickness of the desired change. This gradual transition helps in minimizing stress concentrations and thermal gradients, reducing the risk of warping and defects.

Overhangs and bridges: The orientation of the component in the printing base and in the fixturing during debinding and sintering should consider avoiding or minimizing overhangs and excessive bridging, i.e., extruded feedstock between two points separated by an empty space or gap. Overhangs and bridges without adequate support can lead to sagging and collapse during printing and post-printing stages. As a guideline, overhangs exceeding 45 degrees and spans beyond 5 mm would need support during printing. The aspect ratio (height vs. width) also influences stability. For instance, tall, thin features may need additional support to prevent deformation. Currently, there are systems with rotating printing heads or bases that always maintain the printing perpendicular to the former deposited layer, avoiding the use of supports that otherwise must be removed post-processing or designed-in sacrificial supports. However, even counting with one of these printers, these design considerations should be extended to debinding and sintering. In these steps, weight is also a crucial factor to consider. Heavy parts are more prone to deformation or crumbling under their own weight due to the minimum part integrity during debinding and the high-temperature fluence during sintering. Components exceeding 300 g might exhibit difficulties during both stages. Particularly, unsupported overhangs and bridges are susceptible to deformation.

Support can be integrated directly into the component design. These supports can be removed in a secondary operation during or after sintering if they are made from the same material, or they can be made from a different material designed to be easily separated post-processing. Other strategies can be used to minimize support use that otherwise increases material use and printing times. Designing a flat surface on the component or adequately orienting the geometry during post-printing can facilitate the use of low-cost, standard flat fixtures for sintering. These fixtures are typically made from a high-melting-point porous ceramic that does not react with or adhere to the component material at sintering temperatures. In situations where these solutions are not feasible, a contoured fixture that matches the component's shape can be utilized. These fixtures are designed to fit the as-printed green state of the component and can also accommodate the reduced size after sintering, considering the shrinkage that occurs.

Other considerations: Geometric features like sharp corners, edges, and thin sections can introduce stress concentrations and are particularly susceptible to cracking and warping during both printing and sintering. Rounded edges and fillets are recommended to reduce stress concentrations and improve material flow during extrusion. Additionally, sharp corners may experience excessive shrinkage, leading to dimensional inaccuracies or defects. Designers should consider incorporating small radii in corners and avoiding overly sharp features. The printing direction can significantly impact the entire MEX process, potentially leading to critical defects, such as unwanted shrinkage or porosity, which can compromise shape retention and mechanical properties [16]. To address these challenges, careful assessment of the printing head movement during layer deposition is crucial. One effective strategy is to alternate the printing direction by 90° in successive layers. This cross-layer pattern enhances interlayer bonding, reduces anisotropic behavior, and helps distribute stresses more evenly throughout the part, ultimately improving structural integrity and reducing the likelihood of warping or deformation during debinding and sintering.

4. Available materials and opportunities

4.1 MEX of metal alloys

The precision achieved by MEX in metal alloy additive manufacturing still falls short compared to high-energy direct additive technologies, such as Selective Laser Melting (SLM) or Electron Beam Melting (EBM), and other indirect methods like Binder Jetting (BJ). These technologies are currently preferred for producing high-quality metallic parts due to their superior resolution and accuracy. Additionally, MEX has inherent limitations in part size due to the need for debinding, which poses a barrier for certain applications. However, MEX presents a competitive alternative for producing metal alloys in various industries where the high equipment costs and the need for highly specialized operators associated with other technologies are prohibitive. Near fully dense parts (relative density higher than 98% of the material density) are achievable by MEX when all the processing parameters are well optimized. For example, **Figure 4** shows that just adjusting the layer fan can be critical for attaining good adhesion and defect-free printed parts.

Furthermore, MEX offers significant advantages in terms of microstructure and composition control during the sintering process. Melting-based technologies, such as Selective Laser Melting (SLM) and Electron Beam Melting (EBM), often result in microstructural anisotropy and the development of internal stresses. These issues arise from the rapid heating and cooling cycles inherent to these processes, which cause directional solidification, grain growth, and thermal gradients that create residual stresses in the final part. Additionally, elemental segregation can occur in melting-based technologies; while this is not always a significant problem in many metal alloys, it can be severe in certain materials. For example, high-entropy alloys (HEAs) and alloys with low-solubility elements, such as aluminum-based alloys and superalloys, are prone to elemental segregation and the formation of undesirable phases.

Element vaporization is another critical issue in direct additive manufacturing technologies, particularly for alloys containing elements like manganese (Mn), zinc (Zn), and chromium (Cr), among others. These elements often have high vapor pressures at the elevated temperatures used in AM processes, making them more likely to vaporize and result in compositional changes that degrade material properties. Additionally, the release of these volatile elements can compromise operator safety, creating health risks in the workplace. The need for controlled atmospheres, temperature adjustments, and potentially more expensive material alternatives can increase the overall costs of the AM process. Additionally, there may be limitations in the range of materials that can be effectively processed, thus restricting design options, and complicating the material selection for specific applications.

In contrast, Material Extrusion (MEX), which relies on sintering, facilitates the development of more uniform, isotropic microstructures, as shown in **Figure 5**. This process effectively mitigates the development of residual stresses and minimizes the evaporation

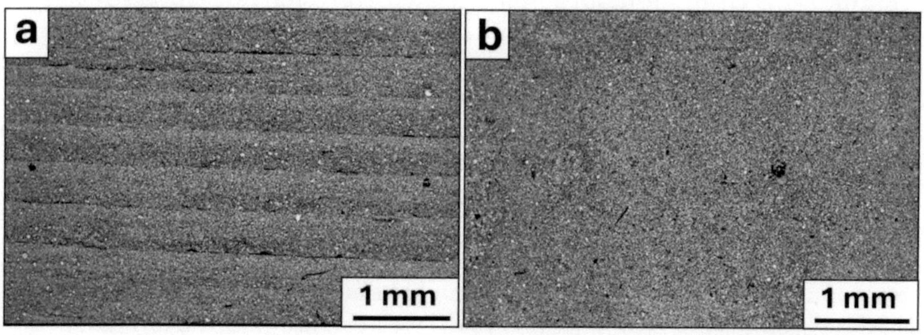

Figure 4.
Scanning electron microscopy (SEM) images of filament printed green 17-4PH specimens, (a) with activated layer fan showing poor layer adhesion and (b) without layer fan showing uniformity and lack of printing defects.

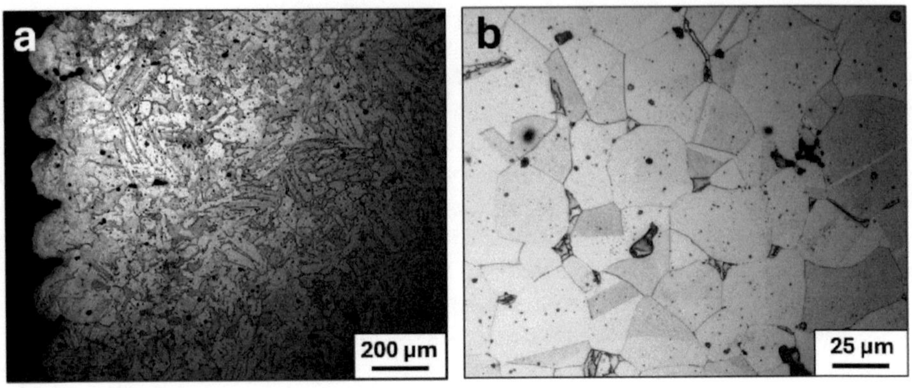

Figure 5.
Light optical microscopy images showing microstructures of some of the most demanded materials processed by filament MEX. (a) 17-4PH stainless steel grade and (b) 316L stainless steel grade.

of alloying elements, thereby preserving the desired microstructure and mechanical properties of the final part. These factors make MEX particularly attractive for manufacturing applications that require materials such as stainless steels and advanced steels, where precise control of elements like chromium and manganese is essential to achieve optimal properties. Unlike melting-based methods, which often result in larger grain sizes due to rapid solidification and localized elevated temperatures, MEX allows for the maintenance of finer grain sizes during the sintering process. This characteristic is especially beneficial for applications requiring fine grain sizes, such as oxide dispersion-strengthened (ODS) steels and mechanically alloyed powders. While other sintering-based technologies, such as Binder Jetting (BJ), also benefit from controlled sintering environments, MEX demonstrates a superior ability to achieve higher densification due to the specific characteristics of the powders it employs—powders that are often unsuitable for BJ. This higher densification can lead to improved material properties, positioning MEX as a valuable option for certain advanced applications.

Regarding composition control in MEX, maintaining precise debinding and sintering conditions are crucial for metal alloys to prevent contamination from carbon or oxidation, which could adversely affect the final part's properties. While this concern is less critical in certain metals, it is particularly important for alloys like stainless steels and light metals, such as titanium, aluminum, and magnesium, where even minor oxidation or unintended carbon incorporation can significantly alter densification during sintering or the desired properties like corrosion resistance, strength, and ductility. Achieving this level of control often requires high-purity protective atmospheres, such as nitrogen or argon, or reductive atmospheres containing certain fractions of hydrogen or methane. Additionally, it might require reformulating the material grade and using powders with exceptionally low levels of carbon and oxygen. All these issues typically increase the overall cost of the process compared to the MEX of other materials like ceramics, which can tolerate air as debinding and sintering atmospheres and do not require special grades. However, if managed properly, debinding and sintering conditions can be strategically engineered to adjust the carbon content in the alloy. This approach could be advantageous in advanced high-strength steels, where controlling the carbon levels can promote the formation of specific phases, such as martensite and austenite, to achieve a balance of high strength and toughness. In tool steels, precise carbon control can enhance wear resistance and hardness by optimizing carbide formation during heat treatment.

To date, MEX has demonstrated success with numerous metal alloys for various applications. One reason is that the typical rough surfaces (see **Figure 5a**) and dimensional inaccuracies of MEX parts can be improved by using secondary finishing processes, such as solvent-assisted smoothing or polishing. Solvents can be applied to soften and smear the surface layer, thereby reducing surface roughness. This can be complemented with mechanical rectification or polishing in the green state, which helps to further refine surface finish and dimensional accuracy while minimizing tool wear compared to finishing processes after sintering [17].

The most significant progress and developments have been principally made in widely demanded alloys, such as 316L stainless steel, 17-4PH steel, Ti-6Al-4V, and copper [18]. Some compelling examples are shown in **Figure 6**. **Figure 6a** illustrates a turbocharger fan for an internal combustion engine, demonstrating the complex geometries achievable through MEX. Made from 17-4PH stainless steel, the part is designed to withstand high axial forces and resist deformation at rotating speeds exceeding 100,000 rpm (revolutions per minute), while also enduring corrosive environments. MEX-produced 17-4PH parts have demonstrated excellent performance in tribocorrosion applications [19, 20], showcasing their durability and resistance to wear and corrosion under challenging conditions. The fan design can be easily modified to optimize compressor performance, making MEX an ideal, cost-effective solution for developing this application in its early stages. **Figure 6b** showcases a face milling tool manufactured from AISI M2 tool steel. A quarter-section cut of the green part reveals internal channels designed to recirculate and dose cutting fluid, preventing overheating due to friction during the milling process [21]. These internal channels are only achievable through additive manufacturing technologies like MEX,

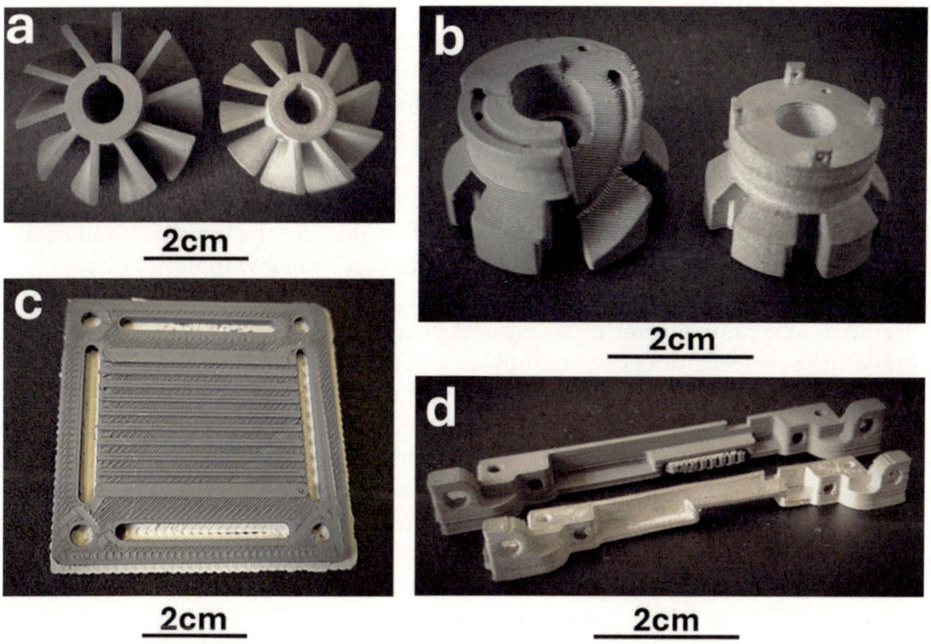

Figure 6.
Green and sintered states of (a) 17-4PH steel turbocharge fan (b) AISI M2 tool steel face milling tool (c) Fe-Cr alloy solid oxide fuel cell interconnector (green state). Courtesy of National Hydrogen Center (CNH2), Spain and DYPAM UCLM (d) Inconel 718 structural ribbon for satellite application. Courtesy of National Institute of Aerospace Technology (INTA), Spain and DYPAM UCLM.

which also offers better microstructure control compared to melting-based alternatives for this challenging alloy. **Figure 6c** presents an interconnector made of a Fe-Cr alloy. The flat, thin, and intricate geometry, including gas-conducting channels in both sides, required a ceramic support structure for the down side channels. One of the main challenges of this part is to avoid warping during part extraction and sintering. The geometric versatility of MEX enables interconnector designs that would be impossible to achieve through conventional lamination, forging, or conventional powder metallurgy methods, offering a cost-effective solution for the fabrication of efficiency-enhanced solid oxide fuel (SOFC) and electrolyzer (SOEC) cells. Lastly, **Figure 6d** features a structural ribbon made from Inconel 718, used in satellite aerospace applications. The multiple holes and bridging make this part particularly demanding to produce, and while post-processing is required to machine the screw threads, MEX offers a more cost-effective and feasible approach compared to traditional machining manufacturing methods, where such complexity and material particularities would result in significantly higher costs and difficulties.

4.2 MEX of ceramics

Many ceramic applications are evolving toward high structural complexity and design personalization, where traditional manufacturing techniques face limitations, offering great opportunities for MEX. Additive manufacturing of ceramics is highly challenging when using high-energy direct technologies, such as laser, electron beam, and wire-arc additive manufacturing. These methods struggle due to ceramics' low absorbance, high melting temperatures, high melt viscosity, and poor thermal shock resistance. As a result, indirect methods involving the use of binders are generally preferred for fabricating ceramic components. Popular ceramic 3D printing techniques are Direct Ink Writing (DIW) and Stereolithography (SLA) but come with challenges, such as achieving proper binder formulation for good powder dispersion, maintaining appropriate rheology, and ensuring shape retention of the printed part. Additionally, the range of raw ceramic powders applicable to these technologies is limited and the printing machines are usually restrained to only employing the inks developed by the corresponding manufacturer. Binder Jetting (BJ) offers another alternative, requiring only a small percentage of binder, but the low flowability and tendency of ceramic powders to agglomerate often result in low integrity of as-printed green parts and undesirable internal porosity. In contrast, MEX is particularly well suited for ceramics, offering significant potential for development. Although formulating new feedstocks can be complex and time-consuming, MEX is compatible with a broad spectrum of ceramics, addressing the primary limitations of competing technologies. For instance, in SLA, ceramic powders must have precisely controlled light absorption and scattering properties to ensure proper resin curing and high printing precision, which limits the range of ceramic powders that can be used and increases the cost of the 3D printers. While ceramic feedstocks developed for powder injection molding have shown high compatibility with pellet-based MEX, producing highly ceramic-loaded filaments remains challenging, which currently restricts the readiness of MEX for printing many ceramic families [22].

To date, MEX has demonstrated versatility in preparing the most demanding advanced ceramics like alumina (Al_2O_3), zirconia (ZrO_2), and silicon carbide (SiC) [15]. These materials are particularly suitable for the aerospace and automotive industries, where high-precision electrical insulators and thermal barriers with custom shapes and complex geometries are required. Additionally, MEX has opened new frontiers in the fabrication of complex ceramic components for electronics, such as dielectric substrates for microwave and radio frequency (RF) applications. The ability to precisely control the geometry and composition of these ceramics allows for the fine-tuning of their dielectric properties, enabling the development of highly efficient and miniaturized devices. In the medical field, MEX has been investigated to create dense zirconia implants with intricate geometries tailored to individual patient anatomy [23]. These implants, particularly in dental and orthopedic applications, offer superior biocompatibility, wear resistance, and mechanical strength compared to their metal counterparts. MEX also shows promise in the fabrication of ceramic components for energy applications, such as SOFC and SOEC components [24]. The ability to produce complex, high-temperature-resistant parts with integrated features, such as internal channels for gas flow, is critical for enhancing the performance and efficiency of these energy devices. In this respect, the component shown in **Figure 7** is a fixation part made of cordierite for an infrared heater and produced by MEX. It serves a dual function: first, it electrically isolates the steel body from the current flowing through the coil, and second, it physically supports the quartz glass tubes, which house the heating element. Traditionally manufactured in steatite ceramics, this combination of electrical insulation and mechanical support makes cordierite an ideal material due to its low thermal expansion, good mechanical strength, and excellent thermal shock resistance. Another promising application of MEX is the processing of advanced ceramic armor components [23], which benefit from the technology's ability to produce lightweight, high-strength materials with complex internal structures. This capability allows for the development of armor systems that offer enhanced protection while reducing the overall weight, a crucial factor in military applications.

4.3 MEX of porosity-controlled materials and metamaterials

Porous materials have gained significant relevance across various industries due to their unique properties and capabilities. Their applications range from lightweight structures and energy absorption to filtration and biomedical implants, making them essential in fields, such as aerospace, automotive, and healthcare. Common fabrication methods for porous materials, including traditional casting of foams and powder metallurgy using space holders, often face challenges in achieving precise control over pore architecture and gradient porosity profiles. These limitations can restrict the performance characteristics attainable, such as impact resistance and fluid dynamics.

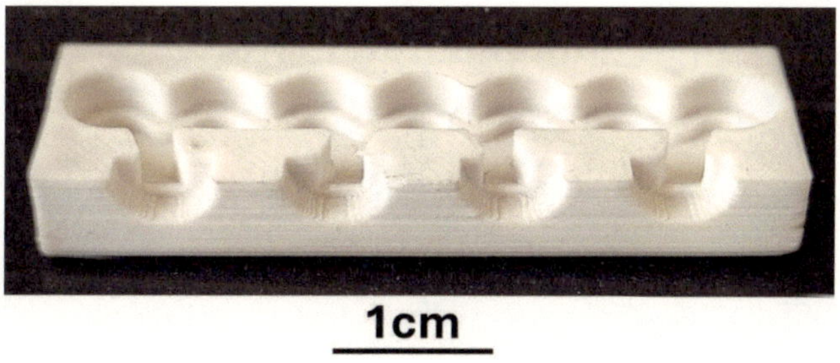

Figure 7.
Fixation part made of cordierite for an infrared heater (sintered state).

In this context, the emergence of AM technologies, particularly MEX, offers exciting opportunities for developing advanced porous materials. MEX enables the precise engineering of pore structures, allowing for tailored densities and mechanical properties essential in lattice materials and metamaterials. The ability of MEX to create intricate lattice structures and varying pore sizes, along with its capability to manipulate infill percentages and patterns during printing, provides a significant advantage over conventional methods that struggle to produce such complex geometries and allows for precise control over pore architecture, size, density, and morphology. **Figure 8** illustrates how these parameters can be adjusted to create porosity gradients or to combine dense and porous structures, effectively engineering the performance of the final product. Moreover, the possibility of incorporation of space holders in the MEX process enhances the control of porosity, enabling the fabrication of parts with hierarchical porosity that optimizes performance for specific applications.

A notable application of MEX technology in porous materials is the filter made of 316L stainless steel shown in **Figure 9**. This component is utilized in filtration systems for chemical processing, oil and gas, food and beverage, and water treatment, leveraging the corrosion resistance of the alloy. The complex porous structure achievable through MEX optimizes filtration efficiency while maintaining structural integrity. Another compelling example is the femoral stem made from biograde CrCoMo alloy, depicted in b. This large component integrates a dense surface with a porous core to reduce weight and control mechanical properties, thereby improving integration with bone. This design allows for the adjustment of the elastic modulus, helping to prevent shielding effects. While the size of the component exceeds typical dimensions for MEX, the internal porosity facilitates binder removal and minimizes the risk of defects during the manufacturing process.

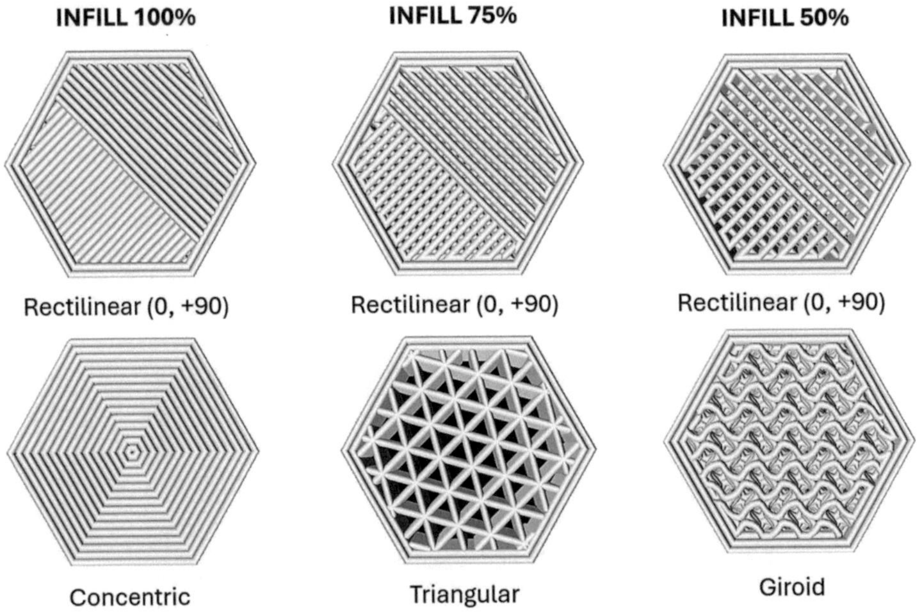

Figure 8.
Different geometric infill patterns and fill ratios available during printing. This hexagonal model was designed to have a length between parallel sizes of 15 mm and with a nozzle size of 0.4 mm.

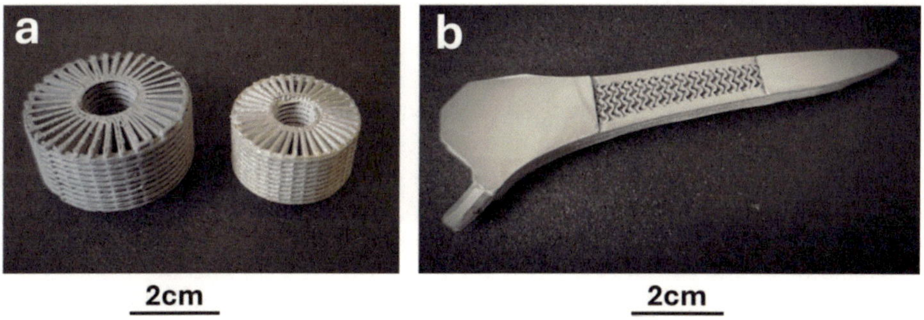

Figure 9.
(a) 316L stainless steel filter with tailored porosity. (b) Biograde CrCoMo alloy femoral stem combining a dense surface and a porous core to adjust the elastic modulus for improved bone-like properties.

The potential for MEX extends beyond metals and it also shines in the fabrication of ceramic-controlled porous structures, which are gaining attention in fields, such as biomedicine, thermal exchangers, and catalysis. For instance, highly porous structures are crucial for catalytic converters, where increased surface area available for catalytic reactions enhances efficiency. Recent advancements in MEX have demonstrated significant potential in this area, such as the printing of controlled porosity cordierite [25], which serves as a structural support material with a catalytic coating in exhaust systems. Cordierite is particularly well suited for high-temperature applications, maintaining durability under the harsh conditions encountered in catalytic converters and reducing the risk of cracking or deforming during thermal cycling, especially due to its thermal expansion characteristics, making it ideal for applications requiring high thermal shock resistance.

Further highlighting the versatility of MEX, yttria-stabilized zirconia (YSZ) and nickel oxide (NiO) porous structures (**Figure 10a**) also can play a crucial role in SOFCs and SOECs due to their electrical conductivity, catalytic activity, and compatibility with other cell components. The pore architecture can be engineered, thanks to MEX versatility to improve the cell performance while retaining the mechanical resistance required in the stack system [24]. The main challenge for MEX of these

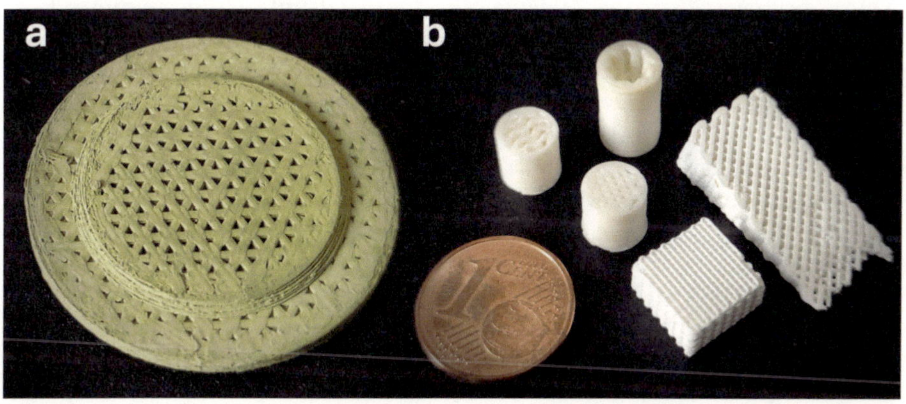

Figure 10.
(a) Porous YSZ-NiO support for planar solid oxide cells either in fuel or in electrolyzer configuration and (b) 3-YSZ implants for bone repair with different porous architectures serving as scaffolds to guide bone growth.

thin components is to avoid warping to perfectly match with other cell components, ensuring good gas sealing. In biomedical applications, ceramic porous grafts and scaffolds for bone injury repair, as those shown in **Figure 10b**, find in MEX a perfect ally to produce structures precisely adjusted to a patient's damaged bone, with engineered gradient porosity that mimics bone, enhancing structural integrity and osseointegration. Custom-designed bone grafts and implants made from bioactive glasses, hydroxyapatite, and biocompatible YSZ [26] have benefited from the design flexibility offered by MEX.

Expanding on the potential of MEX, technology plays a significant role in the creation of metamaterials, which achieve unique characteristics through their structure rather than their composition. These materials typically employ repeated patterns or microstructures that interact with electromagnetic waves, sound, or mechanical forces in unconventional ways, enabling a wide range of applications. Metamaterials can manipulate physical phenomena, such as light, sound, and heat, leading to innovations like invisibility cloaks, superlenses, and advanced antennas. Their behavior is often defined by negative or unusual values of properties, such as refractive index, elasticity, or permeability. For instance, metallic and ceramic metamaterials with negative thermal expansion coefficients have been developed [27], which could be used in advanced thermal management systems where controlling thermal expansion is critical. This capability opens up new possibilities in industries ranging from aerospace to electronics, where the unique properties of metamaterials can be exploited to achieve superior performance in challenging environments.

4.4 MEX of reinforced metal and ceramic composites

The application of MEX in fabricating metal and ceramic-reinforced composites remains underexplored compared to other dominating AM techniques, such as Selective Laser Melting (SLM) and Binder Jetting [28], though the potential of MEX to fabricate metal and ceramic-reinforced composites should not be overlooked, especially considering the technological advancements in related fields like PIM, where impactful examples can be found [29]. Nevertheless, there are pioneering works on the use of MEX for the production of aluminum reinforced with SiC [30] and Al_2O_3 [31], and a nickel-based cermet [32] and in the reinforcement of other ceramics, such as cordierite with two-dimensional (2D) reinforcements like graphene structures. One of the primary advantages of using MEX for metal and ceramic-reinforced composites lies in the binder system used during the extrusion process. Unlike direct high-energy AM methods that often rely on mechanical mixing or premixed powders in a solvent, MEX allows for a more controlled and homogeneous distribution of reinforcement materials within a polymer matrix. The binder in MEX acts as a medium to better disperse and stabilize the reinforcement particles, reducing the risk of agglomeration and ensuring uniformity throughout the material.

Figure 11 presents pioneering results of pellet-based MEX of cordierite reinforced with graphene reduced oxide (rGO). Feedstocks comprising 60 vol.% of a powder mixture of SiO_2, MgO, and Al_2O_3, along with 0.25 wt.% rGO, were successfully processed into pellets and used to print bending test specimens, as shown in **Figure 11a**. Raman spectroscopy confirmed that rGO was predominantly retained in the blue and brown regions of the printed parts. The fracture surface of the sintered specimens (**Figure 11b**) indicates a lack of internal printing defects, although the surface roughness remained high due to the use of a 0.8-mm nozzle diameter and a 0.4-mm layer height. Cordierite primarily formed during the reactive sintering process,

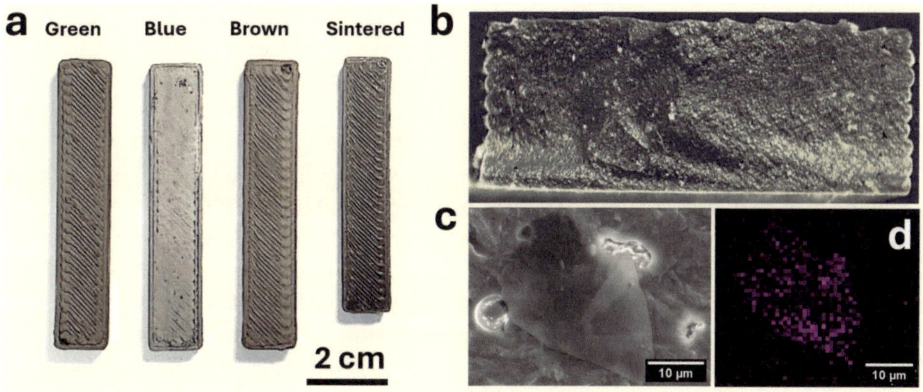

Figure 11.
(a) Bending test specimens in green, blue (after solvent immersion), brown, and sintered states of cordierite reinforced with 0.25 wt.% of rGO. (b) Fracture surface after the bending test. (c) Scanning electron microscopy (SEM) magnified image of the fracture surface, showing a graphene sheet, confirmed by the energy-dispersive X-ray (EDX) map of the carbon element in (d).

leading to approximately 11% linear shrinkage and about 3% residual porosity. The high sintering temperature of 1340°C, necessary for consolidating the part, resulted in a reduction of the rGO content in the final structure. However, the presence of the reinforcement was still confirmed, as illustrated in **Figure 11c** and **d**, leading to a modest 10% increase in bending strength and a 24% improvement in hardness compared to non-reinforced cordierite processed via MEX. To conclude, these promising results demonstrate the potential of MEX combined with reinforced ceramics, such as rGO-reinforced cordierite, to unlock disruptive applications in fields requiring enhanced mechanical performance, such as aerospace, where lightweight structures with tailored coefficient of thermal expansion, strength, and hardness are critical.

4.5 Multi-material and gradient materials by MEX

Additive manufacturing technologies offer unparalleled opportunities for producing multi-material parts and gradient materials, capabilities that are difficult or impossible to achieve with conventional manufacturing methods [33]. They enable the integration of dissimilar materials into a single component, facilitating innovative functionalities and the optimization of key properties, such as mechanical strength, magnetic properties, electrical conductivity, and thermal performance, at strategical spots of the component. Besides, brand-new composite materials filling the properties gap uncovered by current material families can be fabricated by conveniently distributing different metal alloys or ceramic materials at the mesoscale. Another exciting possibility is the creation of gradient materials [34]. These materials exhibit gradual variations in composition and properties within a single component, allowing for customized performance characteristics tailored to specific applications. The particularity of additive technologies is that all these advancements can be integrated into complex shape components with unprecedented properties through advanced design. This opens new avenues for applications in various industries, including aerospace, automotive, biomedical, and electronics, where multi-material and gradient material components can provide significant performance enhancements.

While the application of additive technologies for producing multi-material components is well established, the use of MEX is relatively new and gaining attention as it offers significant advantages over other additive manufacturing technologies. MEX technology has emerged as a competitive option for multi-material manufacturing due to its cost-effectiveness, simplicity, and ability to integrate dissimilar materials.

Melting-based additive manufacturing technologies, although enable high-resolution, face significant challenges when dealing with ceramics and metals multi-material printing. Indeed, to date, they are practically restricted to the combination of different metal alloys. Even in such applications, the control of the microstructure is complicated. The anisotropy developed due to melting-solidification phenomena is difficult to avoid. Besides, the formation of melting pools during the printing of subsequent layers hinders the precise control of the composition and microstructure at material interfaces. Additionally, the melting and solidification phenomena combined with differences in the thermal expansion coefficient of the dissimilar materials create residual stresses that are difficult to manage. In powder bed fusion technologies, another significant limitation is the retrieval and reuse of loose powders, especially when dealing with multiple materials. The recycling process can be complex and costly, affecting the powder's reusability. Moreover, spreading powder to cover a complete layer restricts the simultaneous combination of multiple materials within a single layer, though recent advancements are circumventing this issue. Summed to these technological issues, the high equipment costs and the need for specialized expertise further limit the accessibility of direct additive technologies for processing.

MEX accessibility and materials versatility over other AM technologies are indisputable and currently, there are a bunch of devices that have multiple extruder setups to enable multi-material printing. MEX also offers other advantages to overcome the drawbacks of direct technologies for multi-material printing. It is ideal for combining different ceramics, which is still technologically unviable by its main rival in this material family, VAT photopolymerization, and direct technologies based on melting. MEX also allows for precise control over layer thickness enabling the creation of fine-scale layering in multi-material structures. Microstructure control is more straightforward by adjusting the sintering conditions, resulting in isotropic microstructures like those achieved in wrought metal alloys. This also applies to the interfaces between dissimilar materials, where the extent of compositional gradients is easier to control by the sintering conditions and results in more defined and delimited interfaces. MEX is also advantageous due to high material utilization and reduced waste compared to powder bed AM technologies.

Figure 12 shows a filament-based MEX part successfully combining 17-4PH and 316L stainless steels in a pattern designed to enhance both mechanical strength and corrosion resistance. While MEX offers clear advantages for multi-material part manufacturing, several challenges must be addressed [35]. First, the different feedstocks need to exhibit good chemical compatibility and ensure adequate bonding at the material interfaces. In addition, these feedstocks must display similar debinding and sintering behaviors to avoid issues like warping or differential shrinkage. Achieving precise alignment (position zeroing) of the different print heads and nozzles is critical, as even slight misalignments or material inconsistencies between the two materials can result in catastrophic defects like cracking during sintering. Furthermore, frequent filament changes and retraction to prevent clogging can cause additional wear on the filaments, requiring them to be engineered to withstand such stresses. Finally, closely matching the shrinkage rates during sintering is crucial to minimize residual stresses, which is a particularly challenging aspect to control when working with different materials.

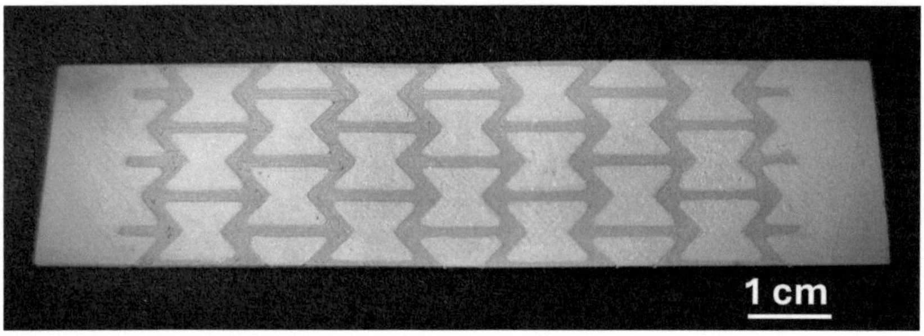

Figure 12.
Sintered part combining 17-4PH (dark gray) and 316L (light gray) stainless steels, designed with a pattern to enhance the overall properties of the part. Dimensions: 7.5 cm × 2.5 cm × 1.5 mm thickness.

5. Conclusions

MEX technologies (pellets and filaments based) can benefit from PIM development and maturity, and they are expected to be close allies to boost new applications and markets. Their compatibility in terms of powder size and distribution, binder, rheology fabrication, and postprocessing (debinding and sintering) has been highlighted. The versatility in the use of prealloyed, mixed-alloyed, and masteralloy powder is interesting to maximize flexibility and processability reducing the final part cost. The morphology affects the extrusion and sintering processes, impacting the quality and mechanical properties of the produced parts, while fine particles improve final density but can cause flow issues due to friction and agglomeration. The particle size distribution (PSD) affects product quality, with a broader PSD being ideal for enhancing packing density and reducing defects. Regarding the binder, similarities and differences between PIM and MEX processes are discussed and the debinding methods are described. Suggestions of binder design are also included for the feedstock fabrication, as well as recommendations for the design of MEX components, considering not only the printing stage but also the debinding and sintering. The chapter emphasizes the advantages of metals and ceramics produced by MEX at some demanding markets, such as new designs with internal channels, combining porous and dense regions or tailoring the final properties according to the infill patterns. Moreover, the opportunities of reinforced and multi-materials manufactured by MEX overtake other additive technologies.

Despite the current limitations of MEX technology, its ability to adapt and find niche applications with innovative designs and materials shows immense promise. With ongoing advancements in feedstock formulation, processing techniques, and multi-material capabilities, MEX is positioned to become a key player in unlocking new possibilities and driving disruptive innovations in additive manufacturing, paving the way for a bright future in both industrial and specialized markets.

Acknowledgements

Part of the results presented in this chapter has been made possible thanks to the financing from the JCCM research project with reference number SBPLY/21/180501/000255 (104), funded by FEDER funds. We also acknowledge support from the Spanish R&D Program UE "Una manera de hacer Europa" (PID2021-125612OB-C21-MCIN/AEI /10.13039/501100011033), co-financed by FEDER.

Author details

Gemma Herranz*, Cristina Berges and Javier Hidalgo
Instituto de Investigaciones Energéticas y Aplicaciones Industriales (INEI), Escuela Técnica Superior de Ingeniería Industrial de Ciudad Real (ETSII), DYPAM Research Group, Castilla-La Mancha University, Ciudad Real, Spain

*Address all correspondence to: gemma.herranz@uclm.es

References

[1] Crump SS. Apparatus and Method for Creating Three-Dimensional Objects. US Patent 5121329A

[2] Lewis JA. Direct ink writing of 3D functional materials. Advanced Functional Materials. 2006;**16**:2193-2204

[3] Gonzalez-Gutierrez J, Cano S, Schuschnigg S, Kukla C, Sapkota J, Holzer C. Additive manufacturing of metallic and ceramic components by the material extrusion of highly-filled polymers: A review and future perspectives. Materials. 2018;**11**:840, 1-36

[4] Altıparmak SC, Yardley VA, Shi Z, Lin J. Extrusion-based additive manufacturing technologies: State of the art and future perspectives. Journal of Manufacturing Processes. 2022;**83**:607-636

[5] Lieberwirth C, Harder A, Seitz H. Extrusion based additive manufacturing of metal parts. Journal of Mechanics Engineering and Automation. 2017;**7**:79-83

[6] Tymrak BM, Kreiger M, Pearce JM. Mechanical properties of components fabricated with open-source 3-D printers under realistic environmental conditions. Materials & Design. 2014;**58**:242-246

[7] Heaney DF, editor. Handbook of Metal Injection Molding. Cambridge, UK, Philadelphia, PA: Woodhead Publishing; 2012

[8] German RM. Powder Metallurgy Science. 2nd ed. Princeton, NJ: Metal Powder Industries Federation; 1994

[9] Bankapalli NK, Gupta V, Saxena P, Bajpai A, Lahoda C, Polte J. Metal fused filament fabricated stainless steel 316L. Composites Part B: Engineering. 2023;**264**:110915

[10] Naranjo JA, Berges C, Campana R, Herranz G. Rheological and mechanical assessment for formulating hybrid feedstock to be used in MIM & FFF. Results in Engineering. 2023;**19**:101258

[11] Limpadapun K, Sukmanee J. 2021 The 11th International Conference on Key Engineering Materials. Vol. 902. Moscow (Russia): Key Engineering Materials; 2021. pp. 101-106. (ISSN: 1662-9795)

[12] Hadian A, Koch L, Koberg P, Sarraf F, Liersch A, Sebastian T, et al. Material extrusion based additive manufacturing of large zirconia structures using filaments with ethylene vinyl acetate based binder composition. Additive Manufacturing. 2021;**47**:102227

[13] Charpentier N, Barrière T, Bernard F, Boudeau N, Gilbin A, Vikner P. Method for material extrusion additive manufacturing process made of bio-based polymer. Application for tool steel powders. Powder Technology. 2024;**438**:119591

[14] Naranjo JA, Perez G, Berges C, Hidalgo J, Herranz G. In: EuroPM2023 Conference Proceedings (2023). Lisbon (Portugal): The European Powder Metallurgy Association (EPMA); 2023

[15] Clemens F, Sarraf F, Borzì A, Neels A, Hadian A. Material extrusion additive manufacturing of advanced ceramics: Towards the production of large components. Journal of the European Ceramic Society. 2023;**43**:2752-2760

[16] Cano S, Lube T, Huber P, Gallego A, Naranjo JA, Berges C, et al. Influence of

the infill orientation on the properties of zirconia parts produced by fused filament fabrication. Materials. 2020;**13**:3158

[17] Dilberoglu UM, Gharehpapagh B, Yaman U, Dolen M. Current trends and research opportunities in hybrid additive manufacturing. International Journal of Advanced Manufacturing Technology. 2021;**113**:623-648

[18] Suwanpreecha C, Manonukul A. A review on material extrusion additive manufacturing of metal and how it compares with metal injection moulding. Metals. 2022;**12**:429

[19] García-Cabezón C, Naranjo JA, García-Hernández C, Berges C, Herranz G, Martín-Pedrosa F, et al. Using fused filament fabrication to improve the tribocorrosion behaviour of 17-4 PH SS in comparison to other metal forming techniques. Friction. 2024;**12**:2325-2343

[20] García-Hernández C, Naranjo JA, Castro-Sastre MÁ, Berges C, Fernandez-Abia AI, Martín-Pedrosa F, et al. Enhancing wear performance: A comparative study of traditional vs. additive manufacturing techniques for 17-4pH SS. Wear. 2024;**540-541**:205258

[21] Naranjo JA, Berges C, Gallego A, Herranz G. A novel printable high-speed steel filament: Towards the solution for wear-resistant customized tools by AM alternative. Journal of Materials Research and Technology. 2021;**11**:1534-1547

[22] Spina R, Morfini L. Material extrusion additive manufacturing of ceramics: A review on filament-based process. Materials. 2024;**17**:2779

[23] Nazari K, Tran P, Tan P, Ghazlan A, Ngo TD, Xie YM. Advanced manufacturing methods for ceramic and bioinspired ceramic composites: A review. Open Ceramics. 2023;**15**:100399

[24] Berges C, Wain A, Andújar R, Naranjo JA, Gallego A, Nieto E, et al. Fused filament fabrication for anode supported SOFC development: Towards advanced, scalable and cost-competitive energetic systems. International Journal of Hydrogen Energy. 2021;**46**:26174-26184

[25] Dorado B, Moreno-Sanabria L, García E, Belmonte M, Miranzo P, Osendi MI. 3D printing of cordierite materials from raw reactive mixtures. Ceramics International. 2023;**49**: 4578-4585

[26] Herranz G, Hidalgo J, Axelrad V, Delgado-Pujol EJ, Berges C, Naranjo JA, et al. Design and manufacturing by fused filament technique of novel YSZ porous grafts infiltrated with PCL/PVA/AgNPs for large bone defects repairing. Journal of Materials Research and Technology. 2024;**29**:3393-3408

[27] Zhang K, Wang K, Chen J, Wei K, Liang B, He R. Design and additive manufacturing of 3D-architected ceramic metamaterials with programmable thermal expansion. Additive Manufacturing. 2021;**47**:102338

[28] Chen L-Y, Qin P, Zhang L, Zhang L-C. An overview of additively manufactured metal matrix composites: Preparation, performance, and challenge. International Journal of Extreme Manufacturing. 2024;**6**:52006

[29] Hidalgo J, González-Velázquez V, Naranjo JA, Berges C, Vázquez E, Herranz G. Towards mass production of graphene-reinforced inconel 718 by powder injection moulding. Results in Engineering. 2024;**22**:102233

[30] Sun Y, Zou H, Yi Z, Xiong H, Kang X, Zhang L, et al. Material extrusion of high-density SiCp/7075Al composite via a high nitrogen-flowing assisted sintering. Materials Characterization. 2024;**214**:114113

[31] Prathumwan R, Subannajui K. Fabrication of a ceramic/metal (Al2O3/Al) composite by 3D printing as an advanced refractory with enhanced electrical conductivity. RSC Advances. 2020;**10**:32301-32308

[32] Yi Z, Shen T, Xiong H, Kang X, Zhang L, Zhou K. Strong and densified 3D metal-ceramic composite with strengthened layer structure by material extrusion additive manufacturing. Additive Manufacturing. 2024;**84**:104136

[33] Sun H, Zou B, Wang X, Chen W, Zhang G, Quan T, et al. Advancements in multi-material additive manufacturing of advanced ceramics: A review of strategies, techniques and equipment. Materials Chemistry and Physics. 2024;**319**:129337

[34] Pelz JS, Ku N, Shoulders WT, Meyers MA, Vargas-Gonzalez LR. Multi-material additive manufacturing of functionally graded carbide ceramics via active, in-line mixing. Additive Manufacturing. 2021;**37**:101647

[35] Hidalgo J, Berges C, Herranz G. Mesostructured emerging 316L-17-4PH composite steel parts by material extrusion for the property maximization through an efficient microstructural design. International Journal of Advanced Manufacturing Technology. 2024;**135**:3343-3362